Pharmaceutical Biotechnology

Edited by
O. Kayser and R.H. Müller

Related Titles

H.-J. Rehm, G. Reed, A. Pühler,
P. Stadler, G. Stephanopoulos

Biotechnology, Second, Completely Revised Edition, Volume 3/Bioprocessing

1993, ISBN 3-527-28313-7

H. Klefenz

Industrial Pharmaceutical Biotechnology

2002, ISBN 3-527-29995-5

G. Walsh

Proteins/Biochemistry and Biotechnology

2001, ISBN 0-471-89906-2

Oliver Kayser, Rainer H. Müller

Pharmaceutical Biotechnology

Drug Discovery and Clinical Applications

WILEY-VCH

WILEY-VCH Verlag GmbH & Co. KGaA

Edited by

Dr. Oliver Kayser
Free University Berlin
Institute of Pharmacy
Pharmaceutical Technology
Biopharmacy & Biotechnology
Kelchstr. 31
12169 Berlin
Germany

Prof. Dr. Rainer H. Müller
Free University Berlin
Institute of Pharmacy
Pharmaceutical Technology
Biopharmacy & Biotechnology
Kelchstr. 31
12169 Berlin
Germany

Library of Congress Card No.: applied for

British Library Cataloguing-in-Publication Data. A catalogue record for this book is available from the British Library.

Bibliographic information published by Die Deutsche Bibliothek Die Deutsche Bibliothek lists this publication in the Deutsche Nationalbibliografie; detailed bibliographic data is available in the Internet at http://dnb.ddb.de.

Printed in the Federal Republic of Germany
Printed on acid-free paper.

Composition: Laserwords Private Ltd, Chennai, India
Printing: betz-druck GmbH, Darmstadt
Bookbinding: Litges & Dopf Buchbinderei GmbH, Heppenheim
ISBN 3-527-30554-8

Preface

Pharmaceutical biotechnology has a long tradition and is rooted in the last century, first exemplified by penicillin and streptomycin as low molecular weight biosynthetic compounds. Today, pharmaceutical biotechnology still has its fundamentals in fermentation and bioprocessing, but the paradigmatic change affected by biotechnology and pharmaceutical sciences has led to an updated definition. Upon a suggestion by the European Association of Pharma Biotechnology (EAPB), pharmaceutical biotechnology is defined as a science covering all technologies required for the production, manufacturing, and registration of biotechnological drugs.

The biopharmaceutical industry has changed dramatically since the first recombinant protein (Humulin®) was approved for marketing in 1982. The range of resources required for the pharmaceutical industry has expanded from its traditional fields. Advances in the field of recombinant genetics allows scientists to routinely clone genes and create genetically modified organisms that can be used in industrial production processes. Also, specific therapeutic proteins can be synthesized in nonbiological ways, and recombinant proteins can be isolated from complex mixtures in commercially viable processes. In contrast to academic research, industrial development and manufacturing is guided by cost and time effectiveness, patent protection, exclusivity periods, and regulatory compliance. There are many critical industry issues that companies have to face; hence there is a need for new pharmaceutical biotechnology textbooks focussing on industrial needs.

Therapeutic proteins and the recently approved antisense oligonucleotide Fomivirsen® represent new and innovative biotech drugs that are different from classical drugs in the development and production process. In this area, pharmaceutical companies are confronted with new challenges to develop new products and to apply new technologies. Industrial needs are particularly different and are either not discussed or are only marginally discussed in existing textbooks, which is why we feel that there is a need for a new pharmaceutical biotechnology textbook.

We asked experts from the pharmaceutical biotech area to present their integrated view to answer questions focussing on industrial needs in the discovery and manufacture of recombinant drugs and new therapies. We are glad that a majority of contributors, active in the pharmaceutical industry, have participated and shared their views on new developments in protein production, production organisms, DNA vaccines, bioinformatics, and legal aspects. Distinct problems related to recombinant proteins that have arisen in recent years, such as drug stability, pharmacokinetics, and metabolization, are discussed in detail. It should be mentioned that for the first time the topic of generic recombinant drugs is presented in this textbook.

Biotechnology is a fast-moving area and crucial topics for future technologies can be recognized today. We wanted to give an insight into these future enterprise technologies and had asked for contributions to highlight new developments in gene therapy, tissue engineering, personalized medicine, and xenotransplantation having a realistic chance of being used in industrial applications.

In this textbook, you will find updated facts and figures about the biotech industry, product approvals, and discussions of how biotechnology is applied in human and animal health care, and in industrial and environmental processes. We address how biotech is being employed in national security efforts as well as the ethical issues that are frequently debated when people discuss the use of biotechnology in health sciences.

We would like to thank all contributors for their contributions, because we know that time was short and most of the papers were written alongside their regular duties. Special thanks to Dr. Andrea Pillmann, Wiley VCH, for her support in the layout, proofreading, and production of this textbook.

We are convinced that this textbook is filling a niche and covering industrial needs and interests in the pharmaceutical biotech area. Our point of view is that this textbook will cater to scientists and decision makers in pharmaceutical and biotechnological companies, venture capitals/finance, and politics.

O. Kayser
R.H. Müllers

Berlin, December 2003

Foreword

Pharmaceutical Biotechnology is a multidisciplinary scientific field undergoing an explosive development. Advances in the understanding of molecular principles and the existence of many regulatory proteins have established biotechnological or therapeutic proteins as promising drugs in medicine and pharmacy. More recent developments in biomedical research highlight the potential of nucleic acids in gene therapy and antisense RNAi technology that may become a medical reality in the future.

The book attempts to provide a balanced view of the biotechnological industry, and the number of experts from the industry sharing their knowledge and experience with the readers gives the book an outstanding value. All contributors provide with each chapter an up-to-date review on key topics in pharmaceutical biotechnology. Section 1 serves as an introduction to basics in protein production and manufacturing. Particular emphasis not only on production organisms like microorganisms and plants but also on industrial bioprocessing will be appreciated by the reader.

The advent and development of recombinant proteins and vaccines is described in detail in Part 2. Biotech drugs have created a number of unique problems because of their mostly protein nature. The production, downstream processing, and characterization is in many aspects different from conventional low molecular weight drugs and is highlighted by selected experts still in touch with the lab bench. Bringing the therapeutic protein to the patient is a major challenge. Protein formulation, biopharmaceutical aspects, and drug regulation are fields that are fast developing and well recognized by their new and innovative techniques. Drug regulation has a major impact on the whole drug manufacturing process, which is why special chapters on the drug approval process in Europe and the United States, and biogenerics are of high interest. Finally, in Part 4, experts provide an outlook on potential drugs and therapeutic strategies like xenotransplantation that are under investigation. Hopefully, some of these concepts will find clinical application in the following years.

I believe that there is a distinct need for a pharmaceutical biotech book focusing on the industrial needs of recombinant drugs and providing detailed insight into industrial processes and clinical use. Therefore, this work is not only a valuable tool for the industrial expert but also for all pharmacists and scientists from related areas who wish to work with biotech drugs. In life-learning courses and the professional environment, this compact book is the basis for a solid understanding for those who wish to gain a better overview of the industry they are working in.

Robert Langer
MIT Boston, November 2003

Contents

List of Contributors

Dr. Albrecht F. Kiderlen
Robert Koch-Institut
Nordufer 20
13353 Berlin
Germany

Prof. Dr. Andreas Bechthold
Albert-Ludwigs-Universität Freiburg
Pharmazeutische Biologie
Stefan-Meier-Straße 19
79104 Freiburg
Germany

Dr. Antonio J. Grillo-López
Neoplastic and Autoimmune Diseases Research Institute
P. O. Box 3797
Rancho Santa Fe, CA 92067
USA

Prof. Dr. Bernd Meibohm
Department of Pharmaceutical Sciences
College of Pharmacy, University of Tennessee, Health Science Center
Memphis, TN 38163
USA

Prof. Dr. David B. Resnik
The Brody School of Medicine
East Carolina University
Greenville, NC 27858
USA

Prof. Dr. Dirk Hoffmeister
The University of Wisconsin
School of Pharmacy
777 Highland Avenue
Madison, WI 53705
USA

Dr. Erno Pungor
Berlex Biosciences
2600 Hilltop Drive
Richmond, CA 94804
USA

Dr. Gary Walsh
Industrial Biochemistry Program
University of Limerick
Limerick City
Ireland

Prof. Dr. Gregory J. Brunn
Transplantation Biology and the Departments of Pharmacology and Experimental Therapeutics
Mayo Clinic
Rochester, MI 55905
USA

Prof. Dr. Hartmut Derendorf
Department of Pharmaceutics, College of Pharmacy
University of Florida
Gainesville, FL 32610
USA

Prof. Dr. Irmgard Merfort
Albert-Ludwigs-Universität Freiburg
Pharmazeutische Biologie
Stefan-Meier-Straße 19
79104 Freiburg
Germany

Dr. Janet Patton
Bone Research Unit
Medical Research Council/
University of the Witwatersrand
7 York Road
Parktown 2193 Johannesburg
South Africa

Prof. Dr. Jeffrey L. Platt
Transplantation Biology and the Departments of Pharmacology and Experimental Surgery, Immunology and Pediatrics
Mayo Clinic
Rochester, MI 55905
USA

Dr. Jeffrey Ulmer
Chiron Corporation
4560 Horton Street
Emeryville, CA 94608-2916
USA

Dr. Jens-Peter Gregersen
Chiron-Behring GmbH
Postfach 1630
35006 Marburg
Germany

Dr. John Donnelly
Chiron Corporation
4560 Horton Street
Emeryville, CA 94608-2916
USA

Dr. Jörg Knäblein
Schering AG
Analytical Development Biologicals
Müllerstraße 178
13342 Berlin
Germany

June Teare
Bone Research Unit
Medical Research Council/
University of the Witwatersrand
7 York Road
Parktown 2193 Johannesburg
South Africa

Dr. Jutta Haunschild
MorphoSys AG
Lena-Christ-Strasse 48
82152 Martinsried
Germany

Prof. Dr. Klaus Cichutek
Paul-Ehrlich-Institut
Paul-Ehrlich-Straße 51–59
63225 Langen
Germany

Dr. Lentsha Nathaniel Ramoshebi
Bone Research Unit
Medical Research Council/
University of the Witwatersrand
7 York Road
Parktown 2193 Johannesburg
South Africa

Louise Renton
Bone Research Unit
Medical Research Council/
University of the Witwatersrand
7 York Road
Parktown 2193 Johannesburg
South Africa

Priv. Doz. Dr. Oliver Kayser
Freie Universität Berlin
Institut für Pharmazie
Pharmazeutische Technologie
Biopharmazie & Biotechnologie
Kelchstraße 31
12169 Berlin
Germany

Prof. Dr. Rainer H. Müller
Freie Universität Berlin
Institut für Pharmazie
Pharmazeutische Technologie
Biopharmazie & Biotechnologie
Kelchstraße 31
12169 Berlin
Germany

Priv. Doz. Dr. Ralf Lipp
Schering AG
Müllerstraße 178
13342 Berlin
Germany

Dr. Stefan Pelzer
Combinature Biopharm AG
Robert-Rössle-Straße 10
13125 Berlin
Germany

Thato Matsaba
Bone Research Unit
Medical Research Council/
University of the Witwatersrand
7 York Road
Parktown 2193 Johannesburg
South Africa

Dr. Titus Kretzschmar
MorphoSys AG
Lena-Christ-Strasse 48
82152 Martinsried
Germany

Dr. Udo Gottschalk
Bayer AG
GB Pharma-Biotechnologie
Friedrich-Ebert-Straße 217
42096 Wuppertal
Germany

Dr. Ugo Ripamonti
Bone Research Unit
Medical Research Council/
University of the Witwatersrand
7 York Road
Parktown 2193 Johannesburg
South Africa

Dr. Walter Hinderer
BioGeneriX AG
Janderstraße 3
68199 Mannheim
Germany

Color Plates

Fig. 2.1 Photography of a sporulated *Streptomyces* strain growing on solid medium. The blue drops indicate the production of an antibiotic (aromatic polyketide).

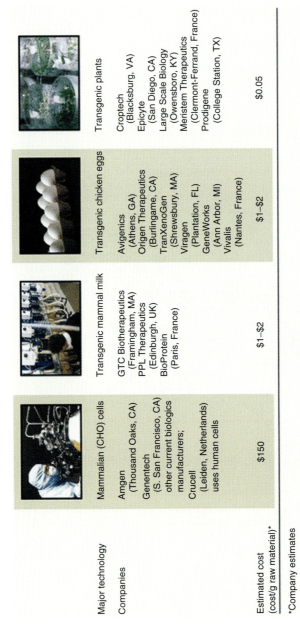

Major technology	Mammalian (CHO) cells	Transgenic mammal milk	Transgenic chicken eggs	Transgenic plants
Companies	Amgen (Thousand Oaks, CA) Genentech (S. San Francisco, CA) other current biologics manufacturers; Crucell (Leiden, Netherlands) uses human cells	GTC Biotherapeutics (Framingham, MA) PPL Therapeutics (Edinburgh, UK) BioProtein (Paris, France)	Avigenics (Athens, GA) Origen Therapeutics (Burlingame, CA) TranXenoGen (Shrewsbury, MA) Viragen (Plantation, FL) GeneWorks (Ann Arbor, MI) Vivalis (Nantes, France)	Croptech (Blacksburg, VA) Epicyte (San Diego, CA) Large Scale Biology (Owensboro, KY) Meristem Therapeutics (Clermont-Ferrand, France) Prodigene (College Station, TX)
Estimated cost (cost/g raw material)*	$150	$1–$2	$1–$2	$0.05

*Company estimates

Fig. 3.2 Companies and technologies in biomanufacturing. A comparison of different expression systems shows the big differences in terms of costs, ranging from US$150 per gram for CHO cells to US$0.05 per gram for transgenic plants [11].

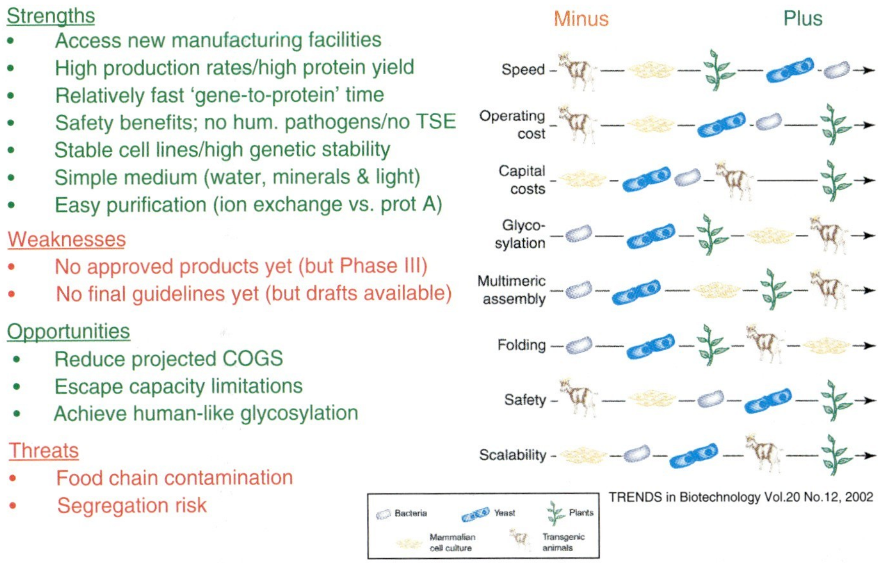
Strengths
• Access new manufacturing facilities
• High production rates/high protein yield
• Relatively fast 'gene-to-protein' time
• Safety benefits; no hum. pathogens/no TSE
• Stable cell lines/high genetic stability
• Simple medium (water, minerals & light)
• Easy purification (ion exchange vs. prot A)
Weaknesses
• No approved products yet (but Phase III)
• No final guidelines yet (but drafts available)
Opportunities
• Reduce projected COGS
• Escape capacity limitations
• Achieve human-like glycosylation
Threats
• Food chain contamination
• Segregation risk
Minus
Plus
Speed
Operating cost
Capital costs
Glyco-sylation
Multimeric assembly
Folding
Safety
Scalability
Bacteria
Yeast
Plants
Mammalian cell culture
Transgenic animals
TRENDS in Biotechnology Vol.20 No.12, 2002

Fig. 3.3 SWOT analysis of plant expression systems. Plant expression systems have a lot of advantages (plus) over other systems and are therefore mostly shown on the right-hand side of the picture (Raskin I et al., Plants and human health in the twenty-first century. *Trends in Biotechnol.* **2002** *20*, 522–531.). Herein different systems (transgenic animals, mammalian cell culture, plants, yeast, and bacteria) are compared in terms of speed (how quickly they can be developed), operating and capital costs and so on, and plants are obviously advantageous. Even for glycosylation, assembly and folding, where plants are not shown on the right-hand side (meaning other systems are advantageous), some plant expression systems are moving in that direction (as will be shown exemplarily in the section for moss). Also, the weaknesses and threats can be dealt with, using the appropriate plant expression system [20].

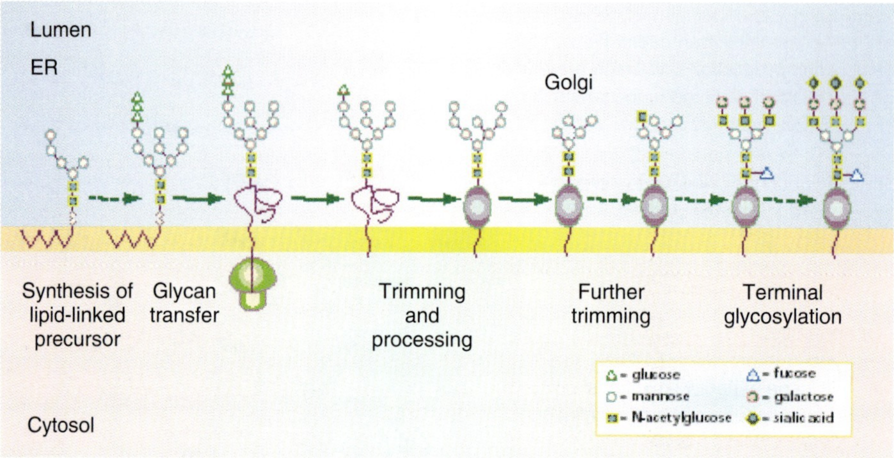

Fig. 3.4 The glycosylation pathway via ER and Golgi apparatus. In the cytosol carbohydrates are attached to a lipid precursor, which is then transported into the lumen of the ER to finish core glycosylation. This glycan is now attached to the nascent, folding polypeptide chain (which is synthesized by ribosomes attached to the cytosolic side of the ER from where it translocates into the lumen) and subsequently trimmed and processed before it is folded and moved to the Golgi apparatus. Capping of the oligosaccharide branches with sialic acid and fucose is the final step on the way to a mature glycoprotein [23].

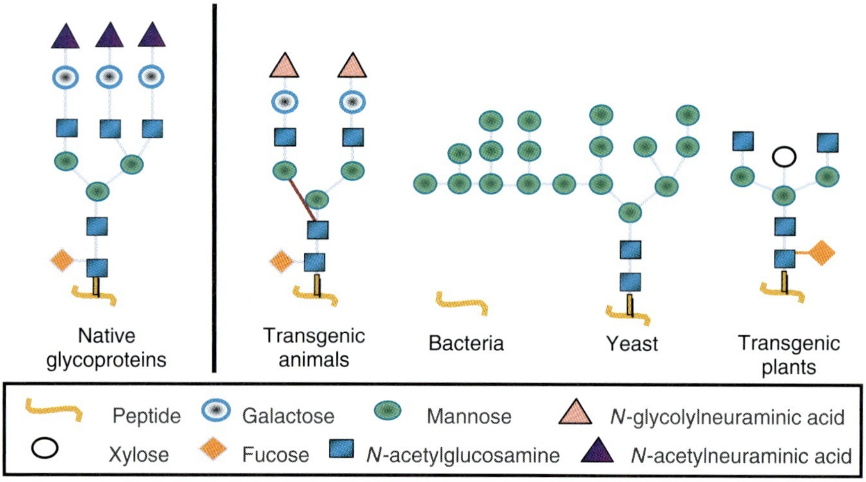

Fig. 3.5 Engineering plants to humanlike glycosylation. The first step to achieve humanlike glycosylation in plants is to eliminate the plant glycosylation pattern, that is, the attachment of β-1-2-linked xylosyl and α-3-linked fucosyl sugars to the protein. Because these two residues have allergenic potential, the corresponding enzymes xylosyl and fucosyl transferase are knocked out. In case galactose is relevant for the final product, galactosyl transferase is inserted into the host genome. Galactose is available in the organism so that this single-gene insertion is sufficient to ensure galactosylation [24].

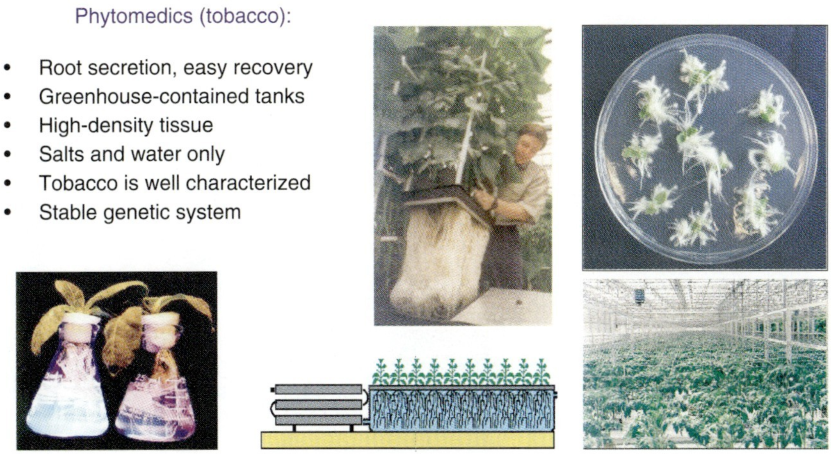
Phytomedics (tobacco):
• Root secretion, easy recovery
• Greenhouse-contained tanks
• High-density tissue
• Salts and water only
• Tobacco is well characterized
• Stable genetic system

Fig. 3.6 Secretion of the biopharmaceuticals via tobacco roots. The tobacco plants are genetically modified in such a way, that the protein is secreted via the roots into the medium ("rhizosecretion"). In this example, the tobacco plant takes up nutrients and water from the medium and releases GFP (green fluorescent protein). Examination of root-cultivation medium by its exposure to near-ultraviolet illumination reveals the bright green-blue fluorescence characteristics of GFP in the hydroponic medium (left flask in panel lower left edge). The picture also shows a schematic drawing of the hydroponic tank, as well as tobacco plants at different growth stages, for example, callus,–fully grown and greenhouse plantation [24].

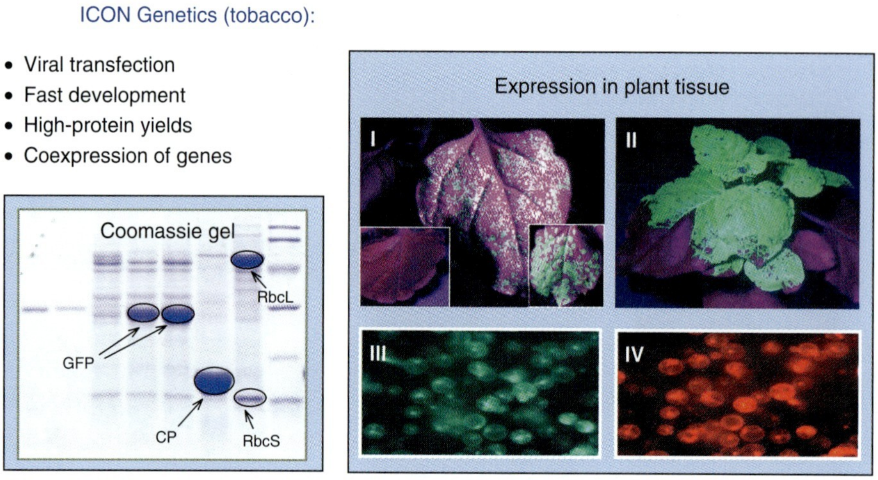

Fig. 3.7 Viral transfection of tobacco plants. This new generation platform for fast (1 to 2 weeks), high-yield (up to 5 g per kilogram of fresh leaf weight) production of biopharmaceuticals is based on proviral gene amplification in a non-food host. Antibodies, antigens, interferons, hormones, and enzymes could successfully be expressed with this system. The picture shows development of initial symptoms on a tobacco following the agrobacterium-mediated infection with viral vector components that contain a *GFP* gene (I); this development eventually leads to a systemic spread of the virus, literally converting the plant into a sack full of protein of interest within two weeks (II). The system allows to coexpress two proteins in the same cell, a feature that allows expression of complex proteins such as full-length monoclonal antibodies. Panel III and IV show the same microscope section with the same cells, expressing green fluorescent protein (III) and red fluorescent protein (IV) at the same time. The yield and total protein concentration achievable are illustrated by a Coomassie gel with proteins in the system: GFP (protein of interest), CP (coat protein from wild-type virus), RbcS and RbcL (small and large subunit of ribulose-1,5-bisphosphate carboxylase) [24].

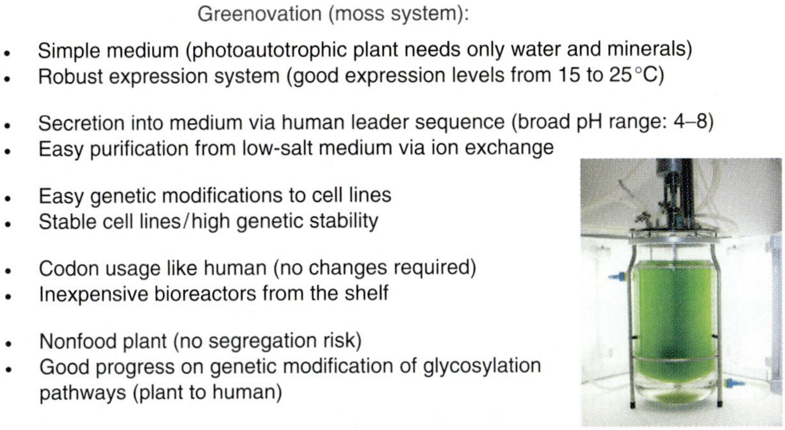
Greenovation (moss system):
• Simple medium (photoautotrophic plant needs only water and minerals)
• Robust expression system (good expression levels from 15 to 25 °C)
• Secretion into medium via human leader sequence (broad pH range: 4–8)
• Easy purification from low-salt medium via ion exchange
• Easy genetic modifications to cell lines
• Stable cell lines/high genetic stability
• Codon usage like human (no changes required)
• Inexpensive bioreactors from the shelf
• Nonfood plant (no segregation risk)
• Good progress on genetic modification of glycosylation pathways (plant to human)

Fig. 3.8 Greenovation use a fully contained moss bioreactor. This company has established an innovative production system for human proteins. The system produces pharmacologically active proteins in a bioreactor, utilizing a moss (*Physcomitrella patens*) cell culture system with unique properties [24].

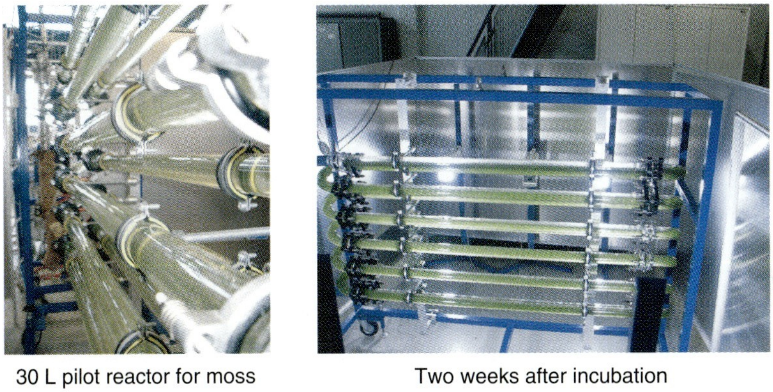
30 L pilot reactor for moss
Two weeks after incubation

Fig. 3.11 Scaling of photobioreactors up to several 1000 L. The moss bioreactor is based on the cultivation of *Physcomitrella patens* in a fermenter. The moss protonema is grown under photoautotrophic conditions in a medium that consists essentially of water and minerals. Light and carbon dioxide serve as the only energy and carbon sources. Cultivation in suspension allows scaling of the photobioreactors up to several 1000 L. Adaptation of existing technology for large-scale cultivation of algae is done in cooperation with the Technical University of Karlsruhe. Courtesy of greenovation Biotech GmbH (Freiburg, Germany) and Professor C. Posten, Technical University (Karlsruhe, Germany).

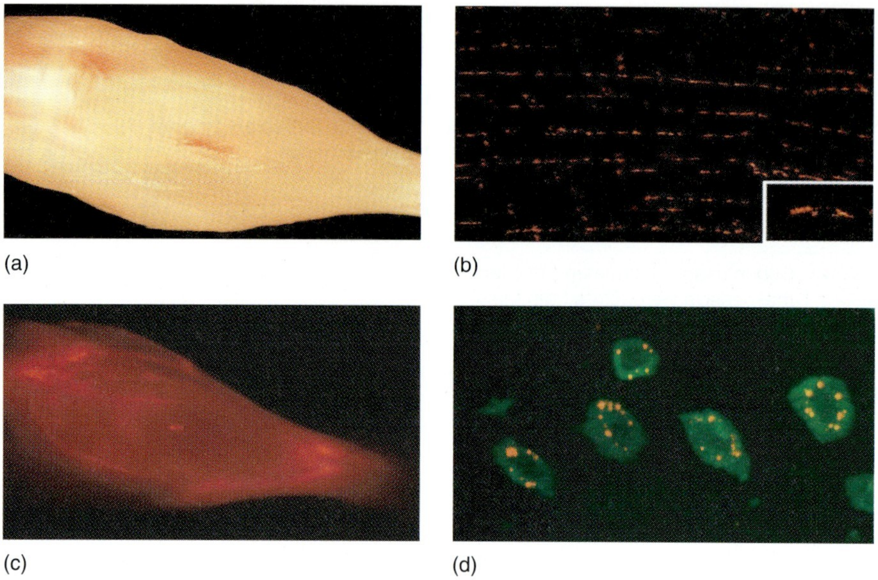

Fig. 5.3 Distribution of injected DNA vaccines. A rhodamine-conjugated DNA vaccine was injected into a tibialis anterior muscle of a mouse shown by light (panel A) and fluorescence (panel C) microscopy (~5× magnification). A longitudinal section of the muscle is shown in panel B (~250× magnification), demonstrating the presence of DNA in cells between the muscle fibers. Panel C shows the phagosomal location of the plasmid DNA (in red) within the cells isolated from the injected tissues (~2500× magnification).

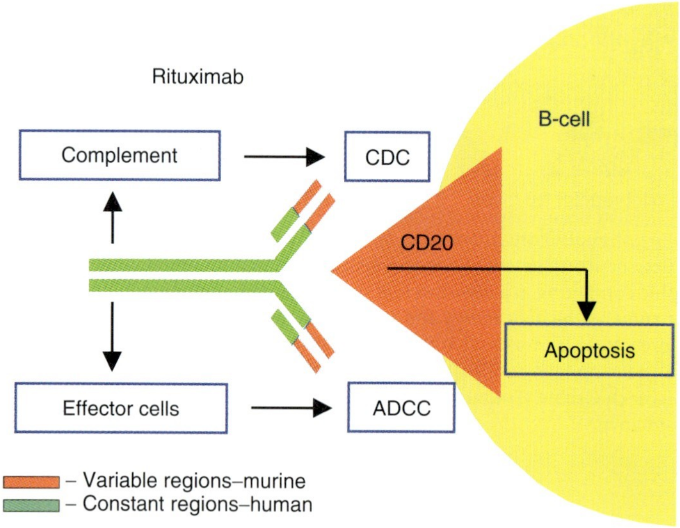

Fig. 12.1 Mechanism of action of rituximab. The chimeric (mouse/human) antibody, rituximab, binds to the CD20 antigen on B-cells and (a) activates complement to effect CDC, (b) attracts effector cells via Fc receptors to effect ADCC, and (c) transmits a signal into the cell to induce apoptosis.

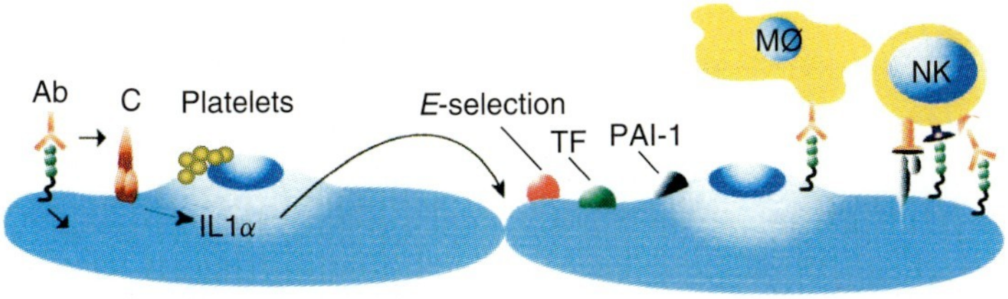

Fig. 15.3 Pathogenesis of acute vascular rejection. Activation of endothelium by xenoreactive antibodies (Ab), complement (C), platelets, and perhaps by inflammatory cells (natural killer (NK) cells and macrophages (M⊘) leads to the expression of new pathophysiologic properties. These new properties, such as the synthesis of tissue factor (TF) and plasminogen activator inhibitor type 1 (PAI-1), promote coagulation; the synthesis of *E*-selectin and cytokines such as IL1α promote inflammation. These changes in turn cause thrombosis, ischemia, and endothelial injury, the hallmarks of acute vascular rejection. (Adapted from *Nature* 1998: **392**(Suppl.) 11–17, with permission.)

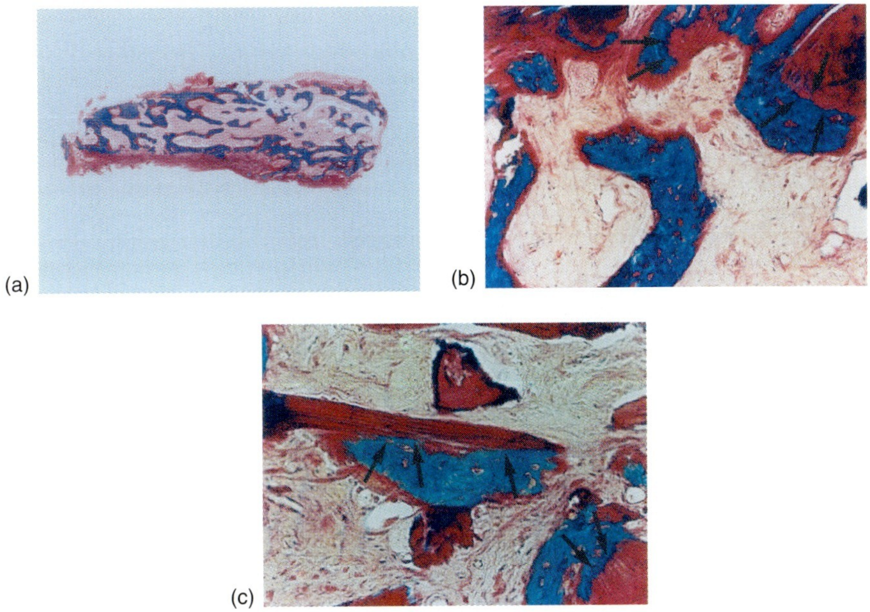

Fig. 16.2 Photomicrographs of tissue induction and morphogenesis in bioptic material 90 days after implantation of naturally derived BMPs/OPs purified from bovine bone matrix in human mandibular defects. (a) Trabeculae of newly formed mineralized bone covered by continuous osteoid seams within highly vascular stroma. (b) and (c) High-power views showing cellular mineralized bone surfaced by osteoid seams. Newly formed and mineralized bone directly opposing the implanted collagenous matrix carrier (arrows) confirms bone formation by induction. Undecalcified sections at 7 μm stained with Goldner's trichrome. Original magnification: (a) ×14; (b) ×40; and (c) ×50.

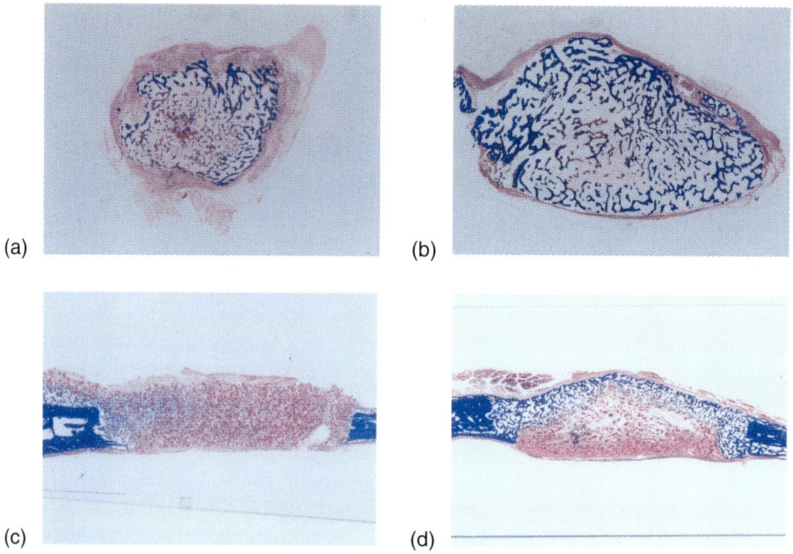

Fig. 16.4 Tissue morphogenesis and site–tissue-specific osteoinductivity of recombinant human-transforming growth factor-$\beta 2$ (hTGF-$\beta 2$) in the adult primate *Papio ursinus*. (a and b) Endochondral bone induction and tissue morphogenesis by hTGF-$\beta 2$ implanted in the *rectus abdominis* muscle and harvested (a) 30 and (b) 90 days after heterotopic implantation. Heterotopic bone induction by a single administration of (a) 5- and (b) 25-μg hTGF-$\beta 2$ delivered by 100 mg of guanidinium-inactivated collagenous matrix. (c and d) Calvarial specimens harvested from the same animals as shown in (a and b). (c) Lack of bone formation in a calvarial defect 30 days after implantation of 10-μg hTGF-$\beta 2$ delivered by collagenous bone matrix. (d) Osteogenesis, albeit limited, is found in a specimen treated with 100-μg hTGF-$\beta 2$ with bone formation only pericranially 90 days after implantation. Note the delicate trabeculae of newly formed bone facing scattered remnants of collagenous matrix particles, embedded in a loose and highly vascular connective tissue matrix. Original magnification: (a and b) ×4.5; (c and d) ×3. Undecalcified sections cut at 4 μm stained with Goldner's trichrome.

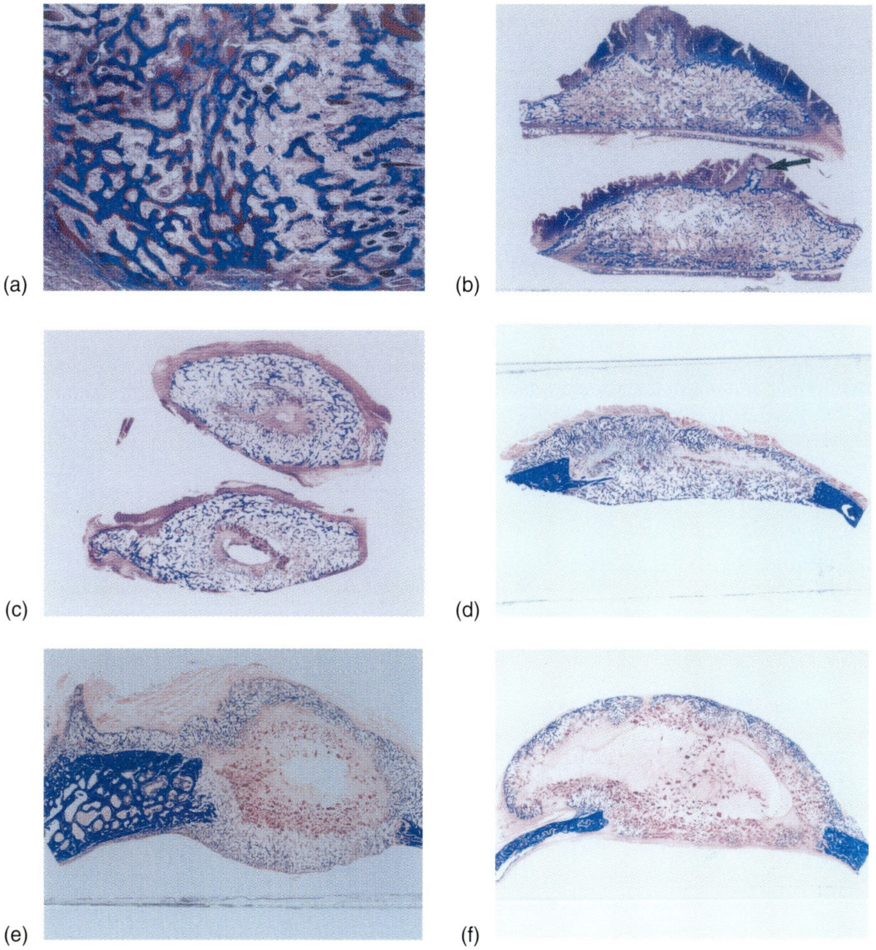

Fig. 16.6 Synergistic tissue morphogenesis and heterotopic bone induction by the combinatorial action of recombinant human osteogenic protein-1 (hOP-1) and transforming growth factor-β1 (hTGF-β1). (a) Rapid and extensive induction of mineralized bone in a specimen generated by 25-μg hOP-1 combined with 0.5-μg hTGF-β1 on day 15. Mineralized trabeculae of newly formed bone are covered by osteoid seams populated by contiguous osteoblasts. (b and c) Photomicrographs of massive ossicles that had formed between the muscle fibers and the posterior fascia of the *rectus abdominis* using binary applications of 25- and 125-μg hOP-1 interposed with 5-μg hTGF-β1 on day 30. Corticalization of the large heterotopic ossicles with displacement of the *rectus abdominis* muscle and extensive bone marrow formation permeating trabeculae of newly formed bone. Arrow in (b) points to a large area of chondrogenesis protruding within the rectus abdominis muscle. (d, e, and f) Low-power photomicrographs of calvarial defects treated by binary applications of 100-μg hOP-1 and 5 μg of naturally derived TGF-β1 purified from porcine platelets as described [55] and harvested on day 30. The calvarial specimens show extensive bone differentiation with pronounced vascular tissue invasion and displacement of the calvarial profile 30 days after implantation of the binary morphogen combinations. Original magnification: (a) ×30; (b, c) ×3.5; (d, e, and f) ×3. Undecalcified sections cut at 4 μm and stained with Goldner's trichrome.

Part I
Introduction to Concepts and Technologies in Pharmaceutical Biotechnology

1
A Primer on Pharmaceutical Biotechnology and Industrial Applications

Oliver Kayser and Raimer H. Müller
Freie Universität Berlin, Berlin, Germany

1.1
Introduction

Today we can mark historic milestones and achievements in the pharmaceutical industry (Table 1). The year 2003 represents the 50th anniversary of the discovery of the double helix structure of DNA and it also marks the 30th anniversary of the discovery of the technique for creating recombinant DNA by Stanley Cohen and Herbert Boyer. This technique still influences modern medicine and the development of new recombinant and therapeutic proteins today. Also, 50 years after Watson and Crick's discovery, the completion of sequencing of the human genome is another milestone in biotechnology leading to new genomic-based drugs [1].

In fact, pharmaceutical biotechnology is one of the key industries today. Recombinant DNA technologies have entered drug discovery and all fields in the development and manufacture of therapeutic proteins and nucleotides. Biotechnology has a major impact on pharmaceutical industry because recent advances in recombinant protein chemistry, vaccine production, and diagnostics have and will revolutionize the treatment paradigms for many serious and unmet diseases. Currently, approximately 150 approved therapeutic proteins and vaccines are available. Recently, the first oligonucleotide for the treatment of cytomegalovirus (CMV) infection of the eyes was approved by the Food and Drug Adiministration (FDA). The drug Fomivirsen (Vitravene®) is an antisense oligonucleotide and represents a new biotechnological group of compounds with new, promising therapeutic purposes [2].

1.2
Actual Status of Biotechnology and its Applications in Pharmaceutical Industry

As mentioned earlier, more than 150 approved biotech drugs or vaccines are on the market and 70% were approved in the last six years (Fig. 1). A recent survey by the Pharmaceutical Research and Manufacturers of America (PhRMA) found 369 drugs in the pipeline meeting the criteria as biotechnological drugs and medicines. These drugs target 200 potential diseases [3] and provide new therapies for autoimmune diseases, asthma, Alzheimer, multiple sclerosis, and cancer,

Tab. 1 Short time line in pharmaceutical biotechnology

Year	*Historic event*
1797	Jenner inoculates child with viral vaccine to protect him from smallpox
1857	Pasteur proposes that microbes cause fermentation
1928	Penicillin is discovered by Fleming
1944	Avery proves DNA as carrier of genetic information
	Waksman isolates streptomycin as antibiotic for tuberculosis
1953	Structure elucidation of double helix of DNA
1967	First protein sequencer is perfected
1970	Discovery of restriction enzymes
1973	Cohen and Boyer produce first recombinant DNA in bacteria with restriction enzymes and ligases
1977	First expression of human protein in bacteria
1980	US Patent for gene cloning to Cohen and Boyer
1981	First transgenic animal
1982	Humulin® as first recombinant biotech drug approved by FDA
1983	Invention of Polymerase Chain Reaction (PCR)
1986	First recombinant vaccine for Hepatitis B (Recombivax HB®)
1988	First US Patent for genetically modified mouse (Onkomouse)
1990	Launching of the Human Genome Project
	First somatic gene therapy to cure ADA-SCID
	First transgenic cow produces human proteins in milk
1994	Approval of DNAse for cystic fibrosis
1997	First animal cloned from adult cell (Dolly)
2000	Rough draft of the human genome is announced
2002	Draft version of the complete map of the human genome is published
	First oligonucleotide drug is approved by FDA

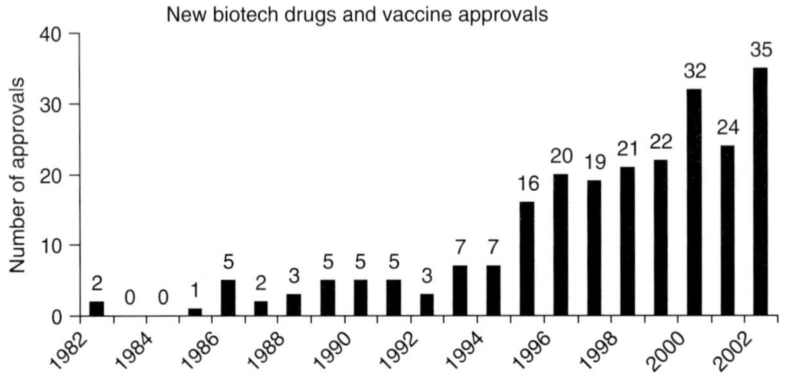

Fig. 1 New biotech drugs and vaccine approval, 1982–2002, *Source:* modified according to L. M. Baron, A. Massey, (Eds.), *Bio – Editors' and Reporters' Guide 2003–2004, 2003.*

including immunization and different infectious diseases (AIDS, Malaria) [3, 4]. Biotechnology-produced pharmaceuticals currently account for 5% of the worldwide pharmaceutical market and are expected to reach approximately 15% by the year 2050. At the same time, the explosive growth of genetic diagnostic techniques will allow personalized genetic profiling of each individual in one hour and for less than US$100.

Not only drugs but also new medical diagnostic tests will be produced and distributed by pharmaceutical biotech industry. Hundreds of tests will be available to increase the safety of blood products. Also, costs for clinical analysis will be reduced. One example is the testing of Low Density Lipoproteins (LDL), cholesterol, and other parameters in one test design. In comparison to conventional tests, cholesterol, total triglycerides, and LDL were determined separately at high costs. In the future, biotechnology-derived tests will be more accurate and quicker than previous tests and will allow earlier diagnosis of the disease. Proteomics may increase sensitivity and may discover today unknown molecular markers that indicate incipient diseases before symptoms appear, helping to prevent diseases and conduct therapies much earlier [5–7].

Xenotransplantation from transgenic animals is a future field in pharmaceutical industry. In general, organ transplantation is an effective and cost-efficient treatment for severe and life-threatening diseases of organs, mostly heart, liver, and kidney. In Europe, there are 35 000, and in the United States, there are 60 000 people on organ-recipient lists. Organ transplantation costs vary from €60 to 120 000 and require a lifelong drug therapy with immunosuppressive drugs to avoid transplant rejection. Genetically modified organs and cells from other organisms like pigs – called as xenotransplantation – are promising sources of donor organs that can be used to overcome the lack of a sufficient number of human organs. But, the spread of infectious pathogens by transplantation of nonhuman organs and the induction of oncogenes is a potential risk and needs close attention [8, 9].

Tissue engineering, in relation to xenotransplantation, is another attractive field in pharmaceutical biotechnology. Tissue engineering combines advances in cell biology and biomaterial science. Tissues consist of scaffolding material (e.g. collagen, biodegradable polymers), which eventually degrades after forming organs or cell implants. Skin tissues and cartilages were the first tissues successfully engineered and tested in vivo; recently, biohybrid systems to maintain patients' liver or kidney function were also successfully tested [10].

Stem cells are considered today as a new avenue in therapy to cure most deadly and debilitating diseases such as Parkinson, Alzheimer, leukemia, and genetic disorders like adenosine deaminase (ADA) deficiency and cystic fibrosis (CF). The potential of embryonic and adult stem cells are intensively discussed, but no major breakthrough can be expected in the next 10 years to turn these cells and techniques into industrial applications. It should also be clear that therapeutic cloning, which is related to stem cell research, will bring ethical questions [11]. Discussions of ethical and social implications are important today to convince the public of potential benefits and to explain the future risks of applied techniques. Significant impediments of diagnostics and therapeutics exist, and deep concerns must be respected before any genetic therapy like somatic gene therapy, stem cell, or cloning will ever be accepted.

1.3 What is the Impact of Biotechnology and Genomics on the Drug Development Process?

The most frequent trends for the pharmaceutical industry in biotechnology are surely new technologies and innovations, especially genomics, and also influences of government regulations, health care legislation affecting own product pricing, changes in demographics, and an aging population [4]. With special emphasis, more topics with minor influence relate to patent protection, e-business, multinational scope of industry, requirements for new drugs, and changes in information technology to address some of them [3]. In the first decade of this century, the biotech industry is likely to show even more explosive growth as progress in computer modeling, automated lab techniques, and knowledge of human genes and proteins continues. As a result, more life-saving therapies will reach the people who most need them. Today, pharma industry seems to be in good shape to work on these challenges, as indicated by some impressive statistics that emphasize the industry's growth [12]:

- During the 1980s, the biotech industry turned out 18 new drugs and vaccines. By comparison, 33 biotech medicines were approved in both 1998 and 1999, and 25 more were approved in the first half of 2000.
- Most of the 1998–1999 approvals were for new products, though a few were for expanding the application of drugs or vaccines to more diseases.
- The number of patents granted to biotech companies has tripled from nearly 3000 per year in the early 1990s to more than 9000 in 1998.
- After a decade of slow, steady growth, biotechnology patent awards began a steep ascent in 1995, when nearly 4000 patents were granted. Since then, the number of patents has skyrocketed at a rate of 25% or more each year.
- Pharmaceutical companies, which traditionally have focused on chemical approaches to treating disease, have become increasingly supportive of biotech R&D – in their own labs, in partnerships with biotech firms, and through acquisitions of biotech firms. Alliances in the biotech industry doubled to nearly 250 between 1998 and 2000.
- Between 1998 and 1999, industry-wide sales and revenues increased by 13% to $16.1 billion and $22.3 billion respectively.

For the future, the biotechnology and genomic way is technology-driven and formed by the integration of high-throughput technologies, genomics, and bioinformatics. Even the genetics wave is data-driven and is an applied new life science field to identify genes that make individuals as their carriers susceptible to particular diseases and allows personalized medicines based on pharmacogenetic facts. So, what is the impact of pharmaceutical biotechnology and genomics on the economics of R&D?

1.3.1 Reducing Costs in R&D

Before biotechnology had been intensively introduced to industrial research, developing costs of each drug had cost companies on average US$880 million and had taken 15 years from start to market authorization. About 75% of these costs were spent on failures. Using genomic technologies, there is a realistic chance of reducing companies' costs to US$500 million, largely as a result of efficacy gains. Significant savings not only of money but also of time by 15% are possible [12, 13].

1.3.2
Increase in Productivity

From trial-and-error approaches and complex biochemical in vitro assays, biotechnology allows industrialized target detection and validation. By the use of micro array technologies and bioinformatics, thousands of genes in diseased and healthy tissues will be analyzed by a single DNA chip. By the use of bioinformatics, results from different assays can be analyzed and linked to an integrated follow-up of information in databases. In total, the potential savings per drug by intelligent information retrieval systems and genetic analytics are estimated at about US$140 million per drug and less than one year of time to market. The Boston Consulting Group (BCG) calculated a sixfold increase in productivity at the same level of investment [12].

1.3.3
Accelerating the Drug Development Process

There is not only an effect on the preclinical development of a drug by biotechnology and genomics but pharmaceutical biotechnology will also help predict drug properties and pharmacokinetic parameters (ADMA/tox) to accelerate the industrial drug development process. Companies will be in the position to pull certain preclinical activities into the chemistry and drug validation part of the value chain. Potential savings are in the order of US$20 million and 0.3 years per drug [12].

1.3.4
Maintaining High Standards in Quality Assurance

Biotechnological drugs have the same high standard in quality and safety as conventional drugs. Of high interest is the question of costs of quality control for recombinant drugs. Boston Consulting Group expects an increase of US$200 million and more than one year per drug [12]. The main reason for this is explained by the extra time needed for unknown chemical and physical properties of recombinant proteins and oligonucleotides. Another time- and cost-consuming aspect is the importance of developing new drug-specific appropriate test assays for drug validation, standardization, activity determination (e.g. biological units), toxicity, and bioanalytical methods.

1.4
Future Outlook

Integrating biotechnology and genomics in the whole drug development process gives companies the opportunity to save up to US$300 million per drug – about one-third of the costs today – and the prospect of bringing the drug two years earlier on the market [3]. Each day lost before market entry will lead to a loss of US$1.5 million per day, indicating the value of recombinant drugs and the need for making manufacturing processes operational and effective.

Any predictions for the near future are challenging. Future reports estimate a significant increase of recombinant drugs replacing up to 30% of commercially used low-molecular drugs up to 2015. For the production of recombinant biotech drugs, bioprocessing in all reactor sizes will be routinely used [14]. From 2010, genetically modified plants and animals – transgenic organisms – will also be routinely used to produce recombinant drugs (Gene Pharming). Somatic gene therapy and the introduction of nanorobotic devices may be expected in the time period between 2010 and 2018 to end up with individual genome profiling for €100 in 2050. Personalized medicine and diagnostics on a biochip may also find industrial interest in the next 10 years. Interestingly, creation of artificial life or complex biochemical networks is expected to be unrealistic in the next 25 years.

References

1. J. A. Miller, V. Nagarajan, *Trend Biotechnol.* **2000**, *18*, 109–191.
2. J. Kurrek, *Eur. J. Biochem.* **2003**, *270*, 1628–1644.
3. A. M. Baron, A. Massey *Editors' and Reporters' Guide to Biotechnology*, URL: http://www.bio.org (24.07.2003), 2001.
4. J. M. Reichert, *New biopharmaceuticals in the USA: trends in development and marketing approvals* 1995–1999, *Trends in Biotechnology* **2000**, *18*, 364–369.
5. K. K. Jain, *Nanodiagnostics: application of nanotechnology in molecular diagnostics Expert Rev Mol Diagn* **2003**, *3*, 153–161.
6. K. K. Jain, Proteomics: delivering new routes to drug discovery – Part 1. *Drug Discov. Today* **2001 Aug 1**; *6*(15): 772–774.
7. K. K. Jain, Proteomics: delivering new routes to drug discovery – Part 2. *Drug Discov. Today* **2001 Aug 15**; *6*(16): 829–832.
8. D. K. Langat, J. M. Mwenda, *Acta Tropica* **2000**, *76*, 147–158.
9. F. H. Bach, *Annu. Rev. Med.* **1998**, *49*, 301–310.
10. C. K. Colton, *Cell Transplantation* **1995**, *4*, 415–436.
11. R. Eisenberg, *Nat. Genet.* **2000**, *1*, 70–74.
12. P. Tollman, P. Guy, J. Altshuler, N. Vrettos, C. Wheeler, *A Revolution in R&D – The Impact of Genomics*, The Boston Consulting Group, Boston, 2001.
13. Anonymous, *Convergence – The Biotechnology Industry Report, Ernst & Young, Score Retrieval Files*, URL: www.ey.com/industry/health (24.07.2003), 2000.
14. Anonymous, *Endurance – The European Biotechnology Report 2003*, URL: www.ey.com/industry/health (24.07.2003), 2000.

2
Procaryotic and Eucaryotic Cells in Biotech Production

Stefan Pelzer
Combinature Biopharm AG, Berlin, Germany

Dirk Hoffmeister
The University of Wisconsin, Madison, WI, USA

Irmgard Merfort and Andreas Bechthold
Albert-Ludwigs-Universität Freiburg, Freiburg, Germany

2.1
Introduction

The production of compounds used in the food and pharmaceutical industries by biotech processes is both an old and a very young business. Over the past 70 years, fermentation of microorganisms or the use of yeast and plants in the production of important pharmaceuticals has been well established. The promises of genomics in drug discovery and drug production, which were eagerly embraced in the mid-1990s, have now been fulfilled in many areas. A systematic integration of technologies results in a superior output of data and information, and thereby enhances our understanding of biological function – drug discovery and development is hence facing a new age.

Bacterial strains, especially Actinomycetes have been used in biotech production and drug discovery for years. Genetic methods now open the field of combinatorial biosynthesis that has improved impressingly in the past couple of years. Also, the productivity of yeast and other fungi in a variety of different processes has improved significantly since genetic methods have been introduced. In addition, a number of recent works considerably widens the potential of plant biotechnology. This review covers examples describing the use of procaryotic cells and plant cells in biotech production. The use of other eucaryotic cells, especially of animal origin, is reviewed in other chapters of this book.

2.2
Actinomycetes in Biotech Production

Soil bacteria of the order Actinomycetes are the most important producers of pharmaceutically relevant bioactive metabolites including antibiotics, antitumor agents, immunosuppressants, antiparasitic agents, herbicides, and enzyme-inhibiting agents. The success story of these bacteria began about 60 years ago with the groundbreaking work of Waksman, who discovered and described streptomycin as the first antibiotic synthesized by an Actinomycete [1]. Ever since, systematic large-scale screens performed by the pharmaceutical industry have revealed numerous therapeutically relevant drugs. More than two-thirds of all naturally derived antibiotics currently used are produced by Actinomycetes strains, underlining their importance to medicine [2].

2.2.1
Actinomycetes: Producer of Commercially Important Drugs

Natural products ("secondary metabolites") have been the largest contributors to drugs in the history of medicine. Before antibiotics were introduced in the 1940s and 1950s (see above), patients with bacteraemia faced low survival chances [3], and the mortality from tuberculosis was 50% [4]. It has been stated that the doubling of our life span in the twentieth century is mainly due to the use of plant and microbial secondary metabolites [5]. Of the 520 new drugs approved between 1983 and 1994, 39% were natural products or those derived from natural products and 60 to 80% of antibacterial and anticancer drugs were derived from natural products [6]. Almost half of the best-selling pharmaceuticals are natural or related to them [7,8]. In 2001, over 100 natural product–derived compounds were in clinical development [9]. Natural products and their derivatives account for annual revenues of about US$30 billion in the antiinfectives market, US$20 billion in the anticancer market, and US$14 billion in the lipid-lowering market [10]. Actinomycetes and, particularly, Streptomycetes (Fig. 1) are the largest antibiotic-producing genus in the microbial world discovered so far. Of the 12 000 or so antibiotics known in 1995, 55% were produced by Streptomycetes and an additional 11% by other Actinomycetes [11]. A compilation of numerous bioactive and commercially important metabolites, which are all synthesized by Actinomycetes strains, is shown in Table 1. This list includes not only very important drugs such as the macrolide erythromycin A synthesized by *Saccharopolyspora (Sac.) erythraea* (in 2000, the annual sales of semisynthetic derivatives reached US$2.6 billion [12]), the glycopeptide vancomycin synthesized by *Amycolatopsis (A.) orientalis* (in 2000, the annual sales of glycopeptides reached US$424 million, [12]) and tetracycline synthesized by *Streptomyces (S.) aureofaciens* (in 2000, the annual sales reached US$217 million [12]) but also anticancer agents like doxorubicin synthesized by *S. peucetius* (Fig. 2). Many compounds produced by Actinomycetes belong to the large family of polyketides. Polyketides are structurally diverse (Fig. 2) and exhibit a wide scope of bioactivities. More than 500 aromatic polyketides have been characterized from Actinomycetes [13]. Polyketides are particularly important for drug discovery, since statistics show that 1 out of 100 polyketides will make its way to commercialization. With an average of as low as 1 out of 5000 compounds, other substances are far less likely to hit the market [14]. Sales of drugs based on polyketides exceed US$15 billion a year [14]. In general, for industrial production, overproducing strains have to be developed. Today, modern processes allow the production of

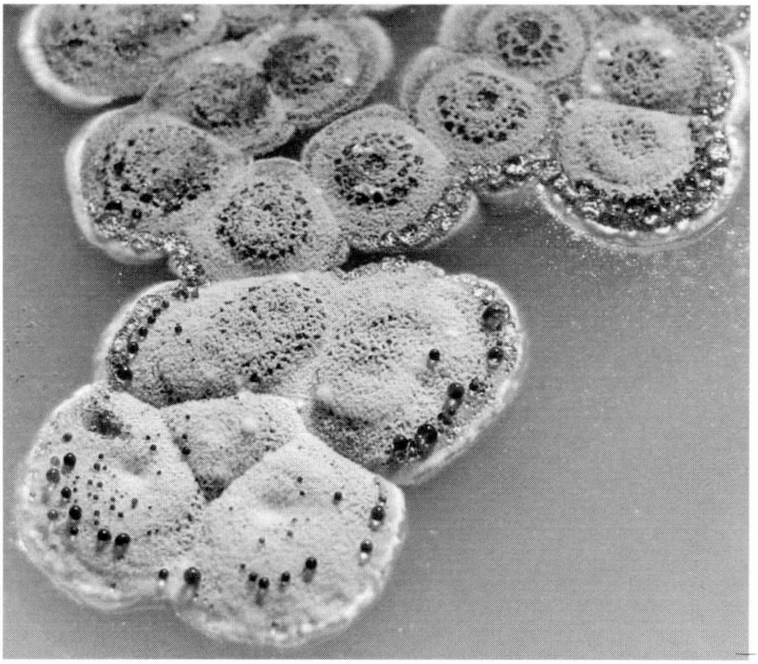

Fig. 1 Photography of a sporulated *Streptomyces* strain growing on solid medium. The blue drops indicate the production of an antibiotic (aromatic polyketide). (See Color Plate p. xv).

compounds at concentrations even higher than 10 g L^{-1} [15–17].

2.2.2
Actinomycetes Genetics: The Basis for Understanding Antibiotic Biosynthesis

Streptomyces coelicolor A3(2) is the genetically best characterized strain among the filamentous Actinomycetes [19]. Actinomycetes genetics has been the subject of research since 1958 when Prof. Dr. Sir D. Hopwood published the first linkage map of *S. coelicolor*, performing the first genetic recombination experiments considering six marker genes [20]. Afterwards, Actinomycetes genetics developed continuously, by the identification of mutants interrupted in the biosynthesis of actinorhodin [21]. After identification and isolation of easily selectable antibiotic resistance genes [22], the first gene cloning in *Streptomyces* was described in 1980 [23]. In 1984, Malpartida and Hopwood demonstrated for the first time that antibiotic biosynthesis genes are usually organized as a gene cluster of structural, regulatory, export, and self-resistant genes [24]. Hence, once a single gene within a cluster has been located, the others may be identified quickly by chromosomal walking. In the course of the last two decades, many molecular tools including vector systems (phage and plasmid-based) have been developed along with DNA transfer and gene inactivation techniques, which were all necessary for targeted manipulation of Actinomycetes [2, 25]. The excellent manual published by the John Innes Institute summarizes all the necessary information

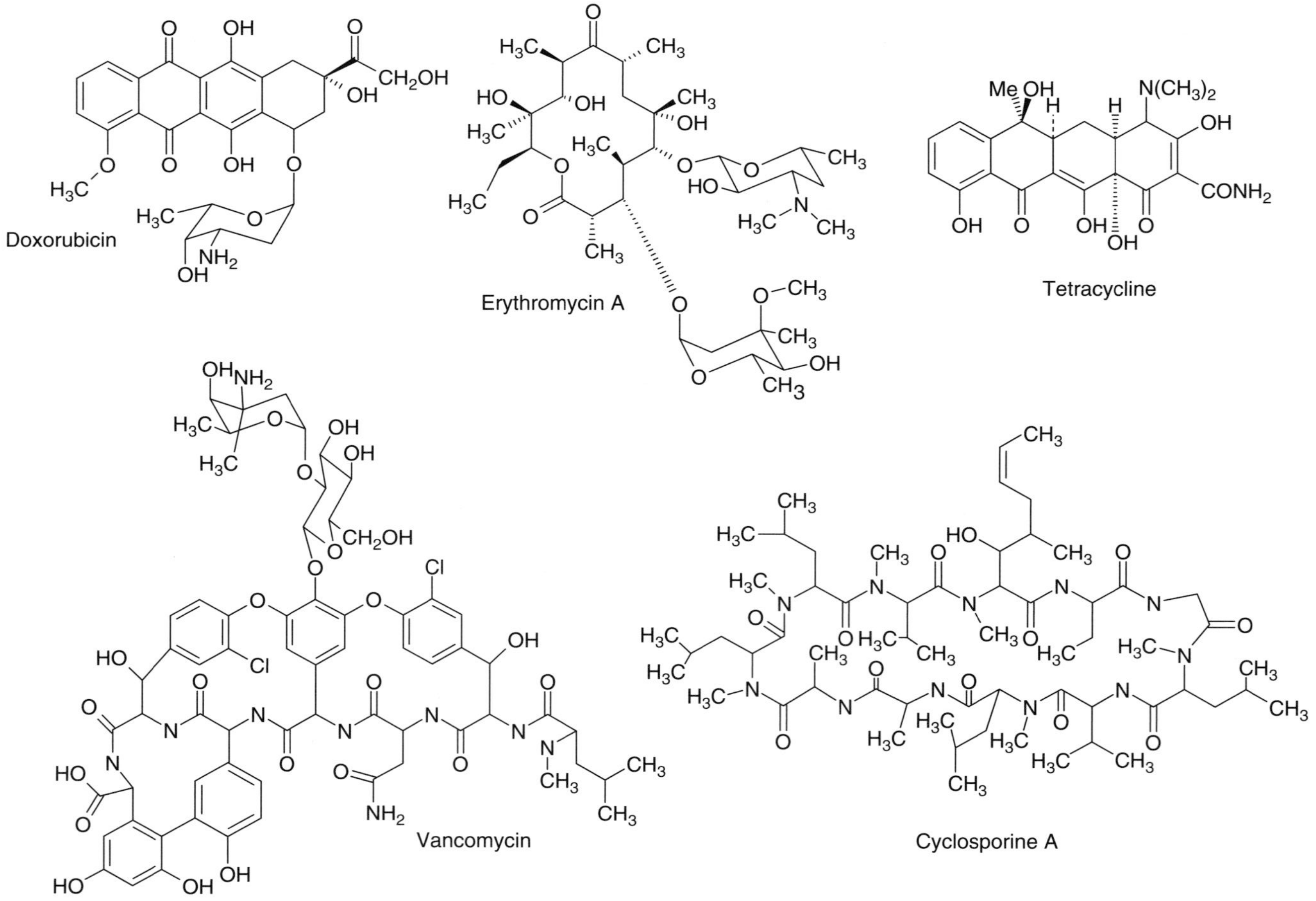

Fig. 2 Structures of natural compounds used as important drugs in pharmacy and medicine.

Tab. 1 Origin, target, and application of commercially important secondary metabolites originating from Actinomycetes [7, 18]

Antibiotic	*Producer*	*Molecular target*	*Application*
A40 926	*Nonomurea* sp.	Cell wall synthesis	Antibacterial
Amphotericin	*S. nodosus*	Membrane (pore-formation)	Antifungal
Ascomycin (FK520)	*S. hygroscopicus*	FKBP12	Immunosuppressive
Avermectin	*S. avermitilis*	Membrane (ion-channel)	Antiparasitic
Avilamycin	*S. viridochromogenes*	Ribosome	Antibacterial
Avoparcin	*A. coloradensis*	Cell wall synthesis	Antibacterial growth promotant
Bleomycin	*S. verticillus*	DNA binding	Antitumor
Bialaphos	*S. hygroscopicus*	Glutamine synthetase	Herbicide
Candicidin	*S. griseus*	Membrane (pore-formation)	Antifungal
Clavulanic acid	*S. clavuligerus*	Beta-lactamases	Combined with β-lactam antibacterial
Chloramphenicol	*S. venezuelae*	Ribosome	Antibacterial
Chlortetracycline	*S. aureofaciens*	Ribosome	Antibacterial
Cyclohexamide	*S. griseus*	Ribosome	Antibiotic
Dactinomycin	*S. parvulus*	DNA intercalation	Antitumor
Daptomycin	*S. roseosporus*	Cell wall synthesis	Antibacterial
Daunorubicin	*S. peucetius*	DNA intercalation	Antitumor
Doxorubicin	*S. peucetius var caesius*	DNA intercalation	Antitumor
Erythromycin A	*Saccharopolyspora erythraea*	Ribosome	Antibacterial
Gentamicin	*Micromonospora purpurea*	Ribosome	Antibacterial
Geldanamycin	*S. hygroscopicus*	Hsp90	Antitumor
Kanamycin	*S. kanamyceticus*	Ribosome	Antibacterial
Lincomycin	*S. lincolnensis*	Ribosome	Antibacterial
Milbemycin	*S. hygroscopicus*	Membrane (ion-channel)	Antiparasitic
Mithramycin	*S. argillaceus*	DNA alkylation	Antitumor
Mitomycin C	*S. lavendulae*	DNA alkylation	Antitumor
Moenomycin	*S. ghanaensis*	Cell wall synthesis	Antibacterial, growth promotant
Monensin	*S. cinnamonensis*	Membrane (ionophore)	Anticoccidial, growth promotant
Natamycin	*S. nataensis*	Membrane (pore-formation)	Antifungal
Neomycin	*S. fradiae*	Ribosome	Antibacterial
Nikkomycin	*S. tendae*	Chitin synthase	Antifungal
Novobiocin	*S. niveus*	DNA gyrase	Antibacterial
Nystatin	*S. noursei*	Membrane (pore-formation)	Antifungal
Oxytetracycline	*S. rimosus*	Ribosome	Antibacterial
Pristinamycin	*S. pristinaespiralis*	Ribosome	Antibacterial

(continued overleaf)

Tab. 1 *(continued)*

Antibiotic	***Producer***	***Molecular target***	***Application***
Ramoplanin	*Actinoplanes spec.*	Cell wall synthesis	Antibacterial
Rapamycin	*S. hygroscopicus*	FKBP	Immunosuppressive
Rifamycin	*A. mediterranei*	RNA polymerase	Antibacterial
Salinomycin	*S. albus*	Membrane (ionophore)	Anticoccidial, growth promotant
Spinosyn	*Sac. spinosa*	unknown	Insecticidal
Spiramycin	*S. ambofaciens*	Ribosome	Antibacterial
Staurosporin	*S. staurosporeus*	Protein kinase C	Antibacterial
Streptomycin	*S. griseus*	Ribosome	Antibacterial
Tacrolimus (FK506)	*Streptomyces spec.*	FKBP	Immunosuppressive
Teicoplanin	*A. teicomyceticus*	Cell wall synthesis	Antibacterial
Tetracycline	*S. aureofaciens*	Ribosome	Antibacterial
Thienamycin	*S. cattleya*	Cell wall synthesis	Antibacterial
Tylosin	*S. fradiae*	Ribosome	Growth promotant
Vancomycin	*A. orientalis*	Cell wall synthesis	Antibacterial
Virginiamycin	*S. virginiae*	Ribosome	Growth promotant

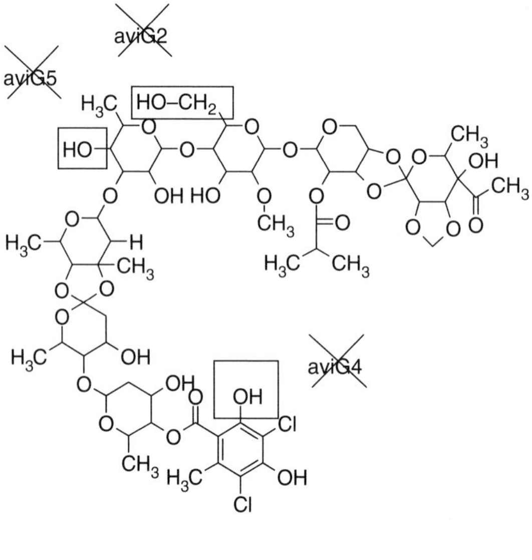

Fig. 3 Production of gavibamycin H1 by a mutant with deletions in three methyltransferase genes.

related to handling and molecular genetic tools essential for working with Actinomycetes [18]. Several hundred biosynthetic gene clusters have so far been identified and genes encoding about 80 pathways for secondary metabolites have been cloned, at least partially sequenced and made available to the public [26–28]. The average size of an antibiotic biosynthetic gene cluster ranges from about 20 kb for a simple aromatic polyketide like actinorhodin to 120 kb for complex polyketides like the antifungal antibiotic nystatin [29].

A highlight in Actinomycetes genetics was the completion of the first genome sequence of the model Actinomycete *Streptomyces coelicolor* A3(2) in July 2001 [30]. Recently, the genome sequence of a second *Streptomyces* strain, the avermectin producer *S. avermitilis*, has been published [31,32], opening up new perspectives for comparative genomics with Actinomycetes. A characteristic feature of Actinomycete chromosomes is their linear structure [33, 34]. The genome of *S. coelicolor* comprises 8 667 507 bp (G/C content of 72.1%), whereas the *S. avermitilis* genome contains 9 025 608 bp (G/C content of 70.7%). Both genomes are densely packed and harbor 7574 ORFs in *S. avermitlis* and 7825 ORFs in *S. coelicolor*, respectively [30]. Comparative analysis of the *S. coelicolor* and *S. avermitilis* chromosomes revealed that both the genomes had an unusual biphasic structure with a core region of 5 Mb and 6.5 Mb, respectively [30, 31]. The most interesting feature in both the completed *Streptomyces* genomes that will impact biotechnology is the abundance of secondary metabolite gene clusters. Before the genome of *S. coelicolor* was sequenced, three antibiotics and a spore pigment were known to be synthesized from this strain. The genome sequence revealed that 23 gene clusters (about 5% of the total genome) are directly dedicated to secondary metabolism including clusters for further putative antibiotics, pigments, complex lipids, signaling molecules and iron-scavenging siderophores [35]. In *S. avermitilis*, 30 gene clusters related to secondary metabolites were identified, corresponding to 6.6% of the genome. From these clusters, 5 out of 30 are putatively involved in pigments and siderophores, 5 in terpenes, 8 in nonribosomal peptides and 12 in polyketide biosynthesis [32]. With avermectin, oligomycin, and a polyene antibiotic, only three complex polyketide clusters have been characterized from this strain before. The completed genome-sequence data can also be used to study the regulatory network of primary metabolism pathways and the cross-talk between primary and secondary metabolism (i.e. the carbon flux). The knowledge gained by these analyses will be useful for the construction of improved strains produced in a rational approach by deleting undesired pathways or adding advantageous pathways, generating precursors and essential cofactors. Moreover, targeted modifications will improve cell growth and fermentation properties (metabolic engineering) [16,17]. Successful metabolic engineering of a strain producing doramectin, a commercial antiparasitic avermectin analog, is an excellent example for the importance of this technology [36].

2.2.3
Urgent Needs for the Development of New Antimicrobial Drugs

Stimulated by the discovery of numerous novel antibacterial agents, which reached a peak in the 1970s [37], US Surgeon General William Stewart declared in 1969 in the US congress that it was "time to close the book on infection diseases" [38]. Today, unfortunately, we know that antibiotics have not won the fight against infectious microorganisms and therefore there is a permanent need for new antibiotics.

One main reason for this development is the problem of emerging resistant forms of pathogens. As an example, according to the WHO more than 95% of *S. aureus* strains worldwide are resistant to penicillin G, and up to 60% are resistant to its derivative methicillin [39]. Several reasons (e.g. use of antibiotics as growth promoters, changes in the spectrum of pathogens) are responsible for this development [40–43].

The past decades witnessed a major decrease in the number of newly discovered compounds. In an almost 40-year period (1962–2000), no new class of antibiotic was introduced to the market (nalidixic acid in 1962, the oxazolidinone antibiotic linezolid in 2000) [44, 45]. The major reason for the decrease in the number of newly discovered compounds might be a decline in screening efforts [37]. Ironically, some of the leading pharmaceutical companies are currently cutting back their antiinfective programs, especially for natural products [46]. They rather focus their activities on the semisynthetic modification of existing antibiotics to produce second- and third-generation antibiotics with improved properties.

Nevertheless, there is no need to resign. According to biomathematical modeling, only 3% of all antibacterial agents synthesized in *Streptomyces* have been reported so far [37]. Additionally, less than 10% of the world's biodiversity has been tested for biological activity, and many more useful natural lead compounds are yet to be discovered [47].

2.2.4 Strategies for the Identification and Development of New Antimicrobial Drugs

2.2.4.1 Approaches to Explore Nature's Chemical Diversity

Vicuron Pharmaceuticals Inc., formerly Biosearch Italia, a company screening for new antibiotics, focuses its activities on a proprietary strain collection of 50 000 microorganisms, including unusual filamentous Actinomycetes and filamentous fungi or strains that are difficult to isolate. The rationale behind this campaign is that these organisms have not been intensively screened in the past and that they may be producers of novel compound classes [48].

Another strategy to reveal the chemical diversity of a single strain is the One strain – many Compounds (OSMAC) approach described by Bode et al. [49]. By systematic alteration of cultivation parameters, the number of secondary metabolites increased tremendously in a single strain. When this method was applied, up to 20 different metabolites with, in some cases, high production titers were detected. Since recent estimates suggest that only 0.1 to 1% of the microbial flora in the environment can be kept in culture [50], the "metagenome" of the unculturable microorganisms should also have a potential to generate novel secondary metabolites. Indeed, several reports demonstrated that it is possible to construct DNA libraries from "soil-DNA" and to use them for the production of novel metabolites in a heterologous *Streptomyces* host [51, 52].

2.2.4.2 Exploiting the Enormous Genotypic Potential of Actinomycetes by "Genome Mining"

The completion of the sequence of the two *Streptomyces* genomes demonstrated that between 5 and 6.6% of the whole genome are directly involved in the biosynthesis of predominantly unknown secondary metabolites (see above). Prior to genome sequencing, a number of reports were published in which cryptic or silent secondary metabolite pathways were identified during the search for gene clusters for known metabolites. Hence, the occurrence of multiple "orphan" gene clusters has been reported for various compound classes like nonribosomal peptides [53, 54], PKSI [55, 56] and PKSII [57, 58]. Combinature Biopharm AG is a Berlin-based company using modern high throughput genomics for the systematic genetic screening of several Actinomycetes genomes to identify known and "orphan" clusters [27]. Recently, Zazopoulos et al. [59] described how a genomics-guided approach can be rewarding for the discovery and expression of cryptic metabolic pathways (genome mining). The genetic information of these biosynthesis clusters is used for the targeted generation and modification of novel compounds in an approach termed "combinatorial biosynthesis."

2.2.4.3 Generation of Novel Antibiotics by Targeted Manipulation of the Biosynthesis (Combinatorial Biosynthesis)

Researchers have started using biosynthetic genes to alter the structure of natural compounds by genetic engineering or to combine genes from different biosynthetic pathways. This new technology named "combinatorial biosynthesis" results in the formation of novel natural products.

New Drugs by Targeted Gene Disruption

Inactivation of specific selected genes is a very common methodology for the generation of novel structural variations of known natural products. Erythromycin is a macrolide antibiotic that is clinically useful in the treatment of infections by Gram-positive bacteria. A hydroxyl group at C6 of the erythronolide macrolactone is responsible for acidic inactivation in the stomach by conversion into anhydroerythromycin. Erythromycin derivatives lacking this hydroxyl group are therefore interesting from the therapeutic and pharmacological point of view. The gene *eryF* that encodes a cytochrome P450 monooxygenase responsible for the introduction of this hydroxyl group into the macrolactone was inactivated and the mutant produced 6-deoxyerythromycin A. This is a much more acid-stable antibiotic and as efficient as erythromycin because of its higher stability [60].

The orthosomycins are a prominent class of antibiotics produced by various Actinomycetes. Members of this class are active against a broad range of Gram-positive pathogenic bacteria. Prominent examples of orthosomycins are the avilamycins and the everninomicins produced by *S. viridochromogenes* Tü57 and *Micromonospora carbonacea*, respectively. Avilamycins and everninomicins are poorly soluble in water, which poses a major obstacle for their use as therapeutics. The avilamycin biosynthetic gene cluster has been cloned and sequenced [61]. Several putative methyltransferase genes have been found in the cluster. Double and triple mutants have been generated by deleting two or three methyltransferase genes in the chromosome of the producer strain (Fig. 3). All mutants produced novel avilamycin derivatives with improved water solubility.

Improved Yield by Expression of Genes

Pristinamycin, produced by *S. pristinaespiralis*, is a mixture of two types of macrocyclic lactone peptolides, pristinamycins I (PI), a branched cyclic hexadepsipeptide of the streptogramin B group, and pristinamycins II (PII), a polyunsaturated cyclic peptolide of the streptogramin A group. Both the compounds inhibit the growth of bacteria. In combination, they display a synergistic bactericidal activity. The PII component of pristinamycin is produced mainly in two forms, called PIIA (80%)

and PIIB (20%). A water-soluble derivate of pristinamycin, now being marketed under the trade name Synercid, was obtained by the chemical modification of PIIA. To generate a PIIA-specific producer strain, two genes, *snaA* and *snaB*, were isolated from the biosynthetic gene cluster. The enzymes encoded by *snaA* and *snaB* catalyze the conversion of PIIB to PIIA. Both genes were placed under the transcription control of a strong promoter and were cloned into an integrative vector. The integration of this vector into the chromosome of the producer strain resulted in the production of 100% PIIA and this was achieved in high concentrations [62].

New Drugs by Expression of Genes Mithramycin is an aromatic polyketide, which is clinically used as an anticancer agent. It possesses a tricyclic chromophore and is glycosylated at two different positions [63]. Urdamycin A is an angucycline polyketide produced by *S. fradiae* Tü2717, which also shows antitumor activity. It consists of the aglycon aquayamycin, which contains a C-glycosidically linked D-olivose, and three additional O-glycosidically linked deoxyhexoses [64, 65]. The UrdGT2 glycosyltransferase catalyzes the C-glycosyl transfer of activated D-olivose as the first glycosylation step during the urdamycin biosynthesis. Landomycins are produced by *S. cyanogenus* S136 and contain an unusual hexasaccharide consisting of four D-olivose and two L-rhodinose units. These polyketides also show antitumor activities, in particular, against prostata cancer cell lines [66]. To generate novel compounds, genes out of the urdamycin and landomycin clusters were expressed in mutants of *S. argillaceus*: coexpression of *urdGT2* (urdamycin biosynthesis) together with *lanGT1*, (landomycin biosynthesis) in a mutant of the mithramycin producer led to the hybrid molecule 9-C-diolivosylpremithramycinone [67]. This example was listed as a highlight in the field of combinatorial biosynthesis as genes from three different organisms yielded a rationally designed product (Fig. 4) [68].

Recently, a plasmid-based strategy has been described that allows the use of deoxysugar biosynthetic genes to produce a variety of deoxysugars in a cell, which can then be attached to an aglycon by the use of different glycosyltransferases. As an example, a plasmid was generated harboring all the genes necessary for the biosynthesis dTDP-D-olivose. This plasmid was coexpressed with the highly substrate-flexible glycosyltransferase gene *elmG* in *S. albus*. When 8-demethyl-tetracenomycin C was fed to this strain, D-olivosyl-tetracenomycin was produced. In a similar way, L-rhamnosyl-tetracenomycin C, L-olivosyl-tetracenomycin C, and L-rhodinosyl-tetracenomycin C were generated depending on the deoxysugar biosynthetic genes used in each case [69].

Polyketides are synthesized by the action of polyketide synthases (PKSs), which have been classified into two types, type I (modular PKSs) and type II (iterative PKSs).

Modular PKSs are large multifunctional enzymes. Active sites (domains) within these enzymes ketosynthases (KS), acyltransferases (AT), dehydratases (DH), enoyl reductases (ER), ketoreductases (KR), acyl carrier proteins (ACP) and thioesterases (TE) are organized into modules such that each module catalyzes the stereospecific addition of a new monomer onto a growing polyketide chain and also sets the reduction level of the carbon atoms of the resulting intermediate [70]. In 1994, the heterologous expression of the complete erythromycin polyketide synthase was accomplished. The recombinant

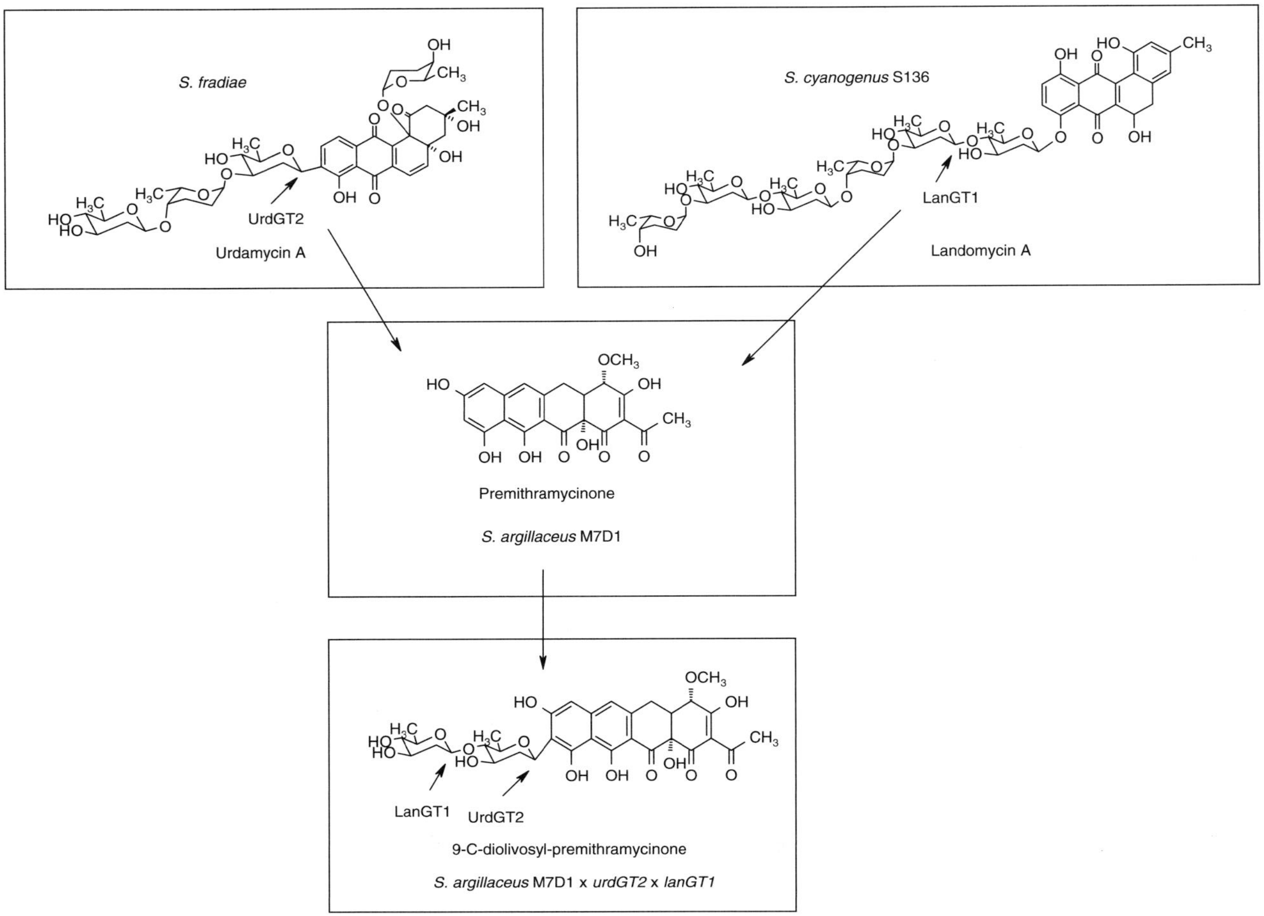

Fig. 4 Generation of a rationally designed hybrid product by coexpression of genes from three different organisms.

strain produced 6-deoxyerythronolide B. This polyketide synthase was then used for a variety of experiments in which modules/domains of the polyketide synthase were exchanged. As an example, compounds produced by the substitution of KR domains are shown in Fig. 5. In these examples, KR domains from the erythromycin PKS have been replaced by domains from the rapamycin producer [71].

New Drugs by Expression of "Artificial" Genes UrdGT1b and UrdGT1c, involved in urdamycin biosynthesis, share 91% identical amino acids. However, the two GTs show different specificities for both nucleotide sugar and acceptor substrate. Targeted amino acid exchanges reduced the number of amino acids, potentially dictating the substrate specificity to 10 in either enzyme. Subsequently, a gene library was created such that only codons of these 10 amino acids from both parental genes were independently combined. Screening of almost 600 library members in vivo revealed 40 active members, acting either like the parental enzymes UrdGT1b and UrdGT1c or displaying a novel specificity. The novel enzymatic specificity is responsible for the biosynthesis of urdamycin P carrying a branched saccharide side chain, hitherto unknown for urdamycins (Fig. 6) [72, 73].

2.2.4.4 Novel Natural Compounds by Glycorandomization

Combinatorial biosynthesis represents an in vivo methodology to diversify natural

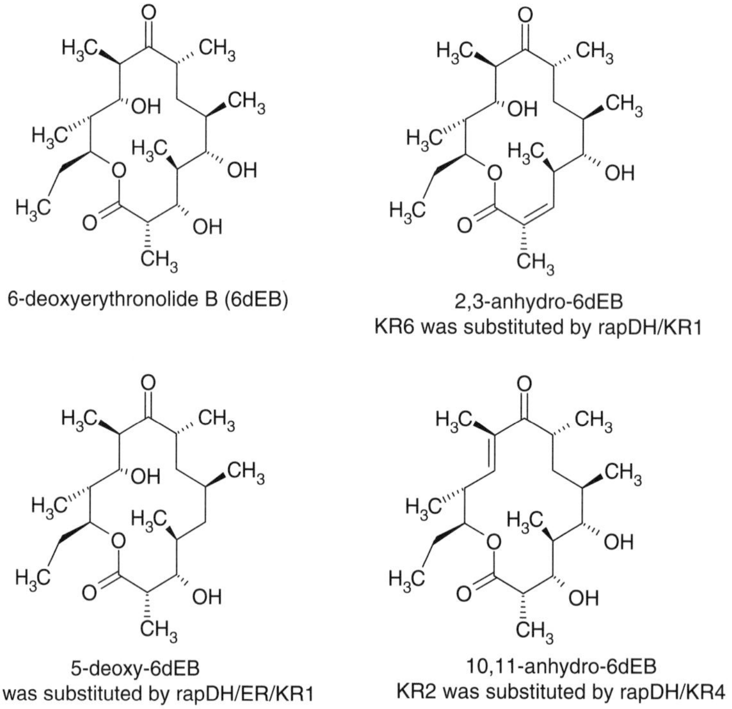

Fig. 5 Production of novel natural compounds by exchanging of modules of polyketide synthases.

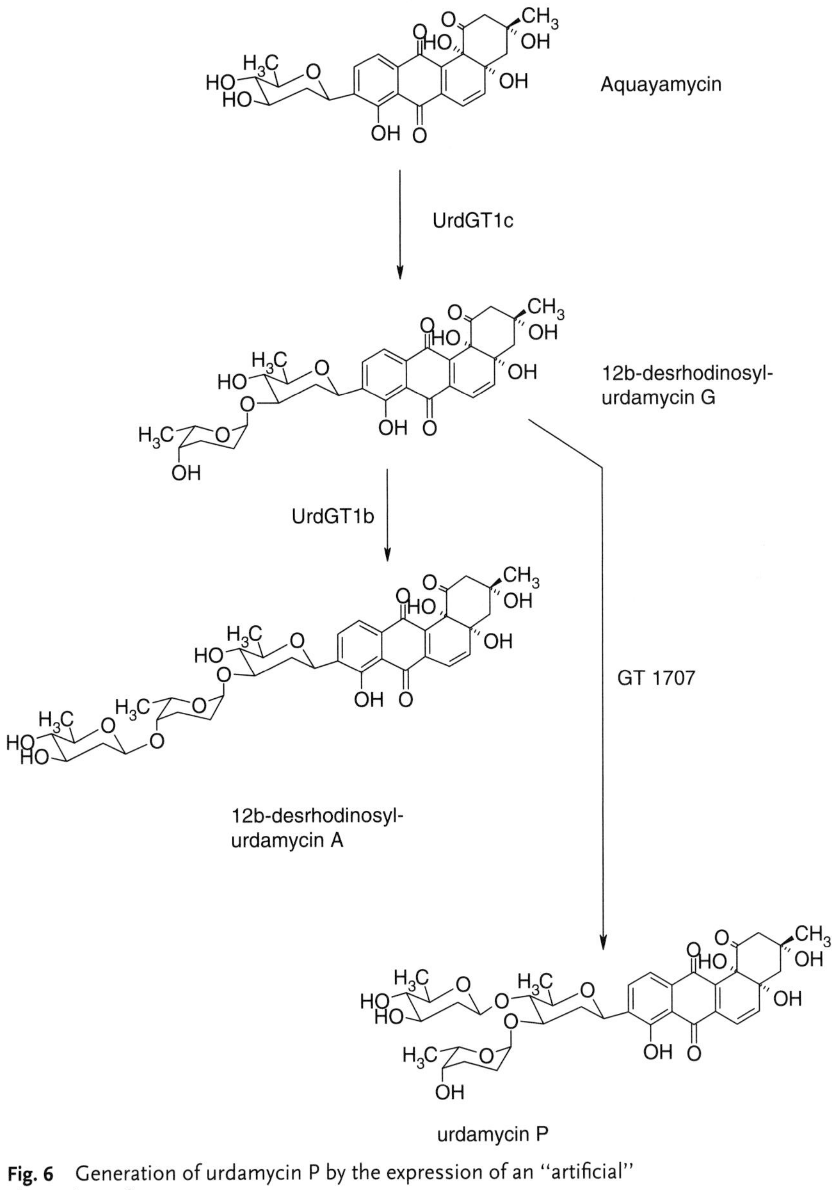

Fig. 6 Generation of urdamycin P by the expression of an "artificial" glycosyltransferase gene.

products, and, among them, a fair amount of glycosides. It is superior to traditional synthetic chemistry in that it makes use of specific enzymes for efficient modifications of complex scaffolds and avoids solubility problems. The reactions are fed out of the host's metabolism, and its cytoplasm is used as an aqueous phase instead of toxic organic solvents that need to be carefully removed from the product and, in most cases, are expensive to dispose of. However, the versatility of combinatorial biosynthesis is somewhat limited. A still unsolved issue is how a novel active compound is dealt with by the host strain. Highly antimicrobial or cytotoxic agents that are not immediately and effectively detoxified by the host's intrinsic mechanisms will kill the host strain long before the compound is detected in a screen. Therefore, especially in case of antimicrobials, scientists run the risk of selecting for structures innocuous to pathogens. In the case of glycosidic structures, true combinatorial approaches suffer from a limited sugar diversity as a second drawback. Only those sugars are available for drug-lead diversification whose biosynthetic routes are understood and the genes involved cloned.

To make use of the possibilities microbial enzymes, particularly natural product glycosyltransferases, offer to pharmaceutical biotechnologists, the glycorandomization paradigm has been proposed and developed. Its key reactions are the bioconversion of sugar phosphates into nucleotide-bound sugars [74, 75] to feed flexible glycosyltransferases. This approach elegantly blends the advantages of structure-based protein engineering with the glycosyltransferases' catalytic potential. Further, it extends the range of available sugars using synthetic chemistry approaches, far beyond what is possible by biosynthetic pathways.

The glycorandomization process starts with a library of chemically synthesized sugar phosphates. The available range includes, for example, deoxy-, amino-, azido-, aminooxy-, methoxy-, and thiosugar phosphates, which are enzymatically converted to their dTDP and, in some cases, UDP derivatives. This reaction is catalyzed by E_P, the *rmlA*-encoded α-D-glucopyranosyl phosphate thymidylyltransferase (E.C. 2.7.7.24) from *Salmonella enterica* LT2. This particular enzyme displays remarkable flexibility toward the sugar donor. Its crystal structure has been solved and some key amino acids have been recognized as hot spots for engineering. A set of targeted amino acid exchanges afforded a pool of even more flexible nucleotidyl-transferases [76, 77]. The E_P-generated sugar nucleotides eventually represent the sugar donor substrates for glycosyltransferases.

Thus, glycorandomization involves only two enzymes (E_P and a glycosyltransferase) to diversify natural product glycosides,

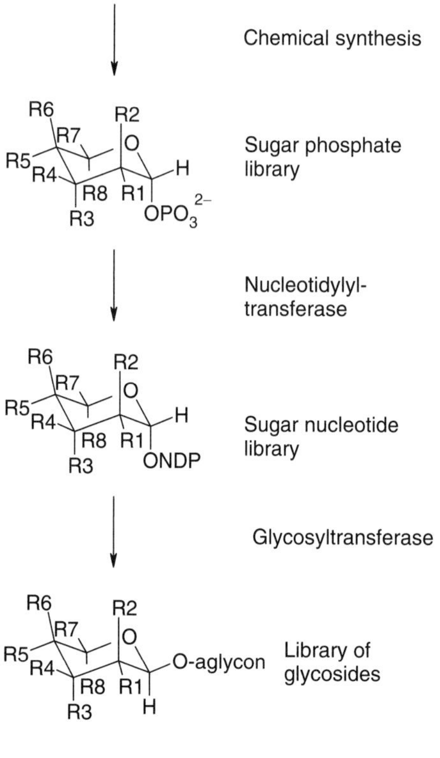

Fig. 7 Schematic of the glycorandomization strategy. Two enzymatic steps convert a chemically synthesized sugar phosphate library into a library of natural product glycosides.

thereby eliminating the need for large sets of biosynthetic genes (Fig. 7).

Although being a very recent technique, glycorandomization has already demonstrated its versatility in its first application toward diversifying the Actinomycete natural products vancomycin and teicoplanin [78], nonribosomally generated sugar-decorated heptapeptides, which are in clinical use as antimicrobial drugs of last resort.

Like conventional combinatorial biosynthesis, glycorandomization requires flexible glycosyltransferases. As recently pointed out in the case of novobiocin [79], a highly specific glycosyltransferase limits the library size. Despite such issues that need to be addressed in future work, glycorandomization is a promising approach to make use of the metabolic potential of procaryotic cells and should promote drug development in the future.

2.2.4.5 **Novel Natural Compounds by Mutasynthesis**

The substrate flexibility of enzymes is also the basis for the "mutasynthesis" approach. During "mutasynthesis," a microorganism containing a defined mutation in an important precursor biosynthesis gene of an interesting metabolite can be fed with alternative or even synthetic precursors. Consequently, derivatives of complex natural products are generated, which may not have been obtained by synthetic methods [80]. This technology was successfully applied to generate the first fluorinated vancomycin-type antibiotics [81].

2.3 Saccharomyces cerevisiae and Other Fungi in Biotech Production

Saccharomyces (*Sa.*) *cerevisiae* might be viewed to be one of the most important fungal organisms used in biotechnology. It has been used in the "old biotechnology" for baking and brewing since prehistory. Yeast genetics, yeast biochemistry, and, finally, yeast molecular biology have substantially contributed to the importance of *Sa. cerevisiae* also in the "new biotechnology" area.

Yeast is a unicellular organism, which, unlike more complex eukaryotes, is amenable to mass production. It can be grown on defined media, giving the investigator a complete control over environmental parameters. The availability of the complete genome sequence of *Sa. cerevisiae* opened the age of "new biotechnology" [82]. In this chapter, we first review how genetic engineering of *Sa. cerevisiae* resulted in improved productivity and yield of important biotech products. Later, three examples of fungal natural products (or their derivatives) are described that have found their way into clinical use.

2.3.1 Generation of Engineered Strains of *Saccharomyces cerevisiae* for the Production of Alcoholic Beverages

Alcohol fermentation is one of the most important processes of biotechnology. Generally, it is initiated by adding yeast to a carbon source and discontinues at a given alcohol concentration. The extension of the substrate range of *Sa. cerevisiae* is of major importance for the large-scale production of several metabolites. *Sa. cerevisiae* is not able to degrade starch and dextrin, since it does not produce starch-decomposing enzymes. Therefore, it is necessary to add starch-decomposing enzymes before fermentation. Attempts have been undertaken to use recombinant strains that contain the decomposing enzymes-encoding genes in order to avoid the preincubation process. The complete assimilation of starch (>98%) was accomplished by coexpression of the *sta2* gene of *Sa. diastaticus* encoding a glucoamylase, the *amy1* gene of *Bacillus amyloliquefaciens* encoding an α-amylase, and the *pulA* gene of *Klebsiella pneumoniae* encoding a pullulanase [83].

Further, genetically engineered strains have been developed, which are able to utilize lactose, melobiose, xylose, and other materials. A thermostable β-galactosidase encoded by *lacA* from *Aspergillus niger* was expressed in *Sa. cerevisiae,* enabling the strain to use lactose as carbon source [84]. A melibiase-producing yeast was constructed by overexpressing the *melI* gene from another *Sa.* strain [85]. Moreover, after the coexpression of *xyl1* and *xyl2,* encoding a xylose reductase and xylitol dehydrogenase from *Pichia stipitis* along with the overexpressed xylulolkinase XKS1 from *Sa. cerevisiae,* xylose was converted to ethanol [86].

Especially in the large-scale production of beer, of highest significance is not ethanol production but a balanced flavor to obtain the desired taste. One unpleasant off-flavor compound is diacetyl, which is a nonenzymatically degraded product of α-acetolactate. Diacetyl is then enzymatically converted to acetoin and subsequently to 2,3-butanediol. The nonenzymatic-degradation step is very slow and requires long lager periods.

One way to avoid the off-flavor is to introduce an alternative route of degradation of α-acetolactate directly to acetoin. α-Acetolactate decarboxylases from different organisms were successfully overexpressed in the beer-producer strains, accelerating the brewing process by diminishing the time of lagering by weeks [87].

Another interesting example is the expression of a β-glucanase of *Bacillus subtilis* in yeast. β-Glucans, the highly viscous side products during fermentation, impede beer filtration, which is still an important separation technique in the brewing industry. The presence of β-glucanase during fermentation did not affect the beer quality and taste but improved the filtration process [88].

2.3.2
Generation of Engineered Strains of *Saccharomyces cerevisiae* for Lactic acid, Xylitol, and Strictosidine Production

Several lactate dehydrogenases (LDHs) were expressed in *Sa. cerevisiae* in order to produce lactic acid. Most successful was the expression of a fungal (*Rhizopus oryzae*) lactate dehydrogenase (LDH). A recombinant strain accumulated approximately 40% more lactic acid with a final concentration of $38\ \mathrm{g\ L^{-1}}$ lactic acid and a yield of 0.44 g of lactic acid per gram of glucose [89]. Xylitol is an attractive sweetener used in the food industry. Xylitol production in yeast was performed by the expression of *xyl1* of *P. stipitis,* encoding a xylitose reductase [90].

A transgenic *Sa. cerevisiae* was constructed harboring the cDNAs that encodes strictosidine synthase (STR) and strictosidine beta-glucosidase (SGD) from the medicinal plant *Catharanthus roseus.* Both enzymes are involved in the biosynthesis of terpenoid indole alkaloids. The yeast culture was found to express high levels of both enzymes. Upon feeding of tryptamine and secologanin, this transgenic yeast culture produced high levels of strictosidine [91].

2.3.3 The Use of Fungi in the Production of Statins, Cyclosporin, and ß-Lactam Antibiotics

2.3.3.1 Statins

Statins are the secondary metabolites of a number of different filamentous fungi. Their medical importance and commercial value stem from their ability to inhibit the enzyme (3*S*)-hydroxy-3-methylglutaryl-CoA (HMG-CoA) reductase. Since this enzyme catalyzes a key step in the endogenous cholesterol biosynthetic pathway, statins have become the widely used antihypercholesterolemic drugs. Along with some synthetic statins, the most prominent examples are lovastatin, mainly from *Aspergillus terreus*, and mevastatin produced by *Penicillium citrinum*, which was the first statin to be discovered [92, 93].

Chemical modification, for example hydroxylation, turned out to be rather unproductive during derivatization efforts. Thus, biotransformation and biotechnological approaches have become the strategy of choice relatively early. For example, a two-step fermentation/biotransformation process has been established for the clinically important pravastatin: Its direct precursor mevastatin is obtained out of a *P. citrinum* culture in the first step and is then subjected to biotransformation, for example, by *S. carbophilus* to complete pravastatin biosynthesis by introducing a hydroxyl group at C6 [94]. Later, improved *Aspergillus* and *Monascus* strains for direct pravastatin production were described [95]. Pioneering work on the genetics and enzymology underlying lovastatin biosynthesis was published in 1999 [96, 97], paving the way for the generation of novel derivatives. Recently, an approach termed "association analysis" [98] was developed to further improve statin producers. The interrelationship between secondary metabolite production levels and genome-wide gene expression was profiled for a minilibrary of *A. terreus* strains engineered to express either wild type or engineered genes that are part of the lovastatin cluster itself or implicated in secondary metabolite regulation. The authors found that multiple genes/proteins attributed to cellular processes as diverse as primary metabolism, secondary metabolism, carbohydrate utilization, sulfur assimilation, transport, proteolysis, and many more correlate with increased (or decreased) lovastatin production levels. This approach revealed multiple points from which to start engineering and may help manipulate statin-producing filamentous fungi for industrial purposes.

2.3.3.2 Cyclosporin

Cyclosporin A (INN: ciclosporin) is a cyclic, nonribosomally synthesized undecapeptide from *Tolypocladium inflatum* (Fig. 2). Apart from its antifungal properties, it represents a potent immunosuppressive drug as it interferes with lymphokine production [99]. Cyclosporin A has been introduced into clinical use to prevent allograft rejection after organ transplants.

The biosynthesis of cyclosporin A has been extensively investigated. A huge 45.8-kb open reading frame was identified as a putative gene coding for the cyclosporine synthetase, a multifunctional peptide synthetase [100]. With a molecular mass of 1689 kDa, it represents one of nature's largest enzymes. Definitive functional evidence arose from targeted gene inactivation, which abolished cyclosporin A biosynthesis [101]. Fungal peptide synthetases are somewhat different from their prokaryotic counterparts in that they usually consist of one huge single enzyme, as is the case for cyclosporin, whereas bacteria generally use multiunit synthetases. Further, fungal peptides quite often include D-configured amino acids. However, these peptide synthetases do not harbor an epimerization domain [102]. In the case of cyclosporin, an external alanin racemase responsible for supplying the synthetase with D-alanine has been purified and characterized [103]. Both cyclosporin synthetase and D-alanine racemase were localized as vacuolar membrane–associated enzymes [104]. Even before the cyclosporin synthetase was characterized, it had become obvious that the biosynthetic pathway tolerates a number of substrate analogs. For example, D-alanine can be replaced by D-serine as was demonstrated by precursor-directed biosynthesis [105], thereby leading to novel cyclosporin derivatives.

For enhanced scaled-up cyclosporin production, a modified sporulation/immobilization method has been proposed, increasing the cyclosporin yield up to tenfold, compared to suspended culture techniques [106]. The immobilization of cells on celite carrier beads decreases the culture's viscosity and therefore increases the mass transfer. It is possible to run the fermentation continuously since the fungal spores can be trapped in the fermenter vessel to populate and germinate on freshly added beads.

2.3.3.3 β-Lactams

Ever since penicillin was discovered and further developed into a drug for use in humans, there has hardly been a natural product that parallels its impact on medicine and pharmacy. Yet, penicillin is only one example of the β-lactam antibiotics class, along with other fungal (and streptomycete) secondary metabolites, for example, cephalosporin C from *Acremonium chrysogenum*. The commercial importance of penicillins and cephalosporins is evident from the worldwide annual sales for these compounds (including their semisynthetic derivatives), which were estimated to reach US$15 billion, and of that, US$4.8 billion derived from sales in the United States [107, 108]. β-lactam biosynthesis has been thoroughly investigated, and been reviewed in a number of compilations [109]. In brief, a nonribosomal peptide synthetase assembles the building blocks, L-α-aminoadipic acid, L-valine (which is epimerized to D-valin), and L-cysteine, into a linear tripeptide, which is then cyclized to isopenicillin N (IPN) by the enzyme IPN synthase. The latter enzyme has been crystallized and, by elegant X-ray diffraction investigations, been used to support the notion that the bicyclic enzymatic product IPN is synthesized in a two-step process, with β-lactam formation preceding the closure of the thiazolidine ring [110, 111]. IPN represents the last common intermediate along the pathways toward penicillin and cephalosporins. While penicillin biosynthesis requires only side-chain modifications, the five-membered penicillin thiazolidine ring is expanded to a dihydrothiazine system on the route to cephalosporins. Gene clusters coding for the enzymatic machinery of β-lactam biosynthesis have been cloned from *Penicillium chrysogenum, Aspergillus nidulans* (Penicillins), and *Acremonium chrysogenum* (Cephalosporin C). In the latter case, the cluster is split up and located on two different chromosomes. Strain development, fermentation, recovery, and purification conditions for β-lactam producers have been subjects of optimization ever since commercial production started. For example, modern fed-batch fermentations yield penicillin titers exceeding 40 g L^{-1} (compared to less than 1 g L^{-1} in 1950). Penicillin G and V are produced in highly automated fermentation vessels with a capacity in the 100 to 400 000 liter range. Although costs for energy and labor have increased, the advances in penicillin production techniques have led to a drastic decrease in bulk prices from ~US$300 per kilogram in the early 1950s to ~US$15 to 20 per kilogram today [108].

2.4 Plants in Biotech Production

From the very roots of humanity, plants have made a crucial contribution to our well-being. Plant products have been used as food and as medicine. Even today, plants are an important source for the discovery of novel pharmacologically active compounds, though the recent competition from combinatorial chemistry and computational drug design has declined the interest. In recent years, gene technology has opened up exciting perspectives and offers tools not only to improve the existing properties of plants, such as the amount of bioactive compounds, but also to create transgenic plants with new properties. Facile transformation and cultivation not only make plants suitable for the production of secondary metabolites but also for recombinant proteins. Plants are capable of carrying out acetylation, phosphorylation, and glycosylation as well as other posttranslational protein modifications required for the biological activity of many eukaryotic proteins. This chapter gives some examples describing the use of plant cells in biotech production by which pharmaceuticals as well as functional and medicinal food are obtained. In this Chapter, we restrict ourselves to a secondary metabolite plant in the low-molecular-weight range.

2.4.1 Transgenic Plants as Functional Foods or Neutraceuticals

A few years ago, industry started the age of engineered functional food. Numerous examples such as the generation of golden rice, the production of healthy plant oils, and engineered plants with increased levels of essential vitamins and nutrients [112] have been published (Table 2). Golden rice was engineered with two

Tab. 2 Examples of new properties in transgenic plants used as neutraceuticals

Property	*Plant*	*References*
Resveratrol	Peanut	(114)
Increased amount of iron and its bioavailability by the reduction of phytic acid	Maize	(115)
Lactoferrin	Rice	(116, 117)
Enriched ferritin leading to the binding of iron and consequently to its accumulation	Lettuce	(118)
Removal of bitter-tasting compounds (glycoalkaloids)	Potatoes	(119)

plant genes from *Narcissus pseudonarcissus* encoding a phytoene synthase and a lycopene ß-cyclase along with one bacterial gene from *Erwinia uredovora* encoding a phytoene desaturase to synthesize ß-carotene, a precursor of vitamin A [113]. This was possible because the transformation of rice was well established and all carotinoid biosynthetic pathway genes had been identified. Despite these promising results, golden rice is not yet on the market. Further work aims to increase the provitamin A amount and to unify high-iron rice lines with provitamin A lines, as it is known that provitamin A potentially increases iron bioavailability.

Wheat is a further target in the area of engineered functional food. A gene encoding a stilbene synthase was expressed in rice that enabled wheat to produce resveratrol. This natural antioxidant possesses positive effects against the development of thrombosis and arteriosclerosis. Moreover, in its glycosidic form it enhances resistance to fungal pathogenes, which also occurs in transgenic plants [120].

The current health-related objective of plant seed engineering is to increase the content of "healthy fatty acids" and reduce "unhealthy fatty acids" in oilseed crops, such as soybean, oil palm, rapeseed, and sunflower. Genetic engineering was successful in reducing the levels of trans-unsaturated fatty acids and in reducing the ratio between omega-6 and omega-3 unsaturated fatty acids in some vegetable oils [121, 122]. Metabolic engineering also succeeded in increasing the vitamin E, vitamin C, and the lycopene content in plants [112, 123], as well as the content of bioflavonoids, known for their antioxidant, anticancer, and estrogenic properties [124]. In addition, human milk proteins like lactoferrin can now be expressed in plants [116]. These proteins are believed to have a multitude of biological activities that benefit the newborn infant. Functional food selected and advertised for its high content of therapeutically active molecules is already offered in the shelves of supermarkets, leaving the determination of their true medical benefit to the consumer.

A further research area is the elimination of natural compounds from a plant to avoid severe side effects. As an example, peanut causes allergies due to several proteins. Researchers are now working on the generation of plants with reduced levels of these proteins.

2.4.2
Transgenic Plants and Plant Cell Culture as Bioreactors of Secondary Metabolites

Biotech methods are also used to increase the amount of pharmaceutically interesting compounds in plants. Leaves of *Atropa belladonna* contain high amounts of L-hyoscyamin, but negligible contents of L-scopolamin due to the low activity of hyoscyamin-6ß-hydroxylase (H6H) in roots. H6H-cDNA was isolated from *Hyoscyamus niger*, cloned, and introduced in *Atropa belladonna* using *Agrobacterium tumefaciens*. Transgenic plants as well as the sexual descendents contain the transgene and accumulate up to 1% of L-scopolamin, but only traces of L-hyoscyamin (Table 3) [125].

Another approach to produce biologically active secondary metabolites is the use of plant cell cultures. Plant cell cultures are advantageous in that they are not limited by environmental, ecological, or climatic conditions. Further, cells can proliferate at higher growth rates than whole plants in cultivation. As shown in Table 4, some metabolites in plant cell cultures have been reported to accumulate

Tab. 3 Transgenic plants with selected improved production of secondary metabolites [125]

Compound	***Target protein***	***Gene donor***	***Gene recipient***
Cadaverin	Lysin-decarboxylase	*Hafnia alvei*	*Nicotiana tabacum*
Sterols	HMG-CoA-reductase	*Hevea brasiliensis*	*Nicotiana tabacum*
Nicotin	Ornithin-decarboxylase	*Saccharomyces cerevisiae*	*Nicotiana tabacum*
Resveratrol	Stilbene-synthase	*Vitis vinifera*	*Nicotiana tabacum*
Scopolamin	Hyoscyamin-6ß-hydoxylase	*Hyoscyamus niger*	*Atropa belladonna*

Tab. 4 Comparison of product yield of secondary metabolites in cell culture and parent plants [126]

Product	***Plant***	***Yield (% DW[a])***	
		Culture	***Plant***
Anthocyanin	*Vitis* sp.	16	10
	Euphorbia milli	4	0.3
	Perilla frutescens	24	1.5
Anthraquinone	*Morinda citrifolia*	18	2.2
Berberine	*Coptis japonica*	13	4
	Thalictrum minor	10	0.01
Rosmarinic acid	*Coleus blumei*	27	3.0
Shikonin	*Lithospermum erythrorhizon*	14	1.5

[a] Dry weight.

with a higher titer compared to those in the parent plants. Some industrial processes harness this potential, for example, for shikonin, ginseng, and paclitaxel production [126]. Especially *Taxus* cell cultures are an interesting alternative to the isolation of paclitaxel from plantation-grown plants as the slow growth of *Taxus* species, the significant variation in taxoid content, and the costly purification of 10-deacetylbaccatin III from co-occurring taxoids are significantly limiting parameters [127–129]. Using cell cultures, taxol production rates up to 23.4 mg L day^{-1} with paclitaxel comprising 13 to 20% of the total taxoid fraction can now be achieved [130].

However, the use of plant cell cultures for the production of interesting molecules has not gained acceptance in industry as yet. Usually, low productivity and high costs are the most important negative parameters [131]. Nevertheless, research is still going on and is well described in the next chapter. Further application of plant-cell-suspension cultures are aimed at the production of ajmalicine, vinblastine, vincristine, podophyllotoxins, and camptothecin [132–134].

2.4.3
Transgenic Plants as Bioreactors of Recombinant Proteins

Nowadays, plants such as tobacco, potato, tomato, banana, legumes, and cereals as well as alfalfa, are used in molecular farming and have emerged as

Tab. 5 Examples of therapeutic antibodies produced in plants [135]

Antibody format	*Antigen*	*Cellular location*	*Transgenic plant*	*Max. expression level*
dAb	Substance P	Apoplast	*Nicotiana benthamiana*	1% TSP leaves
IgG1, Fab	Human creatine kinase	Apoplast	*N. tabacum* *Arabidopsis thaliana*	0.044% TSP leaves 1.3% TSP leaves
IgG1	Streptococcal surface antigen (I/II)	Plasma membrane	*N. tabacum*	1.1% TSP leaves
IgG1	Human IgG	Apoplast	*Medicago sativa*	1% TSP
IgG1	Herplex simplex virus 2	Apoplast	*Glycine max*	Not reported
SigA	Streptococcal surface antigen (I/II)	Apoplast	*N. tabacum*	0.5 mg g^{-1} FW leaves
scFv	Carcinoembryonic antigen	ER	*Pisum sativum*	0.009 mg g^{-1} seed

Note: dAb: single domain antibody; FW: fresh weight; TSP: total soluble protein.

promising biopharming systems for production of pharmaceutical proteins, such as antibodies, vaccines, regulatory proteins and enzymes, [112, 135–138] (Table 5). The advantages offered by plants include low cost of cultivation, high biomass production, relatively fast protein synthesis, low operating costs, excellent scalability, eucaryotic posttranslational modifications, low risk of pathogenicity toward humans and endotoxins, and a relatively high protein expression level [112, 139]. Using transgenic plants as a host is highly attractive in that proteins can be administered in fruits and vegetables as a source of antigens for oral vaccination [135]. Thus, potatoes expressing a synthetic *LT-B* gene, a labile toxin from *Escherichia coli* were successfully used in a clinical study to examine an edible plant vaccine [137]. Further, interesting vaccines that have been tested clinically are directed against viral diarrhea and hepatitis B. No plant-derived protein has still been developed to be used as a drug, but molecular farming has gained attention as plants can be turned into molecular medicine factories.

Recently, plant cells have also been considered to be an alternative host for the production of recombinant proteins since they are able to glycosylate proteins [133, 139]. Of the various systems used for cultivation, such as hairy roots, immobilized cells, and free cell suspensions, the latter is regarded to be most suitable for large-scale applications. Full-size antibodies, antibody fragments, and fusion proteins can be expressed in transgenic-plant-cell systems, such as *Nicotiana tabacum*, pea, wheat, and rice using shake-flask or fermentation cultures [136]. Yet, these systems are still of low commercial importance due to their unadvantageous productivity.

References

1. A. Schatz, E. Bugie, S. Waksman, *Proc. Soc. Exp. Biol. Med.* **1944**, *55*, 66–69.
2. R. H. Baltz, *Trends Microbiol.* **1998**, *6*, 76–83.
3. R. Austrian, J. Gold, *Ann. Intern. Med.* **1964**, *60*, 759–776.
4. P. Dineen, W. P. Homan, W. R. Grafe, *Ann. Surg.* **1976**, *184*, 717–722.
5. G. L. Verdine, *Nature* **1996**, *384*, 11–13.
6. G. M. Cragg, D. J. Newman, K. M. Snader, *J. Nat. Prod.* **1997**, *60*, 52–60.
7. A. L. Demain, *Appl. Microbiol. Biotechnol.* **1999**, *52*, 455–463.
8. T. Henkel, R. M. Brunne, H. Müller, *Angew. Chem., Int. Ed. Engl.* **1999**, *38*, 643–647.
9. A. Harvey, *Scrip Reports*, PJB Publications Ltd, Richmond, Surrey, UK, 2001.
10. H. H. Cowen, *Pharm. Therap. Categories Outlook* **2001**, *10*.
11. J. Berdy, *Proceedings of the 9th International Symposium on The Biology of Actinomycetes*, Part 1, Allerton Press, New York, 1995, pp. 3–23.
12. L. Gray, *Business Communication Company Inc*, Norwalk, CT, 2002, p. 114.
13. B. S. Moore, J. Piel, *Antonie Van Leeuwenhoek* **2000**, *78*, 391–398.
14. J. K. Borchardt, *Modern Drug Discov.* **1999**, *2*, 22–29.
15. S. Donadio, M. Sosio, G. Lancini, *Appl. Microbiol. Biotechnol.* **2002**, *60*, 377–380.
16. W. R. Strohl, *Metab. Eng.* **2001**, *3*, 4–14.
17. H. Liu, K. A. Reynolds, *Metab. Eng.* **2001**, *3*, 40–48.
18. T. Kieser, M. J. Bibb, M. J. Buttner et al., *Practical Streptomyces Genetics*, The John Innes Foundation, Norwich, UK, 2000.
19. D. A. Hopwood, *Microbiology* **1999**, *145*, 2183–2202.
20. D. A. Hopwood, Genetic Recombination in Streptomyces coelicolor, Ph. D. Thesis, University of Cambridge, Cambridge, 1958.
21. B. A. M. Rudd, D. A. Hopwood, *J. Gen. Microbiol.* **1979**, *114*, 35–43.
22. C. J. Thompson, J. M. Ward, D. A. Hopwood, *Nature* **1980**, *286*, 525–527.
23. M. J. Bibb, J. L. Schottel, S. N. Cohen, *Nature* **1980**, *284*, 526–531.
24. F. Malpartida, D. A. Hopwood, *Nature* **1984**, *309*, 462–464.
25. G. Muth, D. F. Brolle, W. Wohlleben in *Manual of Industrial Microbiology and Biotechnology* (Eds.: A. L. Demain, J. E. Davis, R. M. Atlas), ASM Press, Washington, DC, 1999, pp. 353–367.
26. A. Paradkar, A. Trefzer, R. Chakraburtty et al., *Crit. Rev. Biotechnol.* **2003**, *23*, 1–27.
27. T. Weber, K. Welzel, S. Pelzer et al., *J. Biotechnol.* **2003**, in press.
28. P. R. August, T.-W., Yu, H. G. Floss in *Drug Discovery from Nature* (Eds.: S. Grabley, R. Thiericke), Springer, Berlin-Heidelberg, 2000.
29. T. Brautaset, O. N. Sekurova, H. Sletta et al., *Chem. Biol.* **2000**, *7*, 395–403.
30. S. D. Bentley, K. F. Chater, A. M. Cerdeno-Tarraga et al., *Nature* **2002**, *417*, 141–147.
31. S. Omura, H. Ikeda, J. Ishikawa et al., *Proc. Natl. Acad. Sci. U. S. A.* **2001**, *98*, 12215–12220.
32. H. Ikeda, J. Ishikawa, A. Hanamoto et al., *Nat. Biotechnol.* **2003**, *21*, 526–531.
33. Y. S. Lin, H. M. Kieser, D. A. Hopwood et al., *Mol. Microbiol.* **1993**, *10*, 923–933.
34. M. Redenbach, J. Scheel, U. Schmidt, *Antonie Van Leeuwenhoek* **2000**, *78*, 227–235.
35. D. A. Hopwood, *Nat. Biotechnol.* **2003**, *21*, 505–506.
36. T. A. Cropp, D. J. Wilson, K. A. Reynolds, *Nat. Biotechnol.* **2000**, *18*, 980–983.
37. M. G. Watve, R. Tickoo, M. M. Jog et al., *Arch. Microbiol.* **2001**, *176*, 386–390.
38. N. Johnston, *Modern Drug Discov.* **2002**, *6*, 28–32.
39. H. Breithaupt, *Nat. Biotechnol.* **1999**, *17*, 1165–1169.
40. A. Coates, Y. Hu, R. Bax et al., *Nat. Rev. Drug. Discov.* **2002**, *11*, 895–910.
41. H. Zähner, H.-P. Fiedler in *Past Perspectives and Future Friends* (Eds.: P. A. Hunter, G. K. Darby, N. J. Russel), Cambridge University Press, Cambridge, 1995, pp. 67–84.
42. W. R. Strohl in *Biotechnology of Antibiotics* (Eds.: W. R. Strohl), 2nd ed., Marcel Dekker, New York, 1997, pp. 1–47.
43. J. Rosamond, A. Allsop, *Science* **2000**, *287*, 1973–1976.
44. R. P. Bax, R. Anderson, J. Crew et al., *Nat. Med.* **1998**, *4*, 545–546.
45. S. Tsiodras, H. S. Gold, G. Sakoulas et al., *Lancet* **2001**, *358*, 207–208.
46. A. L. Demain, *Nat. Biotechnol.* **2002**, *20*, 331.

47. A. Harvey, *Drug Discov. Today* **2000**, *5*, 294–300.
48. S. Donadio, P. Monciardini, R. Alduina et al., *J. Biotechnol.* **2002**, *99*, 187–198.
49. H. B. Bode, B. Bethe, R. Höfs et al., *ChemBiochem* **2002**, *3*, 619–627.
50. T. V. J. Goksoyr, F. L. Daae, *Appl. Environ. Micobiol.* **1990**, *56*, 782–787.
51. G. Y. Wang, E. Graziani, B. Waters et al., *Org. Lett.* **2000**, *2*, 2401–2404.
52. I. A. MacNeil, C. L. Tiong, C. Minor et al., *J. Mol. Microbiol. Biotechnol.* **2001**, *3*, 301–308.
53. S. Pelzer, W. Reichert, M. Huppert et al., *J. Biotechnol.* **1997**, *56*, 115–128.
54. M. Sosio, E. Bossi, A. Bianchi et al., *Mol. Gen. Genet.* **2000**, *264*, 213–221.
55. X. Ruan, D. Stassi, S. A. Lax et al., *Gene* **1997**, *203*, 1–9.
56. B. Shen, L. Du, C. Sanchez et al., *Bioorg. Chem.* **1999**, *27*, 155–171.
57. C. Mendez, E. Künzel, F. Lipata et al., *J. Nat. Prod.* **2002**, *65*, 779–782.
58. M. Metsä-Ketelä, K. Palmu, T. Kunnari et al., *Antimicrob. Agents Chemother.* **2003**, *47*, 1291–1296.
59. E. Zazopoulos, K. Huang, A. Staffa et al., *Nat. Biotechnol.* **2003**, *21*, 187–190.
60. J. Weber, J. Leung, S. Swanson et al., *Science* **1991**, *252*, 114–117.
61. G. Weitnauer, A. Mühlenweg, A. Trefzer et al., *Chem.Biol.* **2001**, *8*, 569–581.
62. G. Sezonov, V. Blanc, N. Bamas-Jacques et al., *Nat. Biotechnol.* **1997**, *15*, 349–353.
63. J. Skarbek, M. Speedie in *Antitumor Compounds Of Natural Origin* (Ed.: A. Aszalos), CRC Press, Boca Raton, FL, 1981, pp. 191–235, Vol. 1.
64. H. Drautz, H. Zähner, J. Rohr et al., *J. Antibiot.* **1986**, *39*, 1657–1669.
65. A. Trefzer, D. Hoffmeister, E. Künzel et al., *Chem. Biol.* **2000**, *7*, 133–142.
66. R. Crow, B. Rosenbaum, R. Smith et al., *Bioorg. Med. Chem. Lett.* **1999**, *9*, 1663–1666.
67. A. Trefzer, G. Blanco, L. Remsing et al., *J. Am. Chem. Soc.* **2002**, *124*, 6056–6062.
68. U. Rix, C. Fischer, L. Remsing et al., *Nat. Prod. Rep.* **2002**, *19*, 542–580.
69. L. Rodriguez, I. Aguirrezabalaga, N. Allende et al., *Chem. Biol.* **2002**, *9*, 721–729.
70. D. Hopwood, *Chem. Rev.* **1997**, *10*, 2465–2498.
71. R. McDaniel, A. Thamchaipenet, C. Gustafsson et al., *Proc. Natl. Acad. Sci. U. S. A.* **1999**, *96*, 1846–1851.
72. D. Hoffmeister, K. Ichinose, A. Bechthold, *Chem. Biol.* **2001**, *8*, 557–567.
73. D. Hoffmeister, B. Wilkinson, G. Foster et al., *Chem. Biol.* **2002**, *9*, 287–295.
74. J. Jiang, J. B. Biggins, J. S. Thorson, *J. Am. Chem. Soc.* **2000**, *122*, 6803–6804.
75. J. Jiang, J. B. Biggins, J. S. Thorson, *Angew. Chem., Int. Ed.* **2001**, *40*, 1502–1505.
76. W. A. Barton, J. Lesniak, J. B. Biggins et al., *Nat. Struct. Biol.* **2001**, *8*, 545–551.
77. W. A. Barton, J. B. Biggins, J. Jiang et al., *Proc. Natl. Acad. Sci.* **2002**, *99*, 13397–13402.
78. H. C. Losey, J. Jiang, J. B. Biggins et al., *Chem. Biol.* **2002**, *9*, 1305–1314.
79. C. Albermann, A. Soriano, J. Jiang et al., *Org. Lett.* **2003**, *5*, 933–936.
80. W. Wohlleben, S. Pelzer, *Chem. Biol.* **2002**, *9*, 1163–1164.
81. S. Weist, B. Bister, O. Puk et al., *Angew. Chem., Int. Ed. Engl.* **2002**, *41*, 3383–3385.
82. S. Ostergaard, L. Olsson, J. Nielsen, *Micobiol. Mol. Biol. Rev.* **2000**, *64*, 34–50.
83. B. Janse, I. Pretorius, *Appl. Microbiol. Biotechnol.* **1995**, *42*, 873–883.
84. V. Kumar, S. Ramakrishnan, T. Teeri et al., *Biotechnology* **1992**, *10*, 82–85.
85. S. Vincent, P. Bell, P. Bissinger et al., *Lett. Appl. Microbiol.* **1999**, *28*, 148–152.
86. N. Ho, Z. Chen, A. Brainard et al., *Adv. Biochem. Eng. Biotechnol.* **1999**, *65*, 163–192.
87. S. Yamano, K. Kondo, J. Tanaka et al., *J. Biotechnol.* **1994**, *14*, 173–178.
88. B. Cantwell, G. Brazil, N. Murphy et al., *Curr. Genet.* **1986**, *11*, 65–70.
89. C. Skory, *J. Ind. Microbiol. Biotechnol.* **2003**, *30*, 22–27.
90. R. Govinden, B. Pillay, W. Van Zyl et al., *Appl. Microbiol. Biotechnol.* **2001**, *55*, 76–80.
91. A. Geerlings, F. Redondo, A. Contin et al., *Appl. Microbiol. Biotechnol.* **2001**, *56*, 420–424.
92. A. Endo, M. Kuroda, K. Tanzawa, *FEBS Lett.* **1976**, *72*, 323–326.
93. M. Manzoni, M. Rollini, *Appl. Microbiol. Biotechnol.* **2002**, *58*, 555–564.
94. T. Matsuoka, S. Miyakoshi, K. Tanzawa et al., *Eur. J. Biochem.* **1989**, *184*, 707–713.
95. M. Manzoni, S. Bergomi, M. Rollini et al., *Biotechnol. Lett.* **1999**, *21*, 253–257.
96. J. Kennedy, K. Auclair, S. G. Kendrew et al., *Science* **1999**, *284*, 1368–1372.
97. L. Hendrickson, C. R. Davis, C. Roach et al., *Chem. Biol.* **1999**, *6*, 429–439.

98. M. Askenazi, E. M. Driggers, D. A. Holtzman et al., *Nat. Biotechnol.* **2003**, *21*, 150–156.
99. C. Randak, T. Brabletz, M. Hergenrother et al., *EMBO J.* **1990**, *9*, 2529–2536.
100. G. Weber, K. Schörgendorfer, E. Schneider-Scherzer et al., *Curr. Genet.* **1994**, *26*, 120–125.
101. G. Weber, E. Leitner, *Curr. Genet.* **1994**, *26*, 461–467.
102. S. Doekel, M. A. Marahiel, *Metab. Eng.* **2001**, *3*, 64–77.
103. K. Hoffmann, E. Schneider-Scherzer, H. Kleinkauf et al., *J. Biol. Chem.* **1994**, *269*, 12710–12714.
104. M. Hoppert, C. Gentzsch, K. Schörgendorfer, *Arch. Microbiol.* **2001**, *176*, 285–293.
105. R. Traber, H. Hofmann, H. Kobel, *J. Antibiot.* **1989**, *42*, 591–597.
106. T. H. Lee, G.-T. Chun, Y. K. Chang, *Biotechnol. Prog.* **1997**, *13*, 546–550.
107. W. R. Strohl in *Encyclopedia of Bioprocess Technology: Fermentation, Biocatalysis and Bioseparations* (Eds.: M. C. Flickinger, S. W. Drew), Wiley, New York, 1999, pp. 2348–2365, Vol. 5.
108. R. P. Elander, *Appl. Microbiol. Biotechnol.* **2003**, *61*, 385–392.
109. J. F. Martin, *Appl. Microbiol. Biotechnol.* **1998**, *50*, 1–15.
110. P. L. Roach, I. J. Clifton, C. M. H. Hensgens et al., *Nature* **1997**, *387*, 827–830.
111. N. I. Burzlaff, P. J. Rutledge, I. J. Clifton et al., *Nature* **1999**, *401*, 721–724.
112. I. D. Raskin, D. M. Ribnicky, S. Komarnytsky et al., *Trends Biotechnol.* **2002**, *20*, 522–531.
113. P. Beyer, S. Al-Babili, X. Ye et al., *Am. Soc. Nutr. Sci.* **2002**, *132*, 506–510.
114. I. M. Chung, M. R. Park, S. Rehmann et al., *Mol. Cells* **2001**, *12*, 353–359.
115. C. Mendoza, F. E. Viteri, B. Lonnerdal et al., *Am. J. Clin. Nutr.* **2001**, *73*, 80–85.
116. B. Lonnerdal, *J. Am. Coll. Nutr.* **2002**, *21*(Suppl. 3), 218S–221S.
117. Y. A. Suzuki, S. L. Kelleher, D. Yalda et al., *J. Pediatr. Gastroenterol. Nutr.* **2003**, *36*, 190–199.
118. F. Goto, T. Yoshihara, H. Saiki, *Theor. Appl. Gen.* **2000**, *100*, 658–664.
119. L. Arnqvist, P. C. Dutta, L. Jonsson et al., *Plant Physiol.* **2003**, *131*, 1792–1799.
120. S. Fettig, D. Heß, *Transgenic Res.* **1999**, *8*, 179–189.
121. J. Thelen, J. B. Ohlrogge, *Metab. Eng.* **2002**, *4*, 12–21.
122. K. Liu, Proceedings of the World Conference on Oilseed Processing Utilization, Cancun, Mexiko, 2000 pp. 84–89.
123. D. Shintani, D. Della Penna, *Science* **1998**, *282*, 2098–2100.
124. G. Forkman, S. Martens, *Curr. Opin. Biotechnol.* **2001**, *12*, 155–160.
125. W. Kreis, D. Baron, G. Stoll, *Biotechnologie der Arzneistoffe*, Dtsch. Apoth. Verlag, Stuttgart, 2001, pp. 250–260.
126. J. Zhong, *Adv. Biochem. Eng. Biotechnol.* **2001**, *72*, 1–26.
127. S. Jennewein, R. Croteau, *Appl. Microbiol. Biotechnol.* **2001**, *57*, 13–19.
128. Y. Yukimune, H. Tabata, Y. Higashi et al., *Nat. Biotechnol.* **1996**, *14*, 1129–1132.
129. R. Cusido, J. Palazon, M. Bonfill et al., *Biotechnol. Prog.* **2002**, *18*, 418–423.
130. R. Ketchum, D. Gibson, R. Croteau et al., *Biotechnol. Bioeng.* **1999**, *62*, 97–105.
131. J. Choi, G. Cho, S. Byun et al., *Adv. Biochem. Eng. Biotechnol.* **2001**, *72*, 63–102.
132. R. M. Moraes, E. Bedir, H. Barrett et al., *Planta Med.* **2002**, *68*, 341–344.
133. R. Fischer, C. Vaquero-Martin, M. Sack et al., *Biotechnol. Appl. Biochem.* **1999**, *30*, 113–116.
134. S. Hiroshi, T. Yamakawa, M. Yamazaki et al., *Biotechnol. Lett.* **2002**, *24*, 359–363.
135. S. Schillberg, R. Fischer, N. Emans, *Naturwissenschaften* **2003**, *90*, 115–155.
136. R. Fischer, J. Drossard, U. Commandeur et al., *Biotechnol. Appl. Biochem.* **1999**, *30*, 101–108.
137. H. S. Mason, H. Warzecha, T. Mor et al., *Trends Mol. Med.* **2002**, *8*, 324–329.
138. A. M. Walmsley, C. J. Arntzen, *Curr. Opin. Biotechnol.* **2000**, *11*, 126–129.
139. J. Miele, *Trends Biotechnol.* **1997**, *15*, 45–50.

3
Biopharmaceuticals Expressed in Plants

Jörg Knäblein
Analytical Development Biologics Schering AG, 13342 Berlin, Germany

3.1
Introduction

Biopharmaceuticals, which are large molecules produced by living cells, are currently the mainstay products of the biotechnology industry. Indeed, biologics such as Genentech's (Vacaville, CA, USA) human growth factor somatropin or Amgen's (Thousand Oaks, CA, USA) recombinant erythropoietin have shown that biopharmaceuticals can benefit a huge number of patients and also generate big profits for these companies at the same time. But it has also become obvious over the last couple of years that current fermentation capacities will not be sufficient to manufacture all biopharmaceuticals (in the market already or in development), because the market and demand for biologics is continuously and very rapidly growing; for antibodies alone (with at least 10 monoclonal antibodies approved and being marketed), the revenues are predicted to expand to US$3 billion in 2002 [1] and US$8 billion in 2008 [2]. The 10 monoclonal antibodies on the market consume more than 75% of the industry's manufacturing capability. And there are up to 60 more that are expected to reach the market in the next six or seven years [3]. Altogether, there are about 1200 protein-based products in the pipeline with a 20% growth rate and the market for current and late stage (Phase III) is estimated to be US$42 billion in 2005, and even US$100 billion in 2010 [4]. But, there are obvious limitations of large-scale manufacturing resources and production capacities – and pharmaceutical companies are competing [5].

To circumvent this capacity crunch, it is necessary to look into other technologies rather than the established ones, like, for example, *Escherichia coli* or CHO (Chinese hamster ovary) cell expression. One solution to avoid these limitations could be the use of transgenic plants to express recombinant proteins at low cost, in GMP (good manufacturing practice) quality greenhouses (with purification and fill finish in conventional facilities). Plants therefore provide an economically sound source of recombinant proteins [6], such as industrial enzymes [7], and biopharmaceuticals [8, 9]. Furthermore, using the existing infrastructure for crop cultivation, processing, and storage will reduce the amount of capital investment required for commercial production. For example, it was estimated that the production costs of recombinant proteins in plants could be between 10 and 50 times lower than those for producing the same protein in *E. coli* [10] and Alan Dove describes a factor of thousand for cost of protein (US dollar per gram of raw material) expressed in, for example, CHO cells compared to transgenic plants [11]. So, at the dawn of this new millennium, a solution is rising to circumvent expression capacity crunches and to supply mankind with the medicines we need. Providing the right amounts of biopharmaceuticals can now be achieved by applying our knowledge of modern life sciences to systems that were on this planet long time before us – plants.

3.2 Alternative Expression Systems

Currently, CHO cells are the most widely used technology in biomanufacturing because they are capable of expressing eukaryotic proteins (processing, folding, and post-translational modifications) that cannot be provided by *E. coli*. A long track record exists for CHO cells, but unfortunately they bring some problems along when it comes to scale-up for production. Transport of oxygen (and other gases) and nutrients is critical for the fermentation process, as well as heat must diffuse evenly to all cultured cells. According to the Michaelis–Menten equation, the growth rate depends on the oxygen/nutrient supply, therefore good mixing and aeration are a prerequisite for the biomanufacturing process and are usually achieved by different fermentation modes (see Fig. 1). But the laws of physics set strict limits on the size of bioreactors. For example, an agitator achieves good heat flow and aeration, but with increased fermenter size, shear forces also increase and disrupt the cells – and building parallel lines of bioreactors multiplies the costs linearly. A 10 000-L bioreactor costs between US$250 000 to 500 000 and takes five years to build (conceptual planning, engineering, construction, validation, etc.). An error in estimating demand for, or inaccurately predicting the approval of, a new drug can be incredibly costly. To compound the problem, regulators in the United States and Europe demand that drugs have to be

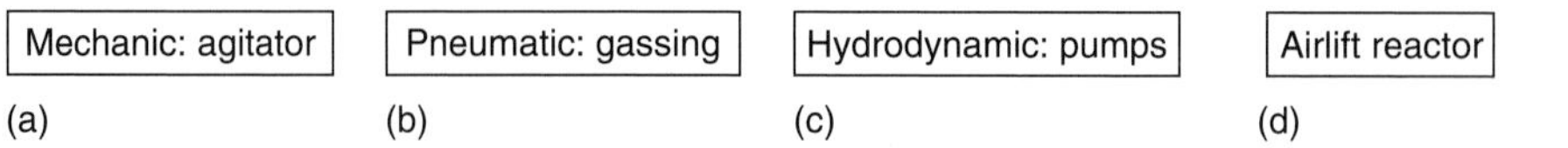

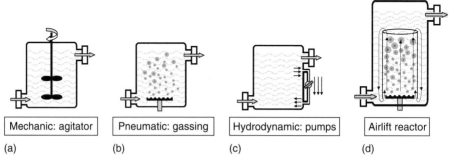

Fig. 1 Different fermentation modes for bioreactors. In order to achieve best aeration and mixing and to avoid high shear forces, different fermentation modes are applied. (a) Mechanical, (b) pneumatical, (c) hydrodynamic pumps, and (d) airlift reactor [12].

produced for the market in the same system used to produce them for the final round of clinical trials, in order to guarantee bioequivalence (e.g. toxicity, bioavailability, pharmacokinetics, pharmacodynamics) of the molecule. So, companies have to choose between launching a product manufactured at a smaller development facility (and struggling to meet market demands) or building larger, dedicated facilities for a drug that might never be approved!

Therefore, alternative technologies are used for the expression of biopharmaceuticals, some of them also at lower costs involved (see Fig. 2). One such alternative is the creation of transgenic animals ("pharming"), but this suffers from the disadvantage that it requires a long time to establish such animals (approximately 2 years). In addition to that, some of the human biopharmaceuticals could be detrimental to the mammal's health, when expressed in the mammary glands. This is why ethical debates sometimes arise from the use of transgenic mammals for production of biopharmaceuticals. Although there are no ethical concerns involved with plants, there are societal ones that will be addressed later. Another expression system (see Fig. 2) utilizes transgenic chicken. The eggs, from which the proteins are harvested, are natural protein production systems. But production of transgenic birds is still several years behind transgenic mammal technology. Intensive animal housing constraints also make them more susceptible to disease (e.g. Asia 1997 or Europe 2003: killing of huge flocks with thousands of chicken suffering from fowl pest). In the light of development time, experience, costs and ethical issues, plants are therefore the favored technology, since such systems usually have short gene-to-protein times (weeks), some are already well established, and as mentioned before, the involved costs are comparatively low. This low cost of goods sold (COGS) for plant-derived proteins is mainly due to low capital costs: greenhouse costs are only US\$10/m^2 versus US\$1000/m^2 for mammalian cells.

3.3
History of Plant Expression

Plants have been a source of medicinal products throughout human evolution. These active pharmaceutical compounds have been primarily small molecules, however. One of the most popular examples is aspirin (acetylsalicylic acid) to relieve pain and reduce fever. A French pharmacist first isolated natural salicin (a chemical relative of the compound used to make aspirin) from white willow bark in 1829. Advances in genetic engineering are now allowing for the production of therapeutic proteins (as opposed to small molecules) in plant tissues. Expression of recombinant proteins in plants has been well documented since the 1970s and has slowly gained credibility in the biotechnology industry and regulatory agencies. The first proof of concept has been the incorporation of insect and pest resistance into grains. For example, "Bt corn" contains genes from *Bacillus thuringensis* and is currently being grown commercially. Genetic engineering techniques are now available for the manipulation of almost all commercially valuable plants. Easy transformation and cultivation make plants suitable for production of virtually any recombinant protein.

Plants have a number of advantages over microbial expression systems, but one of them is of outmost importance: they can produce eukaryotic proteins in

			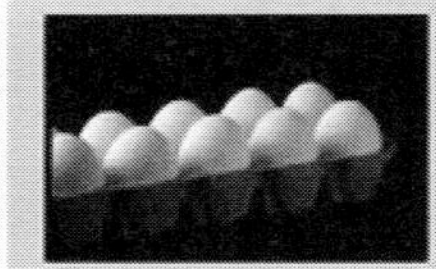	
Major technology	Mammalian (CHO) cells	Transgenic mammal milk	Transgenic chicken eggs	Transgenic plants
Companies	Amgen (Thousand Oaks, CA) Genentech (S. San Francisco, CA) other current biologics manufacturers; Crucell (Leiden, Netherlands) uses human cells	GTC Biotherapeutics (Framingham, MA) PPL Therapeutics (Edinburgh, UK) BioProtein (Paris, France)	Avigenics (Athens, GA) Origen Therapeutics (Burlingame, CA) TranXenoGen (Shrewsbury, MA) Viragen (Plantation, FL) GeneWorks (Ann Arbor, MI) Vivalis (Nantes, France)	Croptech (Blacksburg, VA) Epicyte (San Diego, CA) Large Scale Biology (Owensboro, KY) Meristem Therapeutics (Clermont-Ferrand, France) Prcdigene (College Station, TX)
Estimated cost (cost/g raw material)*	$150	$1–$2	$1–$2	$0.05

*Company estimates

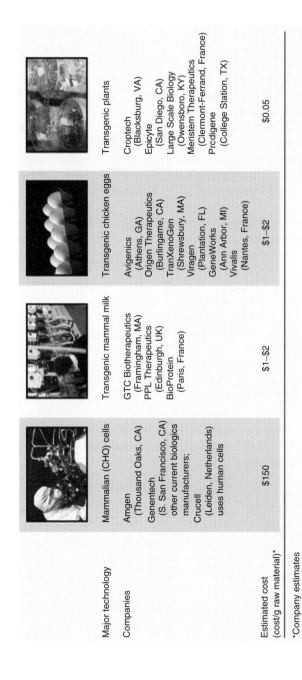

Fig. 2 Companies and technologies in biomanufacturing. A comparison of different expression systems shows the big differences in terms of costs, ranging from US$150 per gram for CHO cells to US$0.05 per gram for transgenic plants [11]. (See Color Plate p. xvi).

their *native* form, as they are capable of carrying out post-translational modifications required for the biological activity of many such proteins. These modifications can be acetylation, phosphorylation, and glycosylation as well as others. Per se, there is no restriction to the kind of proteins that can be expressed in plants: vaccines (e.g. pertussis or tetanus toxins), serum proteins (e.g. albumin), growth factors (e.g. vascular endothelial growth factor (VEGF), erythropoietin), or enzymes (e.g. urokinase, glucose oxidase, or glucocerebrosidase). However, enzymes sometimes have very complex cofactors, which are essential for their catalytic mode of action, but cannot be supplied by most expression systems. This is why, for the expression of some enzymes, expression systems with special features and characteristics need to be developed [13]. Another very important class of proteins are the antibodies (e.g. scFv, Fab, IgG, or IgA). More than 100 antibodies are currently used in clinical trials as therapeutics, drug delivery vehicles, in diagnostics and imaging, and in drug discovery research for both screening and validation of targets [14]. Again, plants are considered as the system of choice for the production of antibodies ("plantibodies") in bulk amounts at low costs. Since the initial demonstration that transgenic tobacco (*Nicotiana tabacum*) is able to produce functional IgG1 from mouse [15], full-length antibodies, hybrid antibodies, antibody fragments (Fab), and single-chain variable fragments (scFv) have been expressed in higher plants for a number of purposes. These antibodies can serve in health care and medicinal applications, either directly by using the plant as a food ingredient or as a pharmaceutical or diagnostic reagent after purification from the plant material. In addition, antibodies may improve plant performance, for example, by controlling plant disease or by modifying regulatory and metabolic pathways [16–19].

3.4
SWOT Analysis Reveals a Ripe Market for Plant Expression Systems

When I analyzed the different expression systems regarding their strengths, weaknesses, opportunities, and threats (SWOT), the advantages of plants and their potential to circumvent the worldwide capacity limitations for protein production became quite obvious (see Fig. 3). Comparison of transgenic animals, mammalian cell culture, plant expression systems, yeast, and bacteria shows certain advantages for each of the systems. In the order in which the systems were just mentioned, we can compare them in terms of their development time (speed). Transgenic animals have the longest cycle time (18 months to develop a goat), followed by mammalian cell culture, plants, yeast, and bacteria (one day to transform *E. coli*). If one looks at operating and capital costs, safety and scalability, the data show that plants are beneficial: therefore, in the comparison (see Fig. 3) they are shown on the right-hand side already. But even for glycosylation, multimeric assembly and folding (where plants are not shown on the right-hand side, meaning other systems are advantageous), some plant expression systems are moving in that direction. An example for this is the moss system from the company greenovation Biotech GmbH (Freiburg, Germany), which will be discussed in detail in the example section. This system performs proper folding and assembly of even such complex proteins like the homodimeric

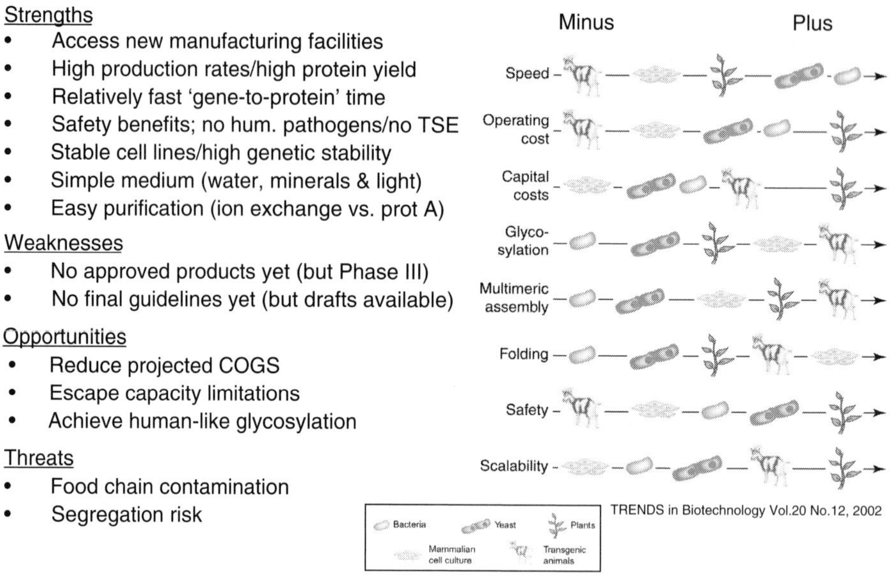
Strengths
• Access new manufacturing facilities
• High production rates/high protein yield
• Relatively fast 'gene-to-protein' time
• Safety benefits; no hum. pathogens/no TSE
• Stable cell lines/high genetic stability
• Simple medium (water, minerals & light)
• Easy purification (ion exchange vs. prot A)
Weaknesses
• No approved products yet (but Phase III)
• No final guidelines yet (but drafts available)
Opportunities
• Reduce projected COGS
• Escape capacity limitations
• Achieve human-like glycosylation
Threats
• Food chain contamination
• Segregation risk
Minus
Plus
Speed
Operating cost
Capital costs
Glyco-sylation
Multimeric assembly
Folding
Safety
Scalability
TRENDS in Biotechnology Vol.20 No.12, 2002
Bacteria
Yeast
Plants
Mammalian cell culture
Transgenic animals

Fig. 3 SWOT analysis of plant expression systems. Plant expression systems have a lot of advantages (plus) over other systems and are therefore mostly shown on the right-hand side of the picture (Raskin I et al., Plants and human health in the twenty-first century. *Trends in Biotechnol.* **2002** *20*, 522–531.). Herein different systems (transgenic animals, mammalian cell culture, plants, yeast, and bacteria) are compared in terms of speed (how quickly they can be developed), operating and capital costs and so on, and plants are obviously advantageous. Even for glycosylation, assembly and folding, where plants are not shown on the right-hand side (meaning other systems are advantageous), some plant expression systems are moving in that direction (as will be shown exemplarily in the section for moss). Also, the weaknesses and threats can be dealt with, using the appropriate plant expression system [20]. (See Color Plate p. xvii).

VEGF. Even the sugar pattern could successfully be reengineered from plant to humanlike glycosylation.

In addition to the potential of performing human glycosylation, plants also enjoy the distinct advantage of not harboring any pathogens, which are known to harm animal cells (as opposed to animal cell cultures and products), nor do the products contain any microbial toxins, TSE (Transmissible Spongiform Encephalopathies), prions, or oncogenic sequences [21]. In fact, humans are exposed to a large, constant dose of living plant viruses in the diet without any known effects/illnesses. Plant production of protein therapeutics also has advantages with regard to their scale and speed of production. Plants can be grown in ton quantities (using existing plant/crop technology, like commercial greenhouses), be extracted with industrial-scale equipment, and produce kilogram-size yields from a single plot of cultivation. These economies of scale are expected to reduce the cost of production of pure pharmaceutical-grade therapeutics by more than two orders of magnitude versus current bacterial fermentation or cell culture reactor systems (plus raw material COGS are estimated to be as low as 10% of conventional cell culture expenses).

Although a growing list of heterologous proteins were successfully produced in a number of plant expression systems with their manifold advantages, there are also obvious downsides. One weakness is that no product has been approved for the market yet (but will be soon, since some are in Phase III clinical trials already, see Table 1). The other weakness is that no final regulatory guidelines exist. But as mentioned before, regulatory authorities (Food and Drug Administration (FDA), European Medicine Evaluation Agency (EMEA), and Biotechnology Regulatory Service (BRS)) and the Biotechnology Industry Organization (BIO) have drafted guidelines on plant-derived biopharmaceuticals (see Table 2) and have asked the community for comments. The FDA has also issued several PTC (Points To Consider) guidelines about plant-based biologics, and review of the July 2002 PTC confirms that the FDA supports this field and highlights the benefits of plant expression systems – including the absence of any pathogens to man from plant extracts. The main concerns of using plant expression systems are societal ones about environmental impacts, segregation risk, and contamination of the food chain. But these threats can be dealt with, using non-edible plants (non-food, non-feed), applying advanced containment technologies (GMP greenhouses, bioreactors) and avoiding open-field production.

Owing to the obvious strengths of plant expression systems, there has been explosive growth in the number of start-up companies. Since the 1990s, a number of promising plant expression systems have been developed, and in response to this "blooming field" big pharmaceutical companies have become more interested. Now, the plant expression field is "ripe" for strategic alliances, and, in fact, the last year has seen several major biotech companies begin partnerships with such plant companies. The selection of several such partnerships shown in Table 1 clearly demonstrates that, in general, there has been sufficient experimentation with various crops to provide the overall proof of concept that transgenic plants can produce biopharmaceuticals. However, and this can be seen in the table as well, the *commercial production* of biopharmaceuticals

Tab. 1 Plant-derived biopharmaceuticals in clinical trials

Company	*Partner*	*Protein /indication*	*Host*	*Stage*
Monsanto	Guy's Hospital London	Anticaries antibody	Corn	Phase III
Large Scale Biology	Own product	scFv (non-Hodgkin)	Tobacco	Phase III
Meristem Therapeutics	Solvay Pharmaceuticals	Gastric lipase	Corn	Phase II
Large Scale Biology	ProdiGene, Plant Bioscience	Anti-idiotype antibody	Tobacco	Phase I
Monsanto	NeoRx	Antitumor antibody	Corn	Phase I
ProdiGene	Own product	TGEV vaccine	Corn	Phase I
Epicyte Pharmaceutical	Dow, Centocor	Anti-HSV antibody	Corn	Phase I
CropTech	Immunex	Enbrel (arthritis)	Tobacco	Preclinical
CropTech	Amgen	Therapeutic antibodies	Tobacco	Preclinical
AltaGen Bioscience Inc.	US Army + 3 biotechs	Antibodies	Potato	Preclinical
Meristem Therapeutics	CNRS	Human lactoferrin	Corn	Preclinical
MPB Cologne GmbH	Aventis CropScience	Confidential	Potato	Preclinical

Tab. 2 Drafted guidelines on plant-derived biopharmaceuticals

Agency	*Guideline*	*Status*
BRS (Biotechnology Regulatory Services)	"Case study on plant-derived biologics" for Office of Science and Technology Policy/Council on Environmental Quality	Released: Mar 5, 2001
BIO (Biotechnology Industry Organization)	"Reference Document for Confinement and Development of Plant-Made Pharmaceuticals in the United States"	Released: May 17, 2002
EMEA (European Medicine Evaluation Agency)	"Concept Paper on the Development of a Committee for Proprietary Medicinal Products (CPMP) Points to Consider on the Use of Transgenic Plants in the Manufacture of Biological Medicinal Products for Human Use"	Released: Mar 01, 2001
FDA (Food and Drug Administration)	"Drugs, Biologics, and Medical Devices Derived from Bioengineered Plants for Use in Humans and Animals"	Issued: Sep 6, 2002
EMEA (European Medicine Evaluation Agency)	"Points To Consider Quality Aspects of Medicinal Products containing active substances produced by stable transgene expression in higher plants"	Issued: Mar 13, 2002

in transgenic plants is still in the early stages of development and yet the most advanced products are in Phase III clinical development.

3.5
Risk Assessment and Contingency Measures

For a number of reasons, including the knowledge base developed on genetically modifying its genome, industrial processes for extracting fractionated products and the potential for large-scale production, the preferred plant expression system has been corn. However, the use of corn touches on a potential risk: some environmental activist groups and trade associations are concerned about the effect on the environment and possible contamination of the food supply. These issues are reflected in the regulatory guidelines and have been the driving force to investigate other plants as well. While many mature and larger companies have been working in this area for many years, there are a number of newcomers that are developing expertise as well. These smaller companies are reacting to the concerns by looking at the use of non-edible plants that can be readily raised in greenhouses. All potential risks have to be assessed and contingency measures need to be established. Understanding the underlying issues is mandatory to make sophisticated decisions about the science and subsequently on the development of appropriate plant expression systems for production of biopharmaceuticals.

Ongoing public fears from the food industry and the public, particularly in Europe ("Franken Food") could have spillover effects on plant-derived pharmaceuticals. Mistakes and misunderstandings have already cost the genetically enhanced grain industry hundreds of millions of dollars [21]. The only way to prevent plant expression systems from suffering the same dilemma is to provide the public with appropriate information on emerging discoveries and newly developed production systems for biopharmaceuticals. Real and theoretical risks involve the spread of engineered genes into wild plants, animals, and bacteria (horizontal transmission). For example, if herbicide-resistance was transmitted to weeds, or antibiotic resistance was to be transmitted to bacteria, superpathogens could result. If these genetic alterations were transmitted to their progeny (vertical transmission), an explosion of the pathogens could cause extensive harm. An example of this occurred several years ago, when it was feared that pest-resistant genes had been transmitted from Bt corn to milkweed – leading to the widespread death of Monarch butterflies. Although this was eventually not found to be the case, the public outcry over the incident was a wake-up call to the possible dangers of transgenic food technology. To avoid the same bad perception for biopharmaceuticals expressed in plants, there is the need for thorough risk assessment and contingency planning. One method is the employment of all feasible safety strategies to prevent spreading of engineered DNA (genetic drift), like a basic containment in a greenhouse environment. Although no practical shelter can totally eradicate insect and rodent intrusion, this type of isolation is very effective for self-pollinators and those plants with small pollen dispersal patterns. The use of species-specific, fragile, or poorly transmissible viral vectors is another strategy. Tobacco mosaic virus (TMV), for example, usually only infects a tobacco host. It requires an injury of the plant to gain entry and cause infection. Destruction of a field of TMV-transformed tobacco requires only plowing under or application of an herbicide. These factors prevent both horizontal and vertical transmission. In addition, there is no known incidence of plant viruses infecting animal or bacterial cells. Another approach is to avoid stable transgenic germ lines and therefore most uses of transforming viruses do not involve the incorporation of genes into the plant cell nucleus. By definition, it is almost impossible for these genes to be transmitted vertically through pollen or seed. The engineered protein product is produced only by the infected generation of plants. Another effective way to reduce the risk of genetic drift is the use of plants that do not reproduce without human aid. The modern corn plant cannot reproduce without cultivation and the purposeful planting of its seeds. If a plant may sprout from grain, it still needs to survive the wintering-over process and gain access to the proper planting depth. This extinction process is so rapid, however, that the errant loss of an ear of corn is very unlikely to grow a new plant. Another very well-known example of self-limited reproduction is the modern banana. It propagates almost exclusively through vegetative cloning (i.e. via cuttings).

Pollination is the natural way for most plants to spread their genetic information, make up new plants, and to deliver their offspring in other locations. The use of plants with limited range of pollen dispersal and limited contact with compatible wild hosts therefore is also very effective to prevent genetic drift. Corn, for example, has pollen, which survives for only 10 to 30 min and, hence, has an effective fertilizing radius of less than 500 m. In North America, it has no wild-type relatives with which it could cross-pollinate. In addition to being spatially isolated from nearby cornfields, transgenic corn can be "temporally isolated" by being planted at least 21 days earlier or 21 days later than the surrounding corn, to ensure that the fields are not producing flowers at the same time [11]. Under recent USDA (U.S. Department of Agriculture) regulations, the field must also be planted with equipment dedicated to the genetically modified crop. For soybeans, the situation is different, since they are virtually 100% self-fertilizers and can be planted in very close proximity to other plants without fear of horizontal spread. Another option is the design of transgenic plants that have only sterile pollen or – more or less only applicable for greenhouses – completely prevent cross-pollination by covering the individual plants. One public fear regards spreading antibiotic resistance from one (transgenic donor) plant to other wild-type plants or bacteria in the environment. Although prokaryotic promoters for antibiotic-resistance are sometimes used in the fabrication and selection of transgenic constructs, once a transgene has been stably incorporated into the plant genome, it is under the control of plant (eukaryotic) promoter elements. Hence, antibiotic-resistance genes are unable to pass from genetically altered plants into bacteria and remain functional. As stated earlier, another common fear is the creation of a "super bug." The chance of creating a supervirulent virus or bacterium from genetic engineering is unlikely, because the construction of expression cassettes from viral or bacterial genomes involves the removal of the majority of genes responsible for the normal function of these organisms. Even if a resultant organism is somewhat functional, it cannot compete for long in nature with normal, wild-type bacteria of the same species.

As one can see from the aforementioned safety strategies, considerable effort is put into the reduction of any potential risk from the transgenic plant for the environment. In general, the scientific risk can be kept at a minimum, if common sense is applied – in accordance with Thomas Huxley (1825–1895) that "Science is simply common sense at its best." For example, protein toxins (for vaccine production) should never be grown in food plants.

Additionally, the following can be employed as a kind of risk management to prevent the inappropriate or unsafe use of genetically engineered plants [21]:

- An easily recognized phenotypic characteristic can be coexpressed in an engineered product (e.g. tomatoes that contain a therapeutic protein can be selected to grow in a colorless variety of fruit).
- Protein expression can be induced only after harvesting or fruit ripening. For example, CropTech's (Blacksburg, VA, USA) inducible expression system in tobacco, MeGA-PharM, leads to very efficient induction upon leaf injury (harvest) and needs no chemical inducers. This system possesses a fast induction response and protein synthesis rate, thus leads to high expression levels with no aged product in the field (no environmental damage accumulation).
- Potentially antigenic or immunomodulatory products can be induced to grow in, or not to grow in, a certain plant tissue (e.g. root, leaf/stem, seed, or pollen). In this way, for example, farmers can be protected from harmful airborne pollen or seed dusts.

- Although no absolute system can prevent vandalism or theft of the transgenic plants, a very effective, cheap solution has been used quietly for many years now in the United States. Plots of these modified plants are being grown with absolutely no indication that they are different from a routine crop. In the Midwest, for example, finding a transgenic corn plot among the millions of acres of concurrently growing grain is virtually impossible. The only question here is, if this approach really helps facilitating a fair and an open discussion with the public. Asking the same question for the EU is not relevant: owing to labeling requirements, this approach would not be feasible, as, in general, it is much more difficult to perform open-field studies with transgenic plants.

3.6 Moving Plants to Humanlike Glycosylation

As discussed earlier, plant production of therapeutic proteins has many advantages over bacterial systems. One very important feature of plant cells is their capability of carrying out post-translational modifications [22]. Since they are eukaryotes (i.e. have a nucleus), plants produce proteins through an ER (Endoplasmatic Reticulum) pathway, adding sugar residues also to the protein – a process called *glycosylation*. These carbohydrates help determine the three-dimensional structures of proteins, which are inherently linked to their function and their efficacy as therapeutics. This glycosylation also affects protein bioavailability and breakdown of the biopharmaceutical; for example, proteins lacking terminal sialic acid residues on their sugar groups are often targeted by the immune system and are rapidly degraded [23]. The glycosylation process begins by targeting the protein to the ER. During translation of mRNA (messenger RNA) into protein, the ribosome is attached to the ER, and the nascent protein fed into the lumen of the ER as translation proceeds. Here, one set of glycosylation enzymes attaches carbohydrates to specific amino acids of the protein. Other glycosylation enzymes either delete or add more sugars to the core structures. This glycosylation process continues into the Golgi apparatus, which sorts the new proteins, and distributes them to their final destinations in the cell (see Fig. 4). Bacteria lack this ability and therefore cannot be used to synthesize proteins that require glycosylation for activity. Although plants have a somewhat different system of protein glycosylation from mammalian cells, the differences are usually not proving to be a problem. Some proteins, however, require humanlike glycosylation (see Fig. 5) – they must have specific sugar structures attached to the correct sites on the molecule to be maximally effective [23]. Therefore, some efforts are being made in modifying host plants in such a way that they provide the protein with human glycosylation patterns. One example of modifying a plant expression system in this way is the transgenic moss, which will be discussed in the next section.

3.7 Three Promising Examples: Tobacco (Rhizosecretion, Transfection) and Moss (Glycosylation)

To further elaborate on improving glycosylation and downstream processing, three interesting plant expression systems will be discussed. All systems share the advantage of utilizing non-edible plants (non-food and non-feed) and can be kept in either a greenhouse or a fermenter to avoid

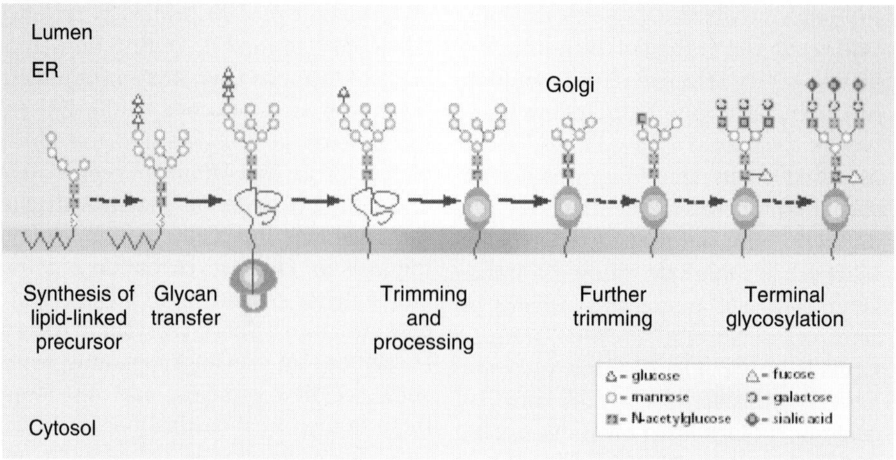

Fig. 4 The glycosylation pathway via ER and Golgi apparatus. In the cytosol carbohydrates are attached to a lipid precursor, which is then transported into the lumen of the ER to finish core glycosylation. This glycan is now attached to the nascent, folding polypeptide chain (which is synthesized by ribosomes attached to the cytosolic side of the ER from where it translocates into the lumen) and subsequently trimmed and processed before it is folded and moved to the Golgi apparatus. Capping of the oligosaccharide branches with sialic acid and fucose is the final step on the way to a mature glycoprotein [23]. (See Color Plate p. xviii).

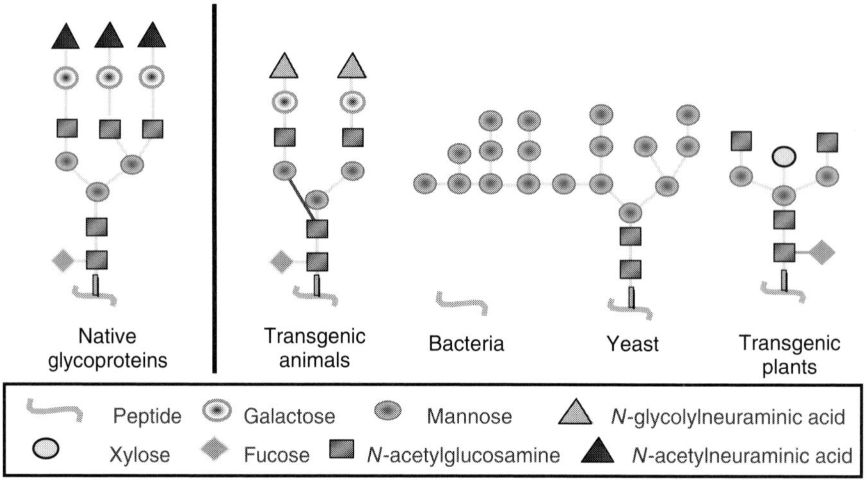

Fig. 5 Engineering plants to humanlike glycosylation. The first step to achieve humanlike glycosylation in plants is to eliminate the plant glycosylation pattern, that is, the attachment of β-1-2-linked xylosyl and α-3-linked fucosyl sugars to the protein. Because these two residues have allergenic potential, the corresponding enzymes xylosyl and fucosyl transferase are knocked out. In case galactose is relevant for the final product, galactosyl transferase is inserted into the host genome. Galactose is available in the organism so that this single-gene insertion is sufficient to ensure galactosylation [24]. (See Color Plate p. xviii).

any segregation risk. Another obvious advantage is secretion of the protein into the medium so that no grinding or extraction is required. This is very important in light of downstream processing: protein purification is often as expensive as the biomanufacturing and should never be underestimated in the total COGS equation [22].

3.8 Harnessing Tobacco Roots to Secrete Proteins

Phytomedics (Dayton, NJ, USA) uses tobacco plants as an expression system for biopharmaceuticals. Besides the advantage of being well characterized and used in agriculture for some time, tobacco has a stable genetic system, provides high-density tissue (high protein production), needs only simple medium, and can be kept in a greenhouse (see Fig. 6). Optimized antibody expression can be rapidly verified using transient expression assays (short development time) in the plants before creation of transgenic suspension cells or stable plant lines (longer development time). Different vector systems, harboring targeting signals for subcellular compartments, are constructed in parallel and used for transient expression. Applying this screening approach, high expressing cell lines can rapidly be identified. For example, transgenic tobacco plants, transformed with an expression cassette containing the GFP (Green Fluorescent Protein) gene fused to an *aps* (amplification-promoting sequence), had greater levels of corresponding mRNAs and expressed proteins compared to transformants lacking *aps* [25]. Usually, downstream processing (isolation/extraction and purification of

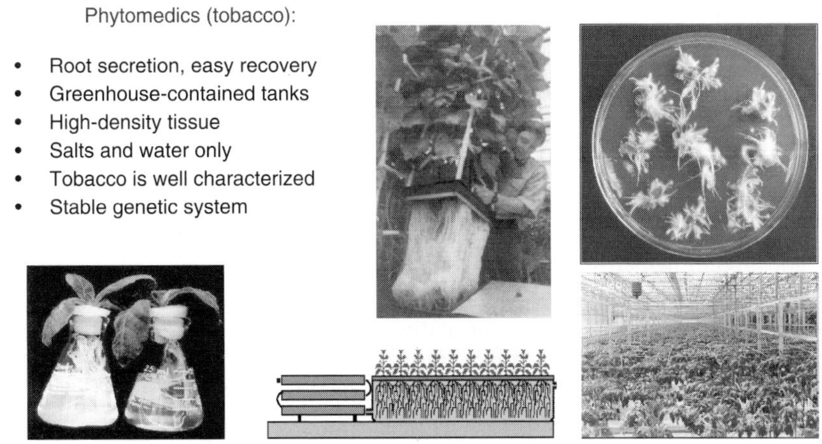
Phytomedics (tobacco):
• Root secretion, easy recovery
• Greenhouse-contained tanks
• High-density tissue
• Salts and water only
• Tobacco is well characterized
• Stable genetic system

Fig. 6 Secretion of the biopharmaceuticals via tobacco roots. The tobacco plants are genetically modified in such a way, that the protein is secreted via the roots into the medium ("rhizosecretion"). In this example, the tobacco plant takes up nutrients and water from the medium and releases GFP (green fluorescent protein). Examination of root-cultivation medium by its exposure to near-ultraviolet illumination reveals the bright green-blue fluorescence characteristics of GFP in the hydroponic medium (left flask in panel lower left edge). The picture also shows a schematic drawing of the hydroponic tank, as well as tobacco plants at different growth stages, for example, callus,–fully grown and greenhouse plantation [24]. (See Color Plate p. xix).

the target protein) is limiting for such a system, for example, if the protein has to be isolated from biochemically complex plant tissues (e.g. leaves), this can be a laborious and expensive process and a major obstacle to large-scale protein manufacturing. To overcome this problem, secretion-based systems utilizing transgenic plant cells or plant organs aseptically cultivated in vitro would be one solution. However, in vitro systems can be expensive, slow growing, unstable, and relatively low yielding. This is why another interesting route was followed. Secretion of molecules is a basic function of plant cells and organs in plants, and is especially developed in plant roots. In order to take up nutrients from the soil, interact with other soil organisms, and defend themselves against numerous pathogens, plant roots have evolved sophisticated mechanisms based on the secretion of different biochemicals (including proteins like toxins) into their neighbourhood (rhizosphere). In fact, Borisjuk and coworkers [26] could demonstrate that root secretion can be successfully exploited for the continuous production of recombinant proteins in a process termed "rhizosecretion." Here, an endoplasmic reticulum signal peptide is fused to the recombinant protein, which is then continuously secreted from the roots into a simple hydroponic medium (based on the natural secretion from roots of the intact plants). The roots of the tobacco plant are sitting in a hydroponic tank (see Fig. 6), taking up water and nutrients and continuously releasing the biopharmaceutical. By this elegant set up, downstream processing becomes easy and cost-effective, and also offers the advantage of continuous protein production that integrates the biosynthetic potential of a plant over its lifetime and might lead to higher protein yields than single-harvest and extraction methods. Rhizosecretion is demonstrated in Fig. 6, showing a transgenic tobacco plant expressing GFP and releasing it into the medium.

3.9 High Protein Yields Utilizing Viral Transfection

ICON Genetics (Halle, Germany) has developed a protein-production system that relies on rapid multiplication of viral vectors in an infected tobacco plant (see Fig. 7). Viral transfection systems offer a number of advantages, such as very rapid (1 to 2 week) expression time, possibility of generating initial milligram quantities within weeks, high expression levels, and so on. However, the existing viral vectors, such as TMV-based vectors used by, for example, Large Scale Biology Corp. (Vacaville, CA, USA) for production of single-chain antibodies for treatment of non-Hodgkin lymphoma (currently in Phase III clinical trials, see Table 1), had numerous shortcomings, such as inability to express genes larger than 1 kb, inability to coexpress two or more proteins (a prerequisite for production of monoclonal antibodies, because they consist of the light and heavy chains, which are expressed independently and are subsequently assembled), low expression level in systemically infected leaves, and so on. ICON has solved many of these problems by designing a process that starts with an assembly of one or more viral vectors inside of a plant after treating the leaves with agrobacteria, which deliver the necessary viral vector components. ICON's proviral vectors provide advantages of fast and high-yield amplification

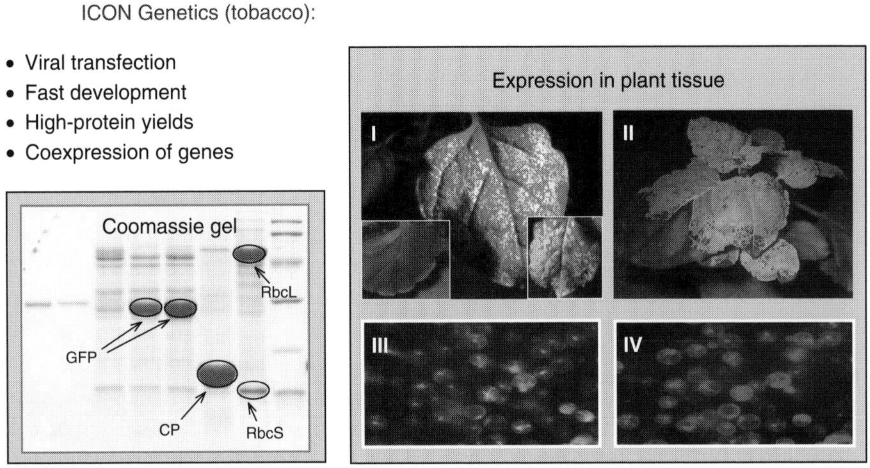

Fig. 7 Viral transfection of tobacco plants. This new generation platform for fast (1 to 2 weeks), high-yield (up to 5 g per kilogram of fresh leaf weight) production of biopharmaceuticals is based on proviral gene amplification in a non-food host. Antibodies, antigens, interferons, hormones, and enzymes could successfully be expressed with this system. The picture shows development of initial symptoms on a tobacco following the agrobacterium-mediated infection with viral vector components that contain a *GFP* gene (I); this development eventually leads to a systemic spread of the virus, literally converting the plant into a sack full of protein of interest within two weeks (II). The system allows to coexpress two proteins in the same cell, a feature that allows expression of complex proteins such as full-length monoclonal antibodies. Panel III and IV show the same microscope section with the same cells, expressing green fluorescent protein (III) and red fluorescent protein (IV) at the same time. The yield and total protein concentration achievable are illustrated by a Coomassie gel with proteins in the system: GFP (protein of interest), CP (coat protein from wild-type virus), RbcS and RbcL (small and large subunit of ribulose-1,5-bisphosphate carboxylase) [24]. (See Color Plate p. xx).

processes in a plant cell, simple and inexpensive assembly of expression cassettes in planta, and full control of the process. The robustness of highly standardized protocols allow to use inherently the same safe protocols for both laboratory-scale as well as industrial production processes. In this system, the plant is modified transiently rather than genetically and reaches the speed and yield of microbial systems while enjoying post-translational capabilities of plant cells. De- and reconstructing of the virus adds some safety features and also increases efficiency. There is no "physiology conflict," because the "growth phase" is separated from the "production phase," so that no competition occurs for nutrients and other components required for growth and also for expression of the biopharmaceutical at the same time.

This transfection-based platform allows to produce proteins in a plant host at a cost of US$1 to 10 per gram of crude protein. The platform is essentially free from limitations (gene insert size limit, inability to express more than one gene) of current viral vector-based platforms. The expression levels reach 5 g per kilogram of fresh leaf tissue (or some 50% of total cellular protein!) in 5 to 14 days after inoculation. Since the virus process (in addition to superhigh production of its own proteins, including the protein of interest) leads to the shut-off of the other cellular protein synthesis, the amount of protein of interest in the initial extract is extremely high (Fig. 7). It thus results in reduced costs of downstream processing. Milligram quantities can be produced within two weeks, gram quantities in 4 to 6 months, and the production system is inherently scalable. A number of high-value proteins have been successfully expressed, including antibodies, antigens, interferons, hormones, and enzymes.

3.10 Simple Moss Performs Complex Glycosylation

Greenovation Biotech GmbH (Freiburg, Germany) has established an innovative production system for human proteins. The system produces pharmacologically active proteins in a bioreactor, utilizing a moss (*Physcomitrella patens*) cell culture system with unique properties (see Fig. 8). It was stated before that post-translational modifications for some proteins are crucial to gain complete pharmacological activity. Since moss is the only known plant system that shows a high frequency of homologous recombination, this is a highly attractive tool for production strain design. By establishing stable integration of foreign genes (gene knockout and new transgene insertion) into the plant genome, it can be programmed to produce proteins with modified glycosylation patterns that are identical to animal cells. The moss is photoautotrophic and therefore only requires simple media for growth, which consist essentially of water and minerals. This reduces costs and also accounts for significantly lower infectious and contamination risks, but in addition to that, the system has some more advantages:

- The transient system allows production of quantities for a feasibility study within weeks – production of a stable expression strain takes 4 to 6 months.
- On the basis of transient expression data, the yield of stable production lines

Greenovation (moss system):

- Simple medium (photoautotrophic plant needs only water and minerals)
- Robust expression system (good expression levels from 15 to 25 °C)

- Secretion into medium via human leader sequence (broad pH range: 4–8)
- Easy purification from low-salt medium via ion exchange

- Easy genetic modifications to cell lines
- Stable cell lines/high genetic stability

- Codon usage like human (no changes required)
- Inexpensive bioreactors from the shelf

- Nonfood plant (no segregation risk)
- Good progress on genetic modification of glycosylation pathways (plant to human)

Fig. 8 Greenovation use a fully contained moss bioreactor. This company has established an innovative production system for human proteins. The system produces pharmacologically active proteins in a bioreactor, utilizing a moss (*Physcomitrella patens*) cell culture system with unique properties [24]. (See Color Plate p. xxi).

is expected to reach 30 mg per liter per day. This corresponds to the yield of a typical fed-batch culture over 20 days of 600 mg per liter.

- Bacterial fermentation usually requires addition of antibiotics (serving as selection marker and to avoid loss of the expression vector). For moss cultivation, no antibiotics are needed – this avoids the risk of traces of antibiotics having a significant allergenic potential in the finished product.
- Genetic stability is provided by the fact that the moss is grown in small plant fragments and not as protoplasts or tissue cultures avoiding somaclonal variation.
- As a contained system, the moss bioreactor can be standardized and validated according to GMP standards mandatory in the pharmaceutical industry.
- Excretion into the simple medium is another major feature of the moss bioreactor, which greatly facilitates downstream processing.

As discussed in detail, the first step to get humanlike glycosylation in plants, is to eliminate the plant glycosylation, for example, the attachment of β-1-2-linked xylosyl and α-1-3-linked fucosyl sugars to the protein, because these two residues have allergenic potential. Greenovation was able to knockout the relevant glycosylation enzymes xylosyl transferase and fucosyl transferase, which was confirmed by RT-PCR (reverse transcriptase PCR). And indeed, xylosyl and fucosyl residues were completely removed from the glycosylation pattern of the expressed protein as confirmed by MALDI-TOF (matrix assisted laser desorption ionization time of flight) mass spectroscopy analysis (see Fig. 9).

A very challenging protein to express is VEGF, because this homodimer consists of two identical monomers linked via

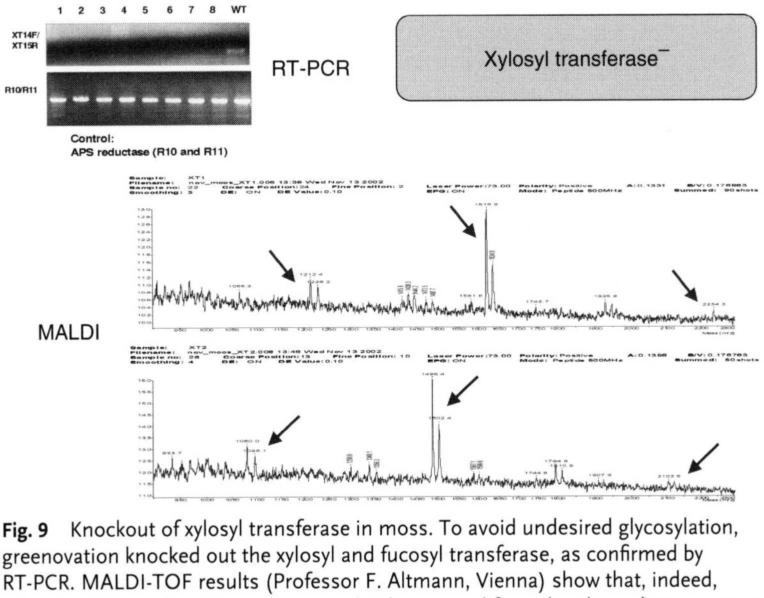

Fig. 9 Knockout of xylosyl transferase in moss. To avoid undesired glycosylation, greenovation knocked out the xylosyl and fucosyl transferase, as confirmed by RT-PCR. MALDI-TOF results (Professor F. Altmann, Vienna) show that, indeed, xylosyl and fucosyl residues were completely removed from the glycosylation pattern of the expressed protein (data for knockout of fucosyl transferase not shown) [24].

a disulfide bond. To produce VEGF in an active form, the following need to be provided:

- Monomers need to be expressed to the right level.
- Monomers need to be correctly folded.
- Homodimer needs to be correctly assembled and linked via a disulfide-bond.
- Complex protein needs to be secreted in its active form.

And in fact, all this could be achieved with the transgenic moss system as shown in Fig. 10. These results are very promising because they demonstrate that this system is capable of expressing even very complex proteins. In addition to that, the moss system adds no plant-specific sugars to the protein – a major step toward humanlike glycosylation. Furthermore, moss is a robust expression system leading to high yields at 15 to 25 °C and the pH can be adjusted from 4 to 8 depending on the optimum for the protein of interest. Adapting existing technology for large-scale cultivation of algae, fermentation of moss in suspension culture allows scaling of the photobioreactors up to several 1000 L (see Fig. 11). Finally, the medium is inexpensive, since only water and minerals are sufficient.

3.11 Other Systems Used for Plant Expression

Several different plants have been used for the expression of proteins in plants. All these systems have certain advantages regarding edibility, growth rate, scalability, gene-to-protein time, yield, downstream processing, ease of use, and so on, which I will not discuss in further detail here. A selection of different expression systems is listed on the next page:

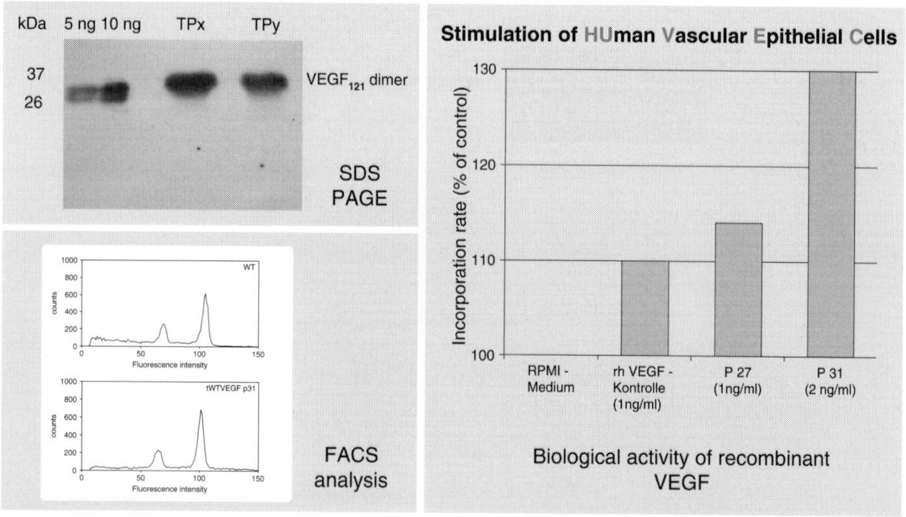

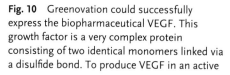

Fig. 10 Greenovation could successfully express the biopharmaceutical VEGF. This growth factor is a very complex protein consisting of two identical monomers linked via a disulfide bond. To produce VEGF in an active form, the monomers need to be expressed to the right level, correctly folded, assembled, and linked via the disulfide bond. The analytical assays clearly show that expression in moss yielded completely active VEGF [20].

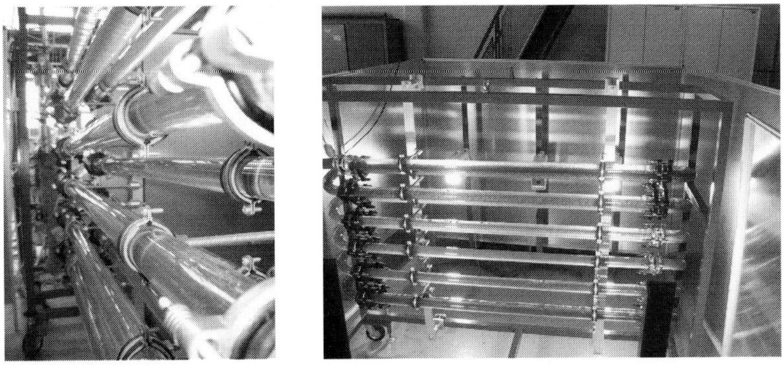

30 L pilot reactor for moss   Two weeks after incubation

Fig. 11 Scaling of photobioreactors up to several 1000 L. The moss bioreactor is based on the cultivation of *Physcomitrella patens* in a fermenter. The moss protonema is grown under photoautotrophic conditions in a medium that consists essentially of water and minerals. Light and carbon dioxide serve as the only energy and carbon sources. Cultivation in suspension allows scaling of the photobioreactors up to several 1000 L. Adaptation of existing technology for large-scale cultivation of algae is done in cooperation with the Technical University of Karlsruhe. Courtesy of greenovation Biotech GmbH (Freiburg, Germany) and Professor C. Posten, Technical University (Karlsruhe, Germany). (See Color Plate p. xxi).

Alfalfa	Ethiopian mustard	Potatoes
Arabidopsis	Lemna	Rice
Banana	Maize	Soyabean
Cauliflower	Moss	Tomatoes
Corn	Oilseeds	Wheat

Some of these systems have been used for research on the basis of their ease of transformation, well-known characterization, and ease to work with. However, they are not necessarily appropriate for *commercial* production. Which crop is ultimately used for full-scale commercial production will depend on a number of factors [21] including

- time to develop an appropriate system (gene-to-protein);
- section of the plant expressing the product/possible secretion;
- cost and potential waste products from extraction;
- "aged" product/ease of storage;
- long-term stability of the storage tissue;
- quantities of protein needed (scale of production).

Depending on the genetic complexity and ease of manipulation, the development time to produce an appropriate transgenic plant for milligram production of the desired protein can vary from 10 to 12 months in corn as compared to only weeks in moss. Estimates for full GMP production in corn are 30 to 36 months and approximately 12 months for moss. Expression of the protein in various tissues of the plant can result in a great variation in yield. Expression in the seed can often lead to higher yields than in the leafy portion of the plant. This is another explanation for the high interest in using corn, which has a relatively high seed-to-leaf ratio. Extraction from leaf can be costly as it contains a high percentage of water, which could result in unavoidable proteolysis during the process. Proteins stored in seeds can be desiccated and remain intact for long periods of time. The purification and extraction of the protein is likely to be done by adaptations of current processes for the extraction and/or fractionation. For these reasons, it is anticipated that large-scale commercial production of recombinant proteins will involve grain and oilseed crops such as maize, rice, wheat, and soybeans. On the basis of permits for open-air test plots issued by the USDA for pharmaceutical proteins and industrial biochemicals, corn is the crop of choice for production with 73% of the permits issued. The other major crops are soybeans (12%), tobacco (10%) and rice (5%).

In general, the use of smaller plants that can be grown in greenhouses is an effective way of producing the biopharmaceuticals and alleviating concerns from environmental activist groups that the transgenic plant might be harmful to the environment (food chain, segregation risk, genetic drift, etc.).

3.12 Analytical Characterization

Validated bioanalytical assays are essential and have to be developed to characterize the biopharmaceuticals during the production process (e.g. in-process control) and to release the final product for use as a drug in humans. These assays are applied to determine characteristics such as purity/impurities, identity, quantity, stability, specificity, and potency of the recombinant protein during drug development. Since the very diverse functions of different proteins heavily depend on their structure [27], one very valuable parameter in protein characterization is the elucidation of their three-dimensional structure. Although over the last couple of years a lot of efforts were put into method improvement for the elucidation of protein structures (during my PhD thesis I was also working in this fascinating field together with my boss Professor Robert Huber [25], Nobel Prize Laureate in 1988 "for the determination of the three-dimensional structure of a photosynthetic reaction centre") it is still very time-consuming to solve the 3D structure of larger proteins. This is why despite the high degree of information that can be obtained from the protein structure, this approach cannot be applied on a routine basis. Therefore, tremendous efforts are put into the development of other assays to guarantee that a potent biopharmaceutical drug is indeed ready for use in humans. A comprehensive overview is given in this textbook in Chapter 6 (Jutta Haunschild and Titus Kretzschmar "Characterization and bioanalytical aspects of recombinant proteins as pharmaceutical drugs").

3.13 Conclusion

The production of protein therapeutics from transgenic plants is becoming a reality. The numerous benefits offered by plants (low cost of cultivation, high biomass production, relatively fast gene-to-protein time, low capital and operating costs, excellent scalability, eukaryotic post-translational modifications, low risk of human pathogens, lack of endotoxins, as well as high protein yields) virtually guarantee that plant-derived proteins will become more and more common for therapeutic uses. Taking advantage of plant expression systems, the availability of cheap protein-based vaccines in underdeveloped countries of the world is in the near future. The cost of very expensive hormone therapies (erythropoietin, human growth hormone, etc.) could fall dramatically within the next decade due to the use of, for example, plant expression systems. Fears about the risks of the plant expression technology are real and well founded, but with a detailed understanding of the technology, it is possible to proactively address these safety issues and create a plant expression industry almost free of mishaps. For this purpose, the entire set up, consisting of the specific plant expression system and the protein being produced, needs to be analyzed and its potential risks assessed on a case-by-case basis. As plant-derived therapeutics begin to demonstrate widespread, tangible benefits to the population, and as the plant expression industry develops a longer safety track record, public acceptance of the technology is likely to improve continuously. Plants are by far the most abundant and cost-effective renewable resource uniquely adapted to complex biochemical synthesis. The increasing cost of energy and chemical raw materials, combined with the environmental concerns associated with conventional pharmaceutical manufacturing, will make plants even more compatible in the future. With the words of Max Planck (1858–1947) "How far advanced Man's scientific knowledge may be, when confronted with Nature's immeasurable richness and capacity for constant renewal, he will be like a marveling child and must always be prepared for new surprises," we will definitely discover more fascinating features of plant expression systems. But there is no need to wait: combining the advantages of some technologies that we have in hand by now could already lead to the ultimate plant expression system. This is what we should focus on, because, then, at the dawn of this new millennium, this would for the first time yield large-enough amounts of biopharmaceuticals to treat everybody on our planet!

Acknowledgments

I would like to thank the companies greenovation Biotech GmbH (Freiburg, Germany), ICON Genetics (Halle, Germany), and Phytomedics (Dayton, NJ, USA) for providing some data and figures to prepare this manuscript.

References

1. K. J. Morrow, *Genet. Eng. News* **2002**, *22*, 34–39.
2. The Context Network, *Biopharmaceutical Production in Plants*, Biopharma Prospectus, West Des Moines, IA, USA, Copyright© 2002.
3. L. Davies, J. Plieth, *Scr. Mag.* **2001**, *10*, 25–29.
4. Arthur D. Little, Inc. (ADL), Cambridge, MA, USA, AgIndustries Research, Copyright© 2002.
5. K. Garber, *Nat. Biotechnol.* **2001**, *19*, 184–185.
6. R. L. Evangelista, A. R. Kusnadi, J. A. Howard et al., *Biotechnol. Prog.* **1988**, *14*, 607–614.
7. A. S. Ponstein, T. C. Verwoerd, J. Pen, Production of enzymes for industrial use in *Engineering Plants for Commercial Products and Applications* (Eds.: G. B. Collins, R. J. Sheperd), New York Academy of Sciences, New York, 1996, pp. 91–98, Vol. 792.
8. G. Giddings, G. Allison, D. Brooks et al., *Nat. Biotechnol.* **2000**, *18*, 1151–1155.
9. R. Fischer, N. Emans, *Transgenic Res.* **2000**, *9*, 279–299.
10. A. Kusnadi, Z. L. Nikolov, J. A. Howard, *Biotechnol. Bioeng.* **1997**, *56*, 473–484.

11. A. Dove, *Nat. Biotechnol.* **2002**, *20*, 777–779.
12. J. Knäblein, *Transport Processes in Bioreactors and Modern Fermentation Technologies*, Lecture at University of Applied Sciences, Emden, Germany, 2002.
13. J. Knäblein, *Isolation, Cloning And Sequencing of the Respiratory Operon of Rhodobacter capsulatus and the Development of a General Applicable System for the Homologue Expression of Difficult-to-Express Proteins*, ISBN 3-933083-23-0, Hieronymus Verlag, München, 1997.
14. Drug & Market Development Publications, *Antibody Engineering: Technologies*, Applications and Business Opportunities, Westborough, MA, USA, Copyright© 2003.
15. A. Hiatt, R. Cafferkey, K. Bowdish, *Nature* **1989**, *342*, 76–78.
16. U. Conrad, U. Fiedler, *Plant Mol. Biol.* **1994**, *26*, 1023–1030.
17. J. K. C. Ma, M. B. Hein, *Plant Physiol.* **1995**, *109*, 341–346.
18. M. D. Smith, *Biotechnol. Adv.* **1996**, *14*, 267–281.
19. G. C. Whitelam, W. Cockburn, *Trends Plant Sci.* **1996**, *1*, 268–272.
20. J. Knäblein, *Biotech: A New Era in the New Millennium – From Plant Fermentation to Plant Expression of Biopharmaceuticals*, PDA International Congress, Prague, Czech Republic, 2003.
21. Technology Catalysts International Corporation, *Biopharmaceutical Farming*, Falls Church, VA, USA, Copyright© 2002.
22. I. Raskin, B. Fridlender et al., *Trends Biotechnol.* **2002**, *20*, 522–531.
23. A. Dove, *Nat. Biotechnol.* **2001**, *19*, 913–917.
24. J. Knäblein, *Biotech: A New Era in the New Millennium – Biopharmaceutic Drugs Manufactured in Novel Expression Systems*, 21. DECHEMA-Jahrestagung der Biotechnologen, Munich, Germany, 2003.
25. N. V. Borisjuk, I. Raskin et al., *Nat. Biotechnol.* **2000**, *18*, 1303–1306.
26. N. V. Borisjuk, I. Raskin et al., *Nat. Biotechnol.* **1999**, *17*, 466–469.
27. M. J. Romao, J. Knäblein, R. Huber et al., *Prog. Biophys. Mol. Biol.* **1998**, *68*, 121–144.
28. J. Knäblein, R. Huber et al., *J. Mol. Biol.* **1997**, *270*, 1–7.

Part II
Industrial Development and Production Process

4
Scientific, Technical and Economic Aspects of Vaccine Research and Development

Jens-Peter Gregersen
Chiron-Behring GmbH, Marburg, Germany

4.1
Introduction

Vaccine research and vaccine development are commonly combined by the term R&D because, in practice, these two different disciplines cannot be easily separated. Vaccine research and development have much in common: they both use the same technical language, apply very similar methods and tools, and have the same ultimate goal, but there is also a fundamental difference, as the underlying motivating factors, working habits, and the final output and results are entirely different. Research is mainly motivated by, and aiming at, scientific publications, which are best achieved by new methods, inventions, and discoveries. As soon as these have been published, a researcher's attention must turn to another and new subject. Developers normally start their work when new discoveries or inventions have been made and may well work on one and the same objective for an entire decade without publishing anything. They are not aiming at inventions; their intention must be to arrive at innovations, that is, products that will have an impact on our daily life. For the researcher, a vaccine could be an antigen or a preparation that has the potential of eliminating or inhibiting microorganisms. In order to convert this into a useful vaccine, developers must then add several other dimensions to the research product, namely quality, safety, a specifically defined clinical efficacy, and practical utility. Building practical utility into a product is probably the most demanding or far-reaching one of those four categories. It encompasses and combines almost any aspect of the product, including its local reactivity, acceptable application schemes with only few vaccinations, a proven and perceived effectiveness, comfortable presentation forms, formulations that guarantee good stability and shelf-lives, and, of course, adequate product prices.

4.2
From the Research Concept to a Development Candidate

Concepts for new vaccines arise from research and are based on combined scientific findings collected over many years and by various scientific institutions and disciplines. New vaccine concepts are regularly presented and proposed in large numbers by scientific publications or patent applications, but these concepts rarely result in new vaccines. After being tested in mice, most concepts slowly fade, since the original results cannot be reproduced under more practical conditions or turn out to be insufficiently effective to justify additional work. On the other hand, there are also organizational and financial aspects that represent serious hurdles. Most academic institutions and scientists simply do not have adequate resources to perform vaccine studies in specific models or even in monkeys or primates. Whereas vaccine antigen candidates can be designed and made by only one or a few individuals, studying these more intensively would normally require other specialists, specific facilities, and, of course, much more money. The initial research project now competes for scarce resources and needs very convincing data to make it to the next stage.

Scientific collaborations across institutional walls are an almost absolute prerequisite for continuing projects beyond testing in small laboratory animals and in order to proceed into a more intensive and application-oriented research phase. During this secondary research phase, promising concepts are taken up, reproduced, and improved until finally – and in only very few cases – a viable product and development concept can be put together. Almost invariably the efficacy of the candidate vaccine needs to be improved and made more reliable. For many new indications, even the tools and models must be established first, by which immunological effects or protection can be adequately measured.

For those few candidates that remain attractive after being studied in a more reliable way or in better models, it will then be important to assess the technical and economic aspects of the vaccine candidate very carefully. As these strongly depend on available facilities, general expertise, and specific experience with certain techniques needed, these aspects are normally evaluated by the developing organization during a project evaluation or "predevelopment" phase. At the end of this phase, a development concept should be available, which at least fulfills the following three criteria:

1. There should be sufficient evidence that the vaccine candidate is effective and protective in humans or in the target animal species. This normally presupposes that meaningful animal models have been established and that the vaccine has been tested successfully.
2. There should be a defined technical base or verified options by which the vaccine can be reliably and safely produced on a large scale. This includes, for example, cell culture or expression systems, purification schemes, and formulations that are qualified for the production of pharmaceutical products and do not contain hazardous components that cannot be removed during later process steps.
3. The expected product cost and the resulting sales prices must be in balance with the envisaged benefit of using the vaccine and expected revenues should be able to recoup the development cost in a reasonable period of time.

Thus, there should be rather clear ideas as to how the vaccine is to be manufactured and how it is characterized in its main qualities. If this base is not yet known or is based only on assumptions, a targeted product development in its strict sense is not possible, as neither the way to go nor the target or end result are known. In this case, the project should still be considered to be a research project. But particularly in the case of vaccines, development projects are frequently started with many uncertainties, assumptions, and compromises, as vaccines are highly complex compositions, which cannot be characterized entirely and completely by analytical methods. Vaccines are products that are defined to a great deal by the process by which they are made, by the analytical tools by which they are tested, and even by the facilities in which they are manufactured. As a consequence, most vaccine development projects have no clear starting point and research and process development activities run in parallel. Although partly impossible, this should be avoided as far as possible, as development activities need many more people and are considerably more expensive than research. No developing organization has sufficient resources to run numerous complex development projects in parallel or to change the direction of a development again and again. By defining adequate criteria and by a proper project organization, critical aspects of a development project can be identified early, so that these are evaluated during the applied research phase prior to the onset of product development.

4.3
Vaccine Research Projects

An excellent overview of ongoing research activities for vaccines is provided by the Jordan Reports issued by the US National Institutes of Health [1]. According to the latest issue of these reports, the number of vaccine R&D projects in the United States in the year 2000 amounted to more than 500 projects. Almost one-third of these were various efforts to develop vaccines against AIDS. A list of the main target indications pursued by recent vaccine research and development efforts is given in Table 1.

The top positions of the vaccine research "hit list" have not changed very much over the past few decades. Well-known viral and bacterial infections continue to occupy the most prominent positions. However, the number of individual projects for many of these vaccine indications has increased considerably. The simple reason for this is that formerly complete microorganisms

Tab. 1 Main infectious agents or targets for new vaccines in advanced R&D

Viruses	*Bacteria*	*Parasites*	*Tumours*
HIV/AIDS	*Streptococcus*	Malaria	B-cell lymphomas
Hepatitis C virus	*Helicobacter pylori*	*Leishmania*	Melanomas
Herpes simplex viruses	*Borrelia/Lyme disease*	*Schistosoma*	Prostata carcinoma
Cytomegalovirus	Salmonella	*Toxoplasma*	CEA-tumors
HRSV	Enterotoxigenic *E. coli*	*Trypanosoma*	
Parainfluenza	*Shigella*		
Rotavirus			

Note: HIV/AIDS: Human immunodeficiency virus/acquired immunodeficiency syndrome; HRSV: Human respiratory syncytialvirus; CEA: Carcino-embryonal antigen: an antigen that is frequently found on colorectal, bronchial, and breast cancer cells.

or subfractions thereof, but rarely purified single antigens, had to be used as vaccine candidates. Modern molecular biology and recombinant techniques result in individual antigens or even single epitope peptides, which may be varied or combined by almost endless options. Of course, this increases the number of candidates significantly and offers many new chances and possibilities, but it does not necessarily increase the chances of success for each individual approach. Molecular biology has not only opened up various new possibilities to approach antiparasite vaccines and tumor vaccines but also, in these particularly complex fields, the number of projects dealing with conventional "whole" organisms or cells is quite remarkable. Antitumor vaccine projects indicate that vaccines should no longer be regarded only as infection prophylaxis. Immunizations can and will in future also be used as therapeutic measures. Vaccine research even covers approaches that attempt to induce temporal infertility by the induction of antihormonal antibodies.

In comparison to current vaccine R&D projects, the number of newly licensed vaccines is extremely small. Most newly licensed products are improvements or combinations of existing vaccines; real vaccine novelties are very rare. Thus, the chances that a vaccine project in advanced research finally ends up as a vaccine product is minimal and is certainly far below 1%. These low success rates in research inevitably lead to long research phases. Short time intervals of around five years between the first publication or patent application of a new vaccine concept and the start of full development are an extremely short, applied research phase for vaccines. These may be applicable to some veterinary vaccines, for which vaccine protection of a candidate vaccine can be measured directly in the target species. For vaccines against human diseases, 10 or more years appear to be a more realistic average estimate for this phase. If one adds those further 10 to 12 years that it takes on average to develop a vaccine product, one must assume that after the basic concept has been published or patented for the first time, about 20 years are needed to successfully develop a new vaccine product. Those who consider these figures as unrealistic estimates are reminded that the average time interval between concept and first appearance on the market for various innovative technical products developed during the past 100 years (including e.g. not only complex products, such as antibiotics, the pace-maker, and radar but also presumably simple products such as the zipper, dry soup mixes, powdered coffee, ball-point pens, and liquid shampoo) was also 20 years [2, 3]. At that time, all innovations had to overcome existing hurdles, such as scientific challenges, technical difficulties, and usually financial limitations also.

4.4 Scientific Challenges of Vaccine R&D

Science and technologies are the driving forces that enable us to develop new vaccines. Regarding the basic technologies, there are few discoveries to be named that had a significant positive influence and resulted in new vaccines. Cultivation of pure bacterial cultures is still the fundamental base for most bacterial vaccines. A remarkable breakthrough came with the invention and development of cell culture techniques in the 1950s, which led to several new or significantly improved antiviral vaccines, including the currently still exclusively used "state-of-the-art" vaccines against poliomyelitis, mumps, measles, rubella, and cell culture rabies vaccines. Compared to these technologies, molecular biology and recombinant techniques up to now had a rather limited success with essentially only one recombinant human vaccine for hepatitis B. DNA vaccines may be regarded as yet another new and basic technology for new vaccines, but only a decade after their discovery they certainly did not yet have enough time to mature to practical applications. Monoclonal antibodies or anti-idiotype antibodies, however, did not lead to new vaccines as expected, although these basic techniques were often quoted as a major breakthrough in vaccine research.

Apart from a few essential technologies, continuous research in virology, microbiology, parasitology, and immunology are the foundations for vaccine research. However, even the most detailed knowledge about cytokines and their regulation of immune responses, or of fundamental genetic mechanisms controlling the growth and replication of microorganisms cannot be expected to bring any direct or immediate success. For the past and for the foreseeable future, it seems that it is more the pragmatic, application-oriented research that primarily fosters vaccine development. Complex immunological hurdles must be overcome in order to arrive at a new vaccine target, and that is mainly done by establishing suitable animal models and by testing all sorts of vaccine candidate antigens in these models in a very pragmatic way for their protective effects.

Current efforts to develop a vaccine against AIDS serve as a good example of illustrating the importance of suitable models for vaccine development. The Jordan Report 2000 [1] lists 135 different AIDS vaccine projects. Only 10% of these were considered to be basic research and development (R&D) projects, that is, they are mainly in a phase of selecting, constructing, and making the desired antigen. The rest of all these projects were allocated to preclinical testing phases in animals or to clinical testing in humans (compare Fig. 1). Less than one-third of these projects seemed to have passed small animal testing successfully and appeared to be worth testing in monkeys. Only 4.4% of the antigen candidates proceeded to trials in chimpanzees. A substantial proportion of 44% of vaccine candidates was tested in humans for safety and efficacy, however, only 1.5% were already in Phase III clinical trials, indicating that these two different vaccine candidates appear interesting enough to go into widespread field-testing for efficacy. The low number of projects in basic R&D shows that after two decades of AIDS research, there are not too many new antigens or entirely new approaches to be discovered. In the absence of reliable animal models, the relatively high number of projects in early human clinical trials and the low number in later stage clinical trials very clearly demonstrate that in this case research is essentially performed in human clinical trials – with all the inherent limitations. Consequently, the chances of success are low, while at the same time the cost of such research is extremely high.

What are the scientific challenges and difficulties to be overcome on the way toward an effective AIDS vaccine? As summarized in Table 2, infectious microorganisms and parasites have developed various mechanisms by which they effectively prevent their elimination by the host's immune response. All of these negative attributes have been found to be

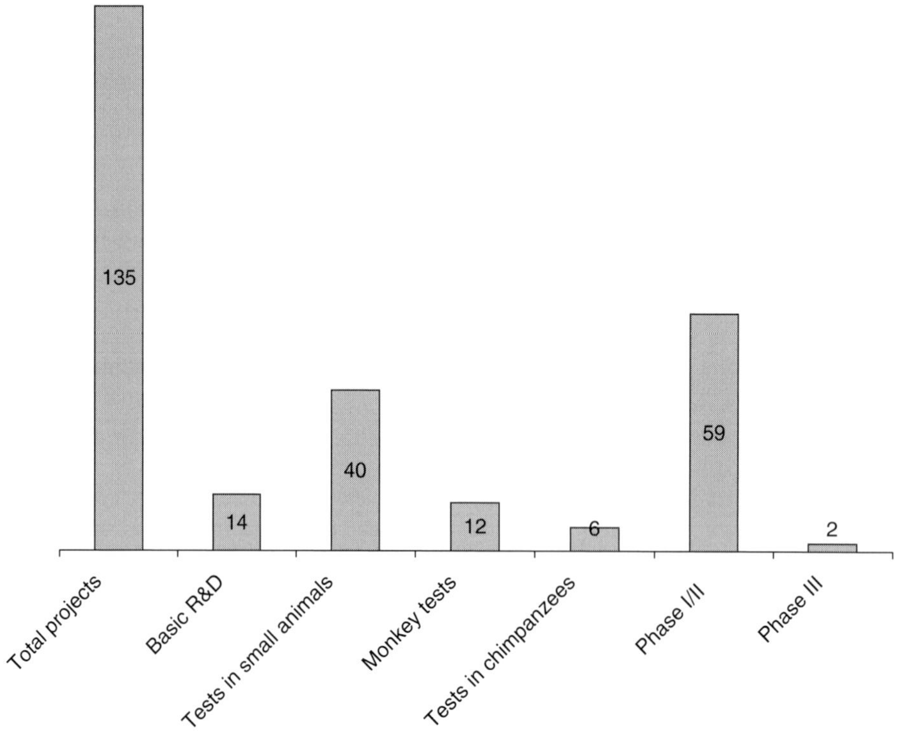

Fig. 1 AIDS vaccine candidates in research and development. The numbers apply to projects identified in the United States in year 2000 [1].

associated with HIV infections. HIV does not only evade the immune responses by presenting itself by different subtypes, by varying its main immunogenic antigens during the protracted course of infection in an infected individual, or by hiding itself in a nonaccessible form by integrating its genome into host cell's genes, it even interferes actively with several important immune functions and modifies these for its own benefit and support. Of particular relevance is the selective preference of HIV for CD4 immune cells, as disturbance of their function can result in numerous deleterious effects. The ability of HIV to persist and replicate in macrophages enables HIV to convert the migrating immune cells into an efficient vehicle across normal barriers. HIV is not only insufficiently neutralized by antibodies, it even uses bound antibodies to get access to immune cells, such as macrophages, which carry receptors for the Fc fragment of antibodies.

Whereas AIDS and HIV was only chosen as examples that contributes any imaginable difficulty to vaccine development, Table 2 also lists many other current vaccine projects and their specific difficulties. A limited number of different serotypes may still be overcome by making and combining several similar vaccines, once a successful vaccine against one of these has been accomplished. Thus, vaccines against parainfluenza infections appear reasonably feasible. Other indications, such as malaria, herpesvirus infections or Lyme

Tab. 2 Immunological challenges on the way toward new vaccines

Attribute	***Examples***	
Different serotypes or subtypes to be covered by the vaccine	Parainfluenza:	3 major pathogenic subtypes
	Dengue:	4 subtypes
	Malaria:	4 major pathogenic plasmodium species
	Borreliosis:	4 genetic and immunological types
	Hepatitis C:	6 major genotypes and >100 subtypes known
	HIV/AIDS:	>10 subtypes known
Antigenic variation of major immunogens	Malaria:	High variance of major antigens within the parasite [4–6]
	HIV:	Antigens vary during the course of infection even within the same patient [7]
	Trypanosoma:	Periodic switching of major surface glycoproteins [8]
Genetic restrictions of immune recognition and immune responses	Malaria:	Multiple HLA restrictions for recognition of Plasmodium falciparum CTL epitopes even within the same individual [5, 6]
	HIV:	HLA-restricted CTL escape mutations associated with viral load and disease progression [9, 10]
Microoganism not accessible to immune responses	HIV:	Virus genetically integrated in host cell genomes [11]
	Herpesviruses:	Virus persists in a latent state in neuronal cells [12]
Microorganism persists in immune cells and may spread with these into tissues or across	HIV:	Persistence and active replication in, for example, macrophages [13]
	Herpesviruses:	Can infect endothelial cells and macrophages [14]
	Borreliosis:	Borrelia survive in macrophages. Complement membrane complexes and macrophages in the endoneutrium of Lyme neuroborreliosis [15, 16]
blood-brain barrier	Hepatitis C:	Macrophages and T-cells found to be infected by Hepatitis C Virus (HCV) [17]
Immune-enhancement and immune-mediated disease	HIV:	Antibody and Fc receptor–mediated enhancement of infection and disease [18, 19]
	Dengue:	Antibody-mediated enhancement of infection [20]
	Borreliosis:	Immune-mediated neuropathology and arthritis [15, 21]
	Respiratory Syncytial Virus:	Inactivated vaccine induced high serum antibodies and aggravated disease upon infection [22]

Note: HIV/AIDS: Human immunodeficiency virus/acquired immunodeficiency syndrome; HLA: Human leucocyte antigens; CTL: Cytotoxic T-lymphocytes.

disease/Borreliosis, however, represent quite significant scientific immunological challenges, because the responsible microorganisms combine many unfavorable immunological characteristics. Finally, the example of a respiratory syncytial virus (RSV) vaccine developed and tested in the late 1960s may serve as an example to illustrate the difficulties and practical effects that some of these imponderable aspects can have. This RSV vaccine turned out to enhance a later disease, rather than preventing it [22]. More than 30 years after those results were published, there is still no real explanation for the underlying mechanism and almost all further efforts to develop a new vaccine were stuck in a preclinical phase.

Another important aspect, which seems to be underrated in many current vaccine research projects, is the fact that most vaccines are not sufficiently effective if these are based on only single antigens. Controlled vaccine studies performed under ideal conditions in genetically homogenous or inbred animals quite often lead to the false assumption that a fully protective vaccine antigen has been identified. But, when the same vaccine is then studied under more practical conditions by fewer numbers of immunizations, in the presence of acceptable and better-tolerated adjuvants, it becomes evident that the selected antigen candidate alone is simply not effective enough. Table 3 summarizes the experiences made with different foot-and-mouth-disease (FMD) experimental vaccines. Results from model studies with this type of vaccine can be correlated reasonably well with protective response in the target species. The FMD virus consists only of three structural proteins and the most relevant virus-neutralizing antigenic epitopes are known to be located on virus protein 1. Thus, FMD vaccines appeared as an excellent target for new vaccines based on recombinant technologies. The standard vaccine, made of inactivated whole virus particles, required a relatively low amount of antigen and only one immunization in order to confer protection. Efforts to make smaller subunit or single-protein vaccines resulted only in a similar protection if several immunizations and/or massively increased antigen doses were given. These results had been established rather early in the molecular biology vaccine era, but despite intensive further research, a commercially viable recombinant FMD vaccine has never been achieved. Meanwhile, conventional whole virus vaccines were successful enough to allow for measures to eradicate the disease completely in those countries where the vaccine has been used intensively.

4.5
Technical Aspects of Vaccine Development

In an ideal situation, vaccine development commences with a proven, protective and well-defined antigenic composition. Successful vaccine development then normally takes 10 or more years, but only a small proportion of all development

Tab. 3 Protectivity of different forms of vaccine antigens

Type of antigen	*Amount of antigen [µg]*	*Vaccinations*
Purified whole virus particles	1	1
Virion subunits (12 S)	10	2
Virus protein 1	200	3
Oligopeptide (N-142-160-C)	200	1

Note: Type of antigen, amount of antigen, and frequency of immunization required to achieve protection against foot-and-mouth-disease virus infection in the guinea pig model [23].

candidates finally end up as a licensed product; the vast majority remains stuck in early development phases or is abandoned [24, 25]. Figure 2 summarizes the essential tasks of a vaccine development project and may give a rough impression of what is to be expected. For the sake of clarity, several time dependencies and overlaps during the preclinical phase have been neglected in this graphic overview.

An extensive range of national and international rules and guidelines exist, covering almost any aspect of pharmaceutical and vaccine development and registration [26–28]. These guidelines describe standards that are not binding in a legal sense, but adherence to these is strongly recommended, as during later registration and licensing, the product will be judged by the same rules. Deviations from guideline recommendations may be inevitable for certain aspects and particularly for vaccines, but these should only be considered if convincing reasons for doing so can be presented. A summary of relevant guideline's requirement along with specific interpretations and applications for biopharmaceuticals and vaccines is given in [29] and may be helpful for prospective developers in order to get a reasonable understanding of the guiding principles.

4.5.1 Preclinical Development

Preclinical development comprises the technical and scientific elaboration of a process to manufacture the desired product on a large scale. Firstly, cell cultures and microorganisms to be used must be established as Master Cell Banks and Working Cell Banks, or as Master and Working Seeds, respectively. These ensure a constant supply of well-characterized, uniform, biological starting materials. Numerous tests in vitro and in vivo are required to guarantee the absence of undesired adventitious agents and to confirm the identity of these cell banks and microbial seeds.

Starting with a single aliquot of the Working Cell Bank and/or the Working Seed material, a process is then established and brought up to a final scale. The term "upstream process" typically means a cell culture or fermentation process up to some 100 L, but for very common vaccines larger scales may be chosen. Downstream processing summarizes activities during the purification process and typically includes recovery and concentration steps, followed by a secondary purification or "polishing" to remove specific impurities and process-related impurities introduced during earlier steps. Inactivation of bacteria or viruses or detoxifying steps for toxoid vaccines is usually included after an initial concentration step.

Formulation development includes the design of adequately buffered and well-tolerated, stable formulations, adjuvantation, the development of specific application forms, combination of vaccines into compatible vaccines, and, particularly for live attenuated vaccines, the development of a lyophilization process. Formulation development also extends to the selection or design of final syringes or other presentation forms and to filling and packing processes. Stability monitoring programs for intermediate and final products are of adamant importance for any development work and should be started as early as possible to avoid difficulties at a late development stage.

Analytical development encompasses all activities to design and use adequate methods to control and specify all parts of

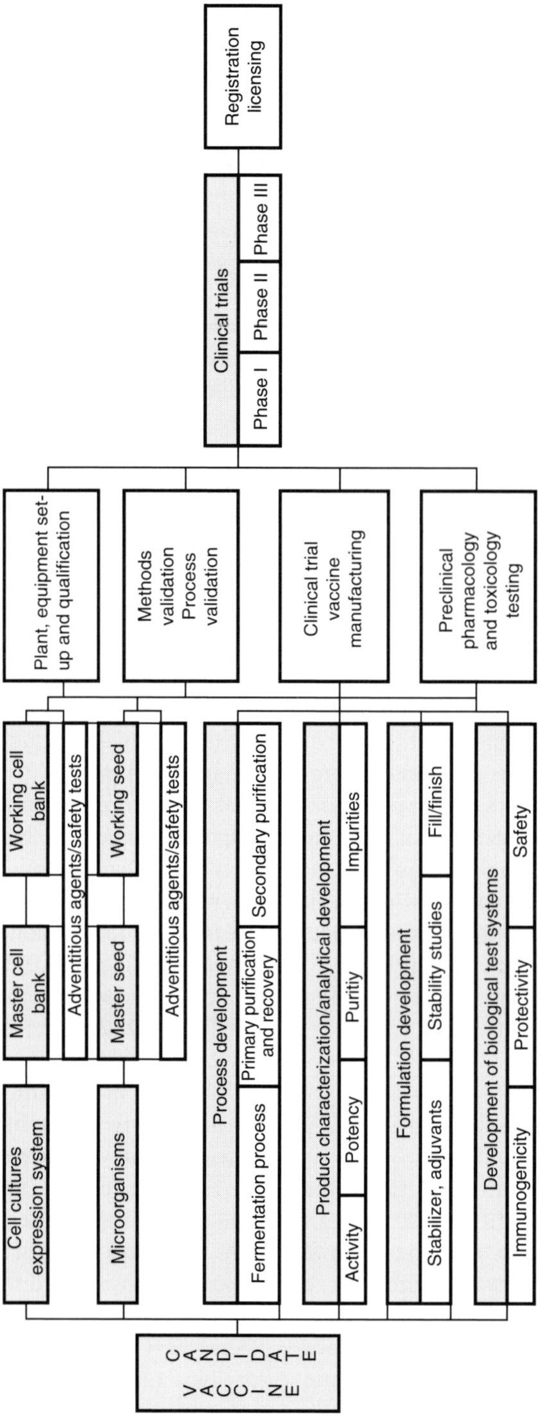

Fig. 2 Schematic diagram of a vaccine development process.

the process and the product. This includes testing of starting materials, intermediate products, and the final product, for example, for identity, specific activity, conformation, purity, and impurities. For a vaccine developed according to today's standards, a range of about 100 different tests and methods will be required. Most of these tests need to be validated for their specific purpose in order to assess the methods specificity, sensitivity, statistical exactness, and its limitations.

In parallel with process development, biological tests and model systems must be available to monitor the vaccine's potency, immunogenicity, or protectivity at any stage, as vaccines are particularly labile products and minor modifications of the process can have a significant – mainly negative – influence on the vaccine antigen. Likewise, biological models must be at hand to study the vaccine's basic pharmacological, immunological, toxicological, and potential immunotoxicological characteristics. As far as this can be adequately studied, these include dose responses, characterization of induced humoral and cellular immune responses or of their major contributing protective mechanisms, longevity of immune responses, and potential immunological side effects. Although vaccines rarely present severe tolerability or toxicological risks, abbreviated classic toxicological testing is mandatory before the onset of studies in humans. Most vaccines need to be tested only in local and systemic tolerance studies and in repeated dose studies in standard toxicology models, but for new adjuvants and certain new excipients, representing a significant part of the vaccine composition, even a complete toxicology program, including two years of carcinogenicity studies, may be needed. Further toxicology and safety studies addressing specific risks, such as embryonal, fetal, or peri- and neonatal toxicity may be required for certain vaccines and applications or if risks are expected or known. The recent withdrawal of a newly licensed Rotavirus vaccine that was suspected to cause intussusceptions and fatal bowl obstructions in vaccinated children may serve as an example that such studies may be required not on entirely hypothetical grounds alone. In this particular case, however, the true reason for the fatalities could also be a mere coincidence and the higher medicinal attention and reporting of fatality cases in vaccinated individuals.

Owing to the biological origin of many starting materials, risks associated with prions and potentially contaminating viruses must be addressed. Organizational measures are to be put in place to avoid risk materials in addition to testing for adventitious agents. Potential risks by starting materials or process contaminants can further be evaluated and assessed by model studies with various viral and microbial agents. If specific risks are identified and if safety margins appear low, specific countermeasures are to be included into the process. As far as possible, within the technical limitations of these safety studies, a residual risk of less than 1 in 1 000 000 cases should be aimed for. In practical terms this means, for example, that an unnoticed contaminating virus is inactivated or eliminated by the process to a degree that no active virus would be found in a vaccine volume equivalent to 1 000 000 doses. For live attenuated vaccines, viral safety must also be assured by assessing the genetic and phenotypic stability of the vaccine virus and by evaluating the chances and consequences of transmissions of the virus to unvaccinated individuals.

4.5.2
Production Facilities

Facilities and equipment for the manufacturing of a vaccine are an immanent part of the registration dossier for the product. Any major change would have to be approved by the regulating authorities. Thus, at least for the later clinical phases, the product should be made in a specific plant and with dedicated equipment. For a development project, this means that after the process has been defined, large investments into buildings, facilities, and equipment are to be expected. Owing to the inherent risks of these investments, pilot plants should be available to produce initial clinical trial vaccine lots on an intermediate scale. A developing organization may even choose to go into Phase III clinical trials with a vaccine that has been produced in a pilot plant and to seek registration for this "preliminary" product. This approach delays the investment decision to a later point of time when all development risks have been abolished, but inevitably requires new clinical trials for the vaccine that is later on made in the final plant and extends the time to the market by several years. The sum to be invested greatly depends upon the scale of operation and dosage volume of the vaccine. For a complete vaccine plant including all auxiliary functions, the total investment may well accumulate to far above € or US$100 million. Vaccine producers who can use their existing infrastructure, such as buildings, filling and packaging facilities, raw material and media production areas, quality control laboratories, and so on, would have to invest significantly less. For small or start-up companies, outsourcing and outlicensing may be chosen to reduce risky capital investments, as only vaccines with high market expectations justify establishing a complete, own manufacturing operation.

4.5.3
Clinical Development

The clinical development of a new vaccine is done in three phases and normally lasts three to seven years. The duration mainly depends not only on the novelty and complexity of the vaccine indication to be explored but also on the availability of measurable immunological surrogate markers of protection. If the vaccine's efficacy must be evaluated by comparing randomly occurring cases of the disease in test groups and in alternatively treated control groups, clinical studies can be extremely long lasting, demanding, and risky.

Prerequisites of all clinical trials are adequate preclinical pharmacological and toxicological safety assessments, including animal studies, to justify tests in humans. On the basis of the available safety data and documentation, approval for clinical trials must be obtained by the relevant ethics committees and health authorities. Trials will only be admitted if these are conducted according to preestablished, systematic, and written procedures for the organization and conduct of the trials for data collection, documentation, and statistical verification of the trial results. The "informed consent" of all participating trial subjects and medical personnel is essential. For trials involving children or mentally handicapped persons, the informed consent must be given by parents or by the person responsible. Clinical trials are to be planned and conducted according to "good clinical practice" standards that require controlled and randomized trials where possible. Control groups are to be treated by established products or treatments. Placebo treatments are only admitted where no alternative treatment exists.

During the initial Phase I, the basic safety features of the vaccine candidate are intensively studied in a limited number (<100) of patients or healthy volunteers. The main purpose of these studies is to confirm the vaccine's local and general tolerance before it is applied in further clinical trial subjects, but Phase I vaccine studies can partly be used for a first dose-finding, and immunological evaluations for adequate immune responses. During Phase I trials, vaccines rarely fail due to safety concerns, but quite frequently due to insufficient or inconsistent immune responses below expected levels.

Phase II clinical studies usually comprise no more than several hundred subjects and are normally done as controlled studies comparing the test vaccine along with an alternative prophylactic or therapeutic treatment. Clinical evaluations are mainly addressing the vaccine's effectiveness and safety, doses, application schemes, and possibly also different target groups selected by age, specific risks, countries, or by epidemiological criteria.

Phase III clinical studies are expanded, controlled, or uncontrolled trials on efficacy and safety in various clinical settings and under practical conditions. Altogether several hundred to several thousand trial subjects are enrolled at various trial sites, which are often distributed over several countries in order to study different epidemiological situations, ethnic populations, and deviating local medical practices. Phase III studies can also be evaluated for risk–benefit relationships and address practicability aspects as well as interactions by other products or concomitantly applied medical treatments. Postmarketing clinical trials of the licensed product, often referred to as Phase IV clinical trials, are nowadays, rather often, also requested as part of a conditioned licensing of pharmaceutical products, mainly in order to specifically investigate those aspects that can only be assessed by large statistical cohorts.

For live attenuated vaccines, specific safety aspects must also be studied clinically. As live viruses or bacteria replicate in the vaccine and may be shed into the environment, the potential transmission of vaccine microorganisms to unvaccinated subjects must be studied. If transmission is possible or likely, the vaccine's genotypic and phenotypic stability must be carefully studied and confirmed.

4.5.4 Licensing and Registration of Vaccine Products

The formal aspects of pharmaceutical product licensing will be dealt with in another chapter of this book [see Chapter 10 in this volume]. On the basis of previous experience and evaluations, the process of getting a vaccine through the evaluation at different national licensing authorities on average takes about two years, which includes time periods for working off and answering questions not adequately covered by the registration dossier.

As vaccines and other biological pharmaceuticals are particularly complex compositions that cannot be adequately characterized by specific quality control methods, the entire process, manufacturing facilities, analytical methods used to specify the product and its starting materials, and all ingredients are considered as being an inherent characterizing part of the product. Any change to these affects the product's license and requires approval by the licensing authority. Changing essential elements, such as production cell substrates or microbial strains, critical test methods such as potency assays, purification methods, or formulations would almost inevitably be seen as a change to the product that needs to be verified by new clinical trials. Furthermore, each individual batch produced must be approved and released by the authorities.

4.6 Economic Aspects of Vaccine Development

Without any doubt, the development of vaccines is a very costly and long-lasting process that bears a significant risk of failure. The following paragraphs intend to provide some deeper insights into the specific risks and chances, cost, and time requirements to develop a new vaccine, as the knowledge of those basics drawn from experience may by helpful in decision making. After all, successful vaccine development depends not only on good science and technical methods but also to a great extent on adequate management decisions.

4.6.1 Vaccine Development Cost

The number of successful vaccine projects is fairly low and retrospective evaluations of the specific cost incurred by these development projects over a time period of 10 or more years are difficult. However, cost evaluations covering developments from the late 1960s to the early 1990s exist, which summarize the development cost of various pharmaceutical developments [30–33]. Although chemical drug products dominated these figures, several vaccine projects were also assessed. With all the inherent variability, we can reasonably assume that these figures also give adequate estimates for vaccine products. These evaluations show that pharmaceutical development cost during those years tended to increase by a factor of about 10 within a decade. As demonstrated by a simple graph (Fig. 3), the rising cost is only in part due to the normal inflation rate, but clearly correlates with the increasing regulatory demands, as exemplified by the number of applicable guidelines.

The latest figure of US$231 million, published in 1991, was based on evaluations of 93 successful product developments. This sum has since been quoted on many occasions and has been willfully and generously projected to later dates. Thus, quoted sums of US$500 to 600 million may be encountered to describe the cost and risks of pharmaceutical development projects. However, these figures are misleading if several important details about the original calculations are not mentioned: those 231 millions include to a great extent, opportunity cost (calculated by an interest rate of 9% of the invested capital over a period of 12 years) and the cost of many unsuccessful or abandoned projects (assuming a success rate of 23%), furthermore, tax credits were not accounted for. All in all, the underlying ex-pocket expenses must be assumed to be only about one-fourth of the total sum and according to today's standards, direct cost between € or US$60 and 100 million may be assumed as a realistic estimate for the development of an new vaccine. If however there is no suitable infrastructure and if investments into completely new production facilities are to be made, this could easily double the cost.

Apart from capital investments, personnel is the most relevant cost factor to be considered. Owing to the high number of

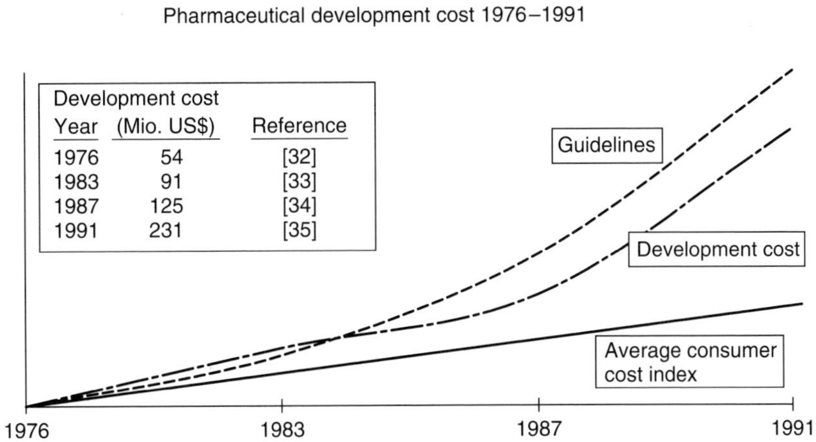

Fig. 3 Pharmaceutical development cost. Development cost denoted for various mainly chemical pharmaceuticals, including vaccine projects. Correlation with rising regulatory requirements as indicated by the counts of existing guidelines at the indicated point of time. Average consumer cost index for a household of four persons taken from figures released by the Feral German Statistics Office (Statistisches Bundesamt). Graphs show relative figures adjusted to a uniform scale.

persons involved in the preclinical development phase and to the long duration of these activities (on average about four years until the start of clinical trials and several years beyond until registration), preclinical activities account for about one half or more of the development cost. Clinical development normally causes about one-fourth of the development cost; the remaining quarter is evenly spread throughout the developing organization and covers overheads, technical support functions, quality control, and quality assurance, as well as various other specialists, for example, for patenting, regulatory affairs, and market research.

Taken together, an average vaccine development project requires about 170 man-years of work with average total expenditures per person and workplace in the pharmaceutical industry being around €180 000 to 200 000 or for the United States around US$220 000 to 240 000. This results in roughly € or US$30 to 40 million for personnel and workplace expenses [34]. External cost of around 20% and highly variable capital investments into plant and facilities are then to be added.

4.6.2
Risks and Chances

The success of a project during and for the entire development process can be estimated by the numbers of projects that make it until the next development phase and finally end up as a commercial product. On the basis of data for products developed during the preceding decades until 1994, one must assume that only 50% of the preclinical vaccine development projects enter the clinical phase and another 50% is abandoned during the clinical trials. Having passed all preceding hurdles, product registration seemed to be uncritical, as only a loss of one out of 100 vaccine projects was noted.

For pharmaceutical drug products, overall success rates of 11% were found, that is, 100 product candidates entering the preclinical development resulted in only 11 licensed products. Vaccine projects appeared to be more successful with an average of 22% licensed products per 100 projects (compare Fig. 4). However, the figures presented in Fig. 1 for AIDS vaccine projects of the recent past show that average figures can also be grossly misleading.

Most current vaccine candidates are dealing with quite "difficult" infectious diseases, which under natural conditions do not induce a lasting protective or sterile immunity, thus doubts about the applicability of those earlier risk evaluations to current vaccine projects are justified. Even if AIDS vaccine projects are not considered, a snapshot view on more recent vaccine developments supports the suspicion that success rates for today's projects and particularly for new vaccine indications are much lower. As shown in Table 4, the success rates of preclinical development in more recent times appeared to be below 50% and only 15% of all projects were found in Phase III clinical trials. Whereas preclinical projects represented a very wide spectrum of entirely new vaccines, the majority of Phase III clinical trials were covering alternatives to already existing vaccines, such as competitor's developments, combinations, or improved formulations. Only 4% of these advanced projects were approaches to develop entirely new vaccines. These figures represent only a static view upon the vaccine development for a certain year, but they clearly indicate that nowadays – and particularly for really new vaccines – development success rates clearly below 5% appear more realistic than earlier estimates that were above 20%.

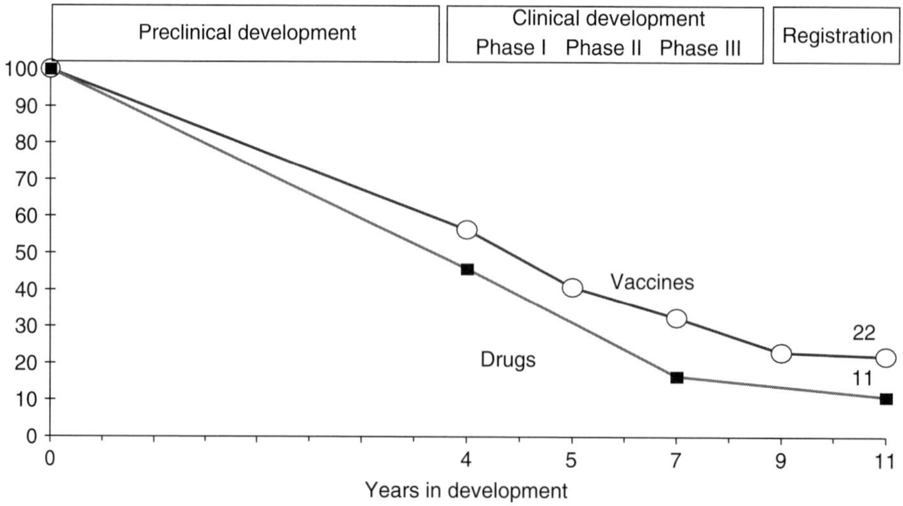

Fig. 4 Success rates of vaccine development projects in comparison with drug development. Data represent the percentage of projects, which successfully completed the respective development phase. Data summarized from [24, 33, 35] and based upon assessments mainly for the 1980s. As discussed in the text, the success rates for vaccines developed today are most likely lower than shown here.

Tab. 4 Human vaccine development projects in the year 2000

Project phase	***No. of vaccine projects***	
Preclinical development	349	100%
Phase I clinical development	158	45%
Phase II clinical development	102	29%
Phase III clinical development	51	15%
Thereof alternatives to existing vaccines in Phase III	37	11%
Thereof new vaccine indications in Phase III	14	4%

Note: Data extracted from listed vaccine projects in the year 2000 [1] without consideration of AIDS vaccine projects. Vaccine projects in Phase III clinical trials for new indications include vaccines against *Coccidioides immitis*, group B *streptococcus*, *Streptococcus pneumoniae*, *Plasmodium falciparum*, *Trypanosomsa cruzi*, *Leishmania major*, *Mycobacterium leprae*, Meningococcus B and C, Rotavirus, and *Vibrio cholerae*.

4.7 Conclusions

Judged by the number of scientific publications in microbiological and biotechnology journals, vaccine R&D appears to have a great attraction for scientists from all pertinent scientific disciplines. Whenever new methods and technologies became available, these have always and immediately triggered a huge number of new vaccine research projects and stimulated research into formerly hopeless vaccines. Along with the good reputation that vaccines enjoy, this scientific enthusiasm is an excellent base for new vaccines, and a good base to attract the required capital as well. But considering the high risks and the long duration of vaccine R&D, there must also be other reasons why investors and pharmaceutical companies invest in this field.

Vaccines represent only a small proportion of the pharmaceutical market, but nevertheless vaccines are extremely successful products. Firstly, vaccines effectively prevent diseases, rather than only curing these. Owing to these advantages, vaccines have often created their own markets and have even defended their market shares against competition by very effective therapeutics or antibiotics. Secondly, most vaccines are recommended by public health authorities and thus enjoy a rather safe position on the market. Furthermore, there are usually only rather limited numbers of competitive products because vaccines are far too complex to become an easy target for producers of generic imitations. And finally, vaccines usually have a very long life span. As long as vaccine products are not neglected and become outdated, but are constantly adapted to a better state of the art, vaccines do not lose their market position, unless they are too successful and by and by eliminate the need to use the vaccine.

Thus, vaccine R&D can be very rewarding for both scientists and investors. Regarding the risks, however, the investor has a quite different perspective than the scientist. The investor may contain risks by putting capital into many different projects and enterprises, thus participating in the statistically very successful "average vaccine". To a limited extent, large companies who develop vaccines can also apply the same strategy. But small enterprises and individual scientists working for only one or a few R&D projects have only few options to manage and reduce risks. They often choose a high-risk approach by aiming only for "block buster" products. In this case, they must be aware that competition in this field will be also very strong, which increases the risks even more. But within the given financial limitations, risks could also be spread over a certain number of projects in early R&D phases, preferably by approaching different indications and concentrating on an attractive new or improved technology.

In any case, vaccine R&D is certainly not a playing ground for those who expect fast success and revenues. Any organization that intends to invest into vaccine R&D should be prepared – both mentally and financially – to endure for at least 10 to 20 years.

References

1. The Jordan Report 2000, Accelerated Development of Vaccines, Division of Microbiology and Infections Diseases, National Institutes of Health. Washington DC, http://www.niaid.nih.gov/publications/pdf/jordan.pdf.
2. S. Rosen, *Future Facts,* Simon & Schuster, New York, 1976, (see also: D. J. Ellis, P. P. Pekar, Planning Basics for Managers, Amacon, New York, 1983).
3. Batelle Memorial Laboratories, Science, Technology and Innovation, Report to the National Science Foundation, USA, 1973.
4. J. D. Smith, C. E. Chitnis, A. G. Craig et al., *Cell* **1995**, *82,* 101–110.
5. R. Wang, D. L. Doolan, T. P. Le et al., *Science* **1998**, *282,* 476–480.
6. T. P. Le, K. M. Coonan, R. C. Hedstrom et al., *Vaccine* **2000**, *18,* 1893–1901.
7. H. C. Lance, J. M. Depper, W. C. Green et al., *N. Engl. J. Med.* **1985**, *313,* 79–84.
8. J. E. Donelson, K. L. Hill, N. N. El-Sayed, *Mol. Biochem. Parasitol.* **1998**, *91,* 51–66.
9. P. J. Goulder, C. Brander, Y. Tang et al., *Nature* **2001**, *412,* 334–338.
10. C. B. Moore, M. John, I. R. James et al., *Science* **2002**, *296,* 1439–1443.
11. P. Luciw, Human immunodeficiency viruses and their replication Chapter 60 in *Fields Virology* (Eds.: B. N. Fields, D. N. Knipe, P. M. Howley), 3rd ed., Lippincott-Raven, 1996, pp. 1881–1952, Vol. 2.
12. B. Roizman, A. E. Sears, Herpes simplex viruses and their replication Chapter 72 in *Fields Virology* (Eds.: B. N. Fields, D. N. Knipe, P. M. Howley), 3rd ed., Lippincott-Raven, 1996, pp. 2231–2296, Vol. 2.
13. A. T. Haase, *Nature* **1986**, *322,* 130–136.
14. J. Odeberg, C. Cerboni, H. Browne et al., *Scand. J. Immunol.* **2002**, *55,* 149–161.
15. D. Maimone, M. Villanova, G. Stanta et al., *Muscle Nerve* **1997**, *20,* 969–975.
16. R. R. Montgomery, M. H. Nathanson, S. E. Malawista, *J. Immunol.* **1993**, *150,* 909–915.
17. A. L. Zignego, M. Decarli, M. Monti et al., *J. Med. Virol.* **1995**, *47,* 58–64.
18. A. Takeda, C. Tuazon, F. A. Ennis et al., *Science* **1988**, *242,* 550–583.
19. J. Homsy, M. Meyer, J. A. Levy, *J. Virol.* **1990**, *64,* 1437–1440.
20. S. B. Halstead, *J. Infect. Dis.* **1979**, *140,* 527–533.
21. B. K. Du Chateau, E. L. Munson, D. M. England et al., *J. Leukoc. Biol.* **1999**, *65,* 162–170.
22. H. W. Kim, J. G. Canchola, C. D. Brandt et al., *Am. J. Epidemiol.* **1969**, *89,* 422–434.
23. M. C. Horzinek, *Kompendium der allgemeinen Virologie,* Enke Verlag, Stuttgart, 1984.
24. M. M. Struck, *Biotechnology* **1994**, *12,* 674–677.
25. M. M. Struck, *Nat. Biotechnol.* **1996**, *14,* 591–593.
26. The Rules Governing Medicinal Products in the European Community, Volume I–IV (1989), Vol. III plus Addendum 1, 2, 3 (1990–1994), as well as numerous newer guidelines, which are not contained in the above mentioned volumes can be obtained from: Office des Publications Officielles des Communantés Européenes, Luxemburg. These guidelines are retrievable via http://www.emea.eu.int/.
27. US-FDA guidelines and Points to Consider (PCT) documents, as well as lists of all available documents about human biological medicines are available via: Congressional,

Consumer and International Affairs Staff, Metro Park North, Building 3, 5600 Fishers Lane, Rockville, MD 20857 and via internet under http://www.fda.gov/cber/guidelines.htm.
28. International regulatory guidelines, harmonized between the EU, USA, and Japan, and covering various aspects are available as ICH (International Conference on Harmonization) documents under: http://www.ich.org/ich5.html.
29. J. P. Gregersen, *Research and Development of Vaccines and Pharmaceuticals from Biotechnology. A Guide to Effective Project Management, Patenting and Product Registration*, VCH, Weinheim, New York, Basel, Cambridge, Tokyo, 1994.
30. R. W. Hansen, The pharmaceutical development process: Estimates of development and the effects of proposed regulatory changes in *Issues in Pharmaceutical Economics* (Ed.: R. J. Chien), Lexington Books, Cambridge, MA, 1979, pp. 151–186.
31. L. Langle, R. Occelli, *J. Econ. Med.* **1983**, *1*, 77–106.
32. S. N. Wiggins, *The Cost of Developing a New Drug*, Pharmaceutical Manufacturers' Association, Washington, DC, 1987.
33. J. A. Di Masi, R. W. Hansen, H. G. Grabowski et al., *J. Health Econ.* **1991**, *10*, 107–142.
34. J. P. Gregersen, Vaccine development: the long road from initial idea to product licensure Chapter 71 in *New Generation Vaccines* (Eds.: M. M. Levine, G. W. Woodrow, J. B. Kaper et al.), 2nd ed., Marcel Dekker, New York, Basel, Hong Kong, 1997, pp. 1165–1177.
35. B. Bienz-Tadmore, P. A. Dicerbo, G. Tadmore et al., *Biotechnology* **1992**, *10*, 521–525.

5
DNA Vaccines: from Research Tools in Mice to Vaccines for Humans

Jeffrey Ulmer and John Donnelly
Chiron Corporation, Emeryville, CA, USA

Jens-Peter Gregersen
Chiron-Behring GmbH, Marburg, Germany

5.1
Introduction

Expression of foreign genes in animals can be achieved through the simple administration of recombinant DNA, as was first demonstrated more than 20 years ago [1, 2], although the impetus for the recent application to vaccines is typically traced to the work of Wolff et al. in 1990 [3]. Shortly thereafter, the induction of antibody responses [4], cytotoxic T-lymphocyte (CTL) responses [5], and protective immunity by DNA vaccines in a lethal animal challenge model [5, 6] were reported. Since then, the field of DNA vaccines (also termed genetic vaccines) has been very active. Over the past decade, the general utility of this approach for prophylaxis and therapy of infectious and noninfectious diseases has been well established (for reviews see [7, 8]), culminating in the human clinical testing of many different DNA vaccines. During this time, an understanding of the mode of action of DNA vaccines has been gained, as well as insights into their limitations. As a consequence, several second-generation DNA vaccine technologies have been developed and some of these are now entering clinical evaluation. This review will address the technological developments that have been achieved, with a look into the issues that will need to be considered if a DNA vaccine approaches registration.

5.2
DNA Vaccine Construction and Immunology

Effective vaccines have three key components: (1) an antigen against which adaptive immune responses are generated, (2) an immune stimulus (or adjuvant) to signal the innate immune system to potentiate the antigen-specific response, and (3) a delivery system to ensure that the antigen and adjuvant are delivered together at the right time and location. For DNA vaccines, the antigen is produced in situ, albeit at very low levels. Thus, the potency of DNA vaccines depends, in part, on effective expression plasmids. With regard to immune stimulation, DNA vaccines appear to contain a built-in adjuvant in the form of immunostimulatory CpG motifs. However, even simple addition of conventional adjuvants can increase DNA vaccine potency, suggesting that there is room for stronger innate immune signaling by DNA vaccines. Finally, DNA vaccines on their own do not have an inherent ability to efficiently enter cells in a functional way (i.e. to transfect them) and thus require means of delivery. Hence, for DNA vaccines to be optimally effective, enabling technologies in all of these three areas must be developed.

5.2.1 DNA Vaccine Expression Plasmids

Most DNA vaccines tested over the past decade have consisted of conventional plasmids with a eukaryotic expression cassette. The important elements of such plasmids are the promoter, the gene insert, the polyadenylation termination sequence, a bacterial origin of replication for production in *Escherichia coli*, and an antibiotic resistance gene for selection (see Fig. 1). Typically, strong viral promoters, such as the intermediate early promoter of cytomegalovirus with the intron A, are used. This ensures constitutive, high levels of antigen production in many tissue

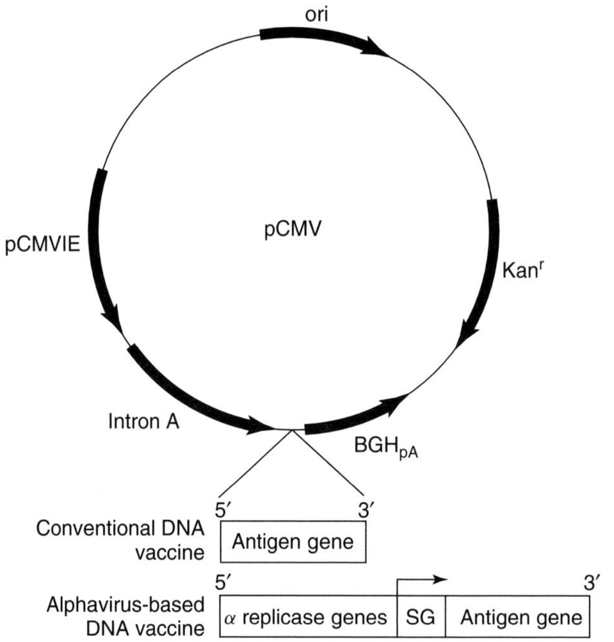

Fig. 1 Schematic representation of a typical DNA vaccine plasmid. Shown are the promoter (pCMVIE), transcription terminator (BGHpa), bacterial origin of replication (ori), and antibiotic resistance gene (Kan^r). Several different types of inserts may be included in a DNA vaccine. Shown are ones containing a discrete open reading frame and those containing an alphavirus RNA replicon. SG = subgenomic promoter.

types and thus increases the likelihood of inducing an immune response. Certain other types of promoters, including those whose expression may be limited to specific types of tissues, have also been used with success. These include the muscle creatine kinase [9, 10], major histocompatibility complex (MHC) class I [11], desmin [12], and elongation factor $1-\alpha$ [13] promoters. A potential advantage of tissue-specific promoters over viral promoters is the additional measure of safety they may provide because of a more limited distribution of antigen production after vaccination. Also, certain viral promoters can be downregulated by cytokines [14], which are produced in situ after DNA vaccination.

Another type of plasmid DNA vaccine introduced more recently encodes an alphavirus RNA replicon. Alphavirus plasmid replicons based on Sindbis virus incorporate the nonstructural "replicase" protein genes and cis replication signals, such that primary transcription from an RNA polymerase II promoter (e.g. CMV) gives rise to an RNA vector replicon capable of directing its own cytoplasmic amplification and expressing an encoded heterologous gene [15, 16]. Similar plasmids based on Semliki Forest virus replicons have also been generated [17–19]. These plasmids have been shown to be more potent than conventional CMV-based DNA vaccines, particularly at low DNA doses. Possible explanations for this increased effectiveness include: (1) amplification of mRNA in the cytoplasm by the RNA replicon that may enhance expression levels, (2) the presence of a dsRNA intermediate that may act to stimulate the innate immune system, (3) expression of other alphavirus nonstructural proteins that may provide additional helper T-cell epitopes, and (4) induction of apoptotic cell death in cells transfected with pSIN, which may facilitate cross-priming of T-cell responses.

A third type of DNA vaccine consists of linear DNA sequences containing simply the promoter, gene, and polyadenylation site. These DNA vaccines can be in the form of a contiguous DNA sequence containing all of the above elements [20] or a gene hybridized to the promoter and termination sequence [21]. The latter version provides the opportunity to rapidly generate DNA vaccines by PCR amplification without the need for bacterial transformation, thereby facilitating the screening of large numbers of vaccine candidates (see Chapter 4).

Irrespective of the type of DNA vaccine vector, antigens expressed from it can be in one of several possible forms, ranging from short peptides of defined T-cell epitopes (as small as 8 amino acids in length) to large polyproteins (>1000 amino acids in length). Several reports have demonstrated that single T-cell epitopes expressed in minigene DNA vaccines can induce potent T-cell responses [22]. In some cases, potency was increased by the addition of an N-terminal signal sequence to facilitate targeting of the epitope to the endoplasmic reticulum [23]. Multiple epitopes expressed end to end in a "string-of-beads" fashion have the ability to elicit multiple T-cell responses [24], and this approach is currently being tested in human clinical trials for malaria and HIV [25] (A. Hill, unpublished). This approach can focus the response on defined dominant epitopes, but requires prior knowledge of these epitopes and may not be sufficient for complete coverage of the diversity of human leukocyte antigen (HLA) haplotypes.

A means around this issue is to express whole antigens to allow determinant selection of epitopes by the host. Expression of whole antigens in their native state is also important for the induction of neutralizing antibodies. Most DNA vaccines reported so far have encoded whole antigens. For antigens whose expression levels are limited by the use of suboptimal codons, synthetic genes containing appropriate codons commonly utilized in eukaryotic cells are often very effective [26, 27]. The nature of the antigen can also affect the quantity and quality of immune responses. Secreted antigens are most effective for induction of antibodies and often for T-cell responses, as well [28, 29]. Targeting antigens for processing and presentation by MHC class I and II molecules can be achieved by the use of ubiquitin [30, 31] and lysosomal [32, 33] targeting signals, respectively. Alternatively, extracellular delivery of expressed antigens to antigen-presenting cells (APCs) through the use of fusion proteins containing ligands for receptors on these cells has been used effectively [34–37].

For many vaccine targets, a single antigen is not sufficient to provide optimal protection. This is particularly true for viruses with variant surface glycoproteins, such as HIV, where multiple clades must also be represented in a vaccine. Thus, the vaccine may require several components. In theory, this can be accomplished by simple mixtures of plasmids, as DNA vaccines are amenable to combinations. However, this adds to the complexity and cost of the vaccine. A potential means to minimize these issues is to express several antigens simultaneously from a minimum number of plasmids, through the use of multicistronic vectors [38, 39], polyprotein gene cassettes [40], or partial genomic constructs [41, 42].

5.2.2
Antigen Presentation and Stimulation of Immune Responses

After DNA vaccination or during viral infection, antigens may be produced directly within APCs, such as Langerhans cells (LC) and dendritic cells (DC), or they may be acquired by APCs in a process termed "cross-presentation," where antigens are produced by one cell type and then transferred to APCs (see Fig. 2). Either way, this is sufficient for the priming of antigen-specific lymphocytes. In practice, DNA vaccination results primarily in the transfection of non-APCs, either myocytes after im injection [3, 43] or fibroblasts and keratinocytes after gene-gun administration [44]. For B-cell responses, non-APCs can be considered simply as factories for the production of antigens, which can then be presented to B-lymphocytes for priming of antibody responses. In this regard, antigens that are secreted or those that spontaneously form higher-order structures, such as virus-like particles, are particularly immunogenic.

Both the gene-gun and im injection methods of DNA vaccination effectively prime T-cell responses, including $CD8^+$ CTL. Because the gene gun can propel DNA-coated gold beads directly into LC resident in the skin, the primary means of CTL induction appears to be by direct priming of LC [45]. In contrast, DNA injected into the muscle has no inherent ability to efficiently enter and transfect DC, thus the predominant mode of CTL priming is by cross-priming [46]. In support of direct priming are the following observations: (1) plasmid DNA, antigen-encoding mRNA, and expressed antigen have been shown to be present in DC and LC after DNA immunization by im injection and gene gun, respectively [45, 47, 48], (2) DC

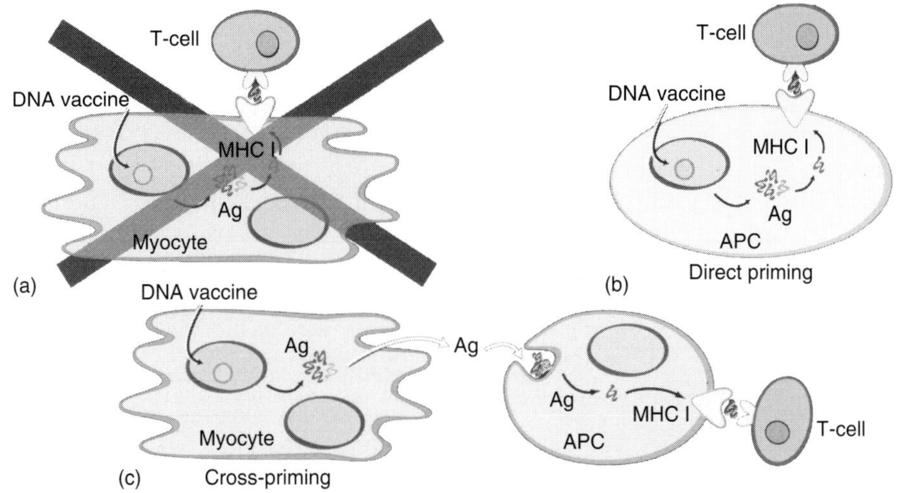

Fig. 2 Possible means of inducing a CTL response by a DNA vaccine. Upon injection of a DNA vaccine, there are three possible modes of CTL induction: (1) direct activation of na ï ve T-cells by transfected muscle cells (a), (2) direct activation of T-cells by transfected antigen-presenting cells (b), and (3) indirect activation of T-cells involving transfer of antigen produced by muscle cells to a professional antigen-presenting cell (c). A current working hypothesis is that CTLs are induced by DNA vaccines utilizing modes (b) and (c), but not (a).

isolated from DNA vaccine–injected tissue can present antigen to T-cells in vitro [49, 50], and (3) passive transfer of DC transfected in vitro with DNA vaccines induces strong T-cell responses in naïve recipients [51]. In support of cross-priming are the following data: (1) transplantation of DNA vaccine-transfected myocytes or tumor cells to naïve or bone marrow chimeric mice induced strong T-cell responses [52–54], (2) DNA encoding FAS (CD95) [55] or mutant caspases [56] to facilitate apoptotic cell death increased induction of T-cell responses by DNA vaccines, (3) in vitro studies suggested that a heat-shock protein chaperone is involved in cross-priming induced by DNA vaccination [57], (4) T-cell responses to DNA vaccines can be increased by targeting the expressed antigen to APCs by utilizing fusion proteins encoding CTLA4 [37] or chemokines [35], or extracellular transfer by HIV tat [58] or herpes simplex vims (HSV) VP22 [59], and (5) in vivo electroporation, which facilitates uptake of DNA vaccines by myocytes, enhanced CTL responses [60, 61]. Thus, taken collectively, the data indicate that both means of priming are sufficient for the induction of CTL.

DNA vaccines appear to have a built-in adjuvant for signaling the innate immune system, in the form of immunostimulatory CpG motifs. It has been well established that oligonucleotides containing unmethylated CpG motifs signal through TLR9 present on plasmacytoid DC and B-lymphocytes, resulting in the activation of DC and the production of cytokines [62]. Oligonucleotides derived from DNA vaccines presumably can be generated in vivo by nuclease digestion and thereby provide an immune stimulus or adjuvant effect for antigens expressed by DNA vaccines. Indeed, the potency of DNA vaccines can, in some instances, be enhanced by the inclusion of additional CpG motifs [63, 64]. However, simple mixtures of plasmid DNA vaccines with immunostimulatory oligonucleotides have not been effective, in contrast to the potent adjuvant effect that CpG can have with protein-based vaccines [65]. This appears to be due to interference in the uptake or expression of DNA vaccines. Thus, appropriate formulation or delivery will be required to take advantage of CpG oligonucleotides as adjuvants for DNA vaccines.

5.2.3
Delivery of DNA Vaccines

After im injection, plasmid DNA is rapidly degraded by nucleases present in the tissues and by macrophages resident in the muscle that phagocytose DNA, with very little of the injected DNA ultimately causing transfection of cells [66]. In addition, injected DNA has a limited distribution in the muscle, being concentrated, for example, at the injection site and the periphery of the tissues in mice (see Fig. 3). Thus, simple needle injection is an inefficient means to deliver DNA vaccines. As a consequence, various methods to facilitate delivery of DNA vaccines have been explored.

The most common alternative to needle injection is the gene gun, which propels gold beads coated with DNA directly into cells in the skin. As described above, resident LC are transfected, then migrate to the draining lymph node whereupon naïve lymphocytes are primed. Because of the greater efficiency with which DNA

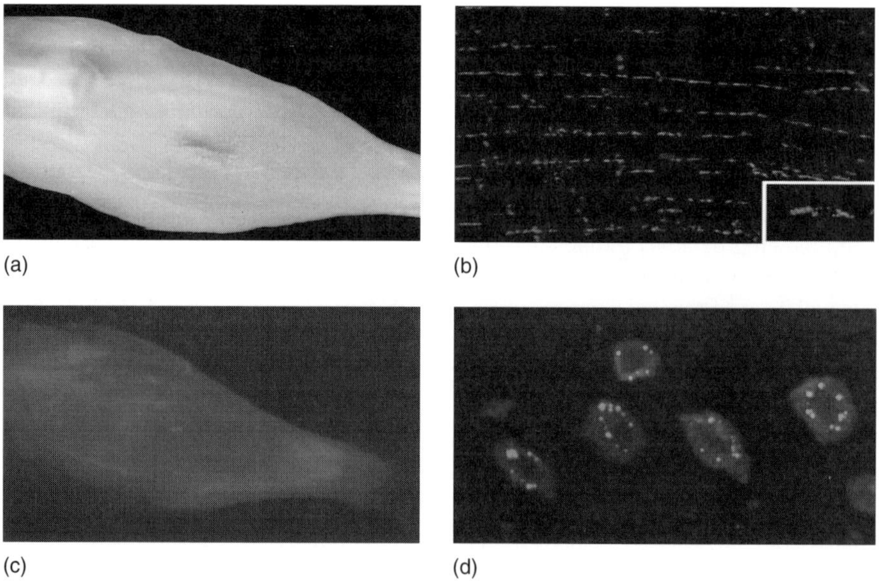

Fig. 3 Distribution of injected DNA vaccines. A rhodamine-conjugated DNA vaccine was injected into a tibialis anterior muscle of a mouse shown by light (panel A) and fluorescence (panel C) microscopy (~5× magnification). A longitudinal section of the muscle is shown in panel B (~250× magnification), demonstrating the presence of DNA in cells between the muscle fibers. Panel C shows the phagosomal location of the plasmid DNA (in red) within the cells isolated from the injected tissues (~2500× magnification). (See Color Plate p. xxii).

is delivered into cells by this technique, less DNA is required for induction of immune responses than by needle injection. However, as this technology is currently practiced, a significant limitation is imposed on the quantity of DNA that can be delivered at one time. Nevertheless, this technology has been shown to induce both humoral and cellular immune responses in human clinical trials [67]. Various noninvasive routes of DNA delivery have also been evaluated. These include intranasal, oral, intravaginal, and topical administration onto the skin. In many cases, particularly by the oral route, naked DNA was not effective due to rapid degradation by hydrolytic enzymes. Thus, formulations designed to protect DNA from digestion are required, such as encapsulation into chitosan particles [68], polylactide coglycolide (PLG) microspheres [69], or liposomes [70].

For parenteral injection of DNA vaccines, naked DNA has been effective in small animal models, but, as mentioned above, there is much room for improvement in the efficiency of DNA delivery. To this end, two basic approaches have been taken: (1) to increase the efficiency of uptake of DNA by cells in the injected tissue (e.g. myocytes) to facilitate cross-priming of immune responses and (2) to target DNA to APCs to facilitate direct priming of immune responses. First, to increase DNA distribution and uptake in the injected tissues, physical techniques generally have been most effective. These include the aforementioned gene-gun approach, needle-free devices (such as the Biojector™) designed to produce better distribution of vaccine [71], in vivo electroporation [60] or sonoporation [72] to induce transient discontinuities in the plasma membranes of cells, and the use of large-volume inoculation to induce high hydrostatic pressure locally in the tissues [73]. These techniques require devices and some are cumbersome, involving invasive procedures that may not be appropriate or practical for widespread use with prophylactic vaccines. Second, to target APCs for DNA vaccine uptake, formulations have generally been used. Liposomes [74] and microparticles based on PLG [75] and chitosan [76] have been a particularly effective strategy, in theory because of their similar size to pathogens. Work with DNA vaccines adsorbed onto the surface of PLG microparticles has shown efficient delivery into DC in vitro, enhancement of transfection of cells in the draining lymph nodes of injected mice, and marked increases in DNA vaccine potency, with a dependence on the size of the microparticles [77]. These observations are consistent with the hypothesis that the microparticles target DNA vaccines to APCs in vivo. These techniques for increasing DNA uptake by cells address the first step in facilitating transfection. Once in the endosome, though, the DNA plasmids must find their way into the cytoplasm and then the nucleus. Thus, further improvements in DNA delivery may be achieved through the inclusion of components that destabilize the endosomal membrane [78] and target DNA to the nucleus [79]. The latter may be particularly important for transfection of muscle cells, which are terminally differentiated; thus the nuclear membrane remains intact.

5.3
Screening for Protective Antigen Candidates

Various efforts have been made to harness the ease of preparation of DNA vaccines for mass screening of candidate antigens for inclusion in vaccines against infectious diseases. The ability of gene-gun immunization to evoke immune responses with small quantities of DNA led to the hypothesis that a small amount of DNA encoding a protective immunogen could still provide protection when mixed with many other DNAs that encoded weak immunogens or no antigen at all. This hypothesis was tested initially using mycoplasma pulmonis [80]. Immunization of mice by gene gun with a random library of M. pulmonis DNA fragments did confer protection from challenge. A comparable library approach to immunization has also been used against murine malaria [81] and murine cysticercosis [82]. Immunization with random libraries has also been used in studies of HIV-2 infection in Hamadryas baboons [83] and simian immunodeficiency vims (SIV) in rhesus macaques [84]. Refinements of this technique include the use of cDNA libraries to ensure that each DNA plasmid encodes a gene [85, 86] or simply to use unligated PCR products that have been hybridized with a synthetic promoter and terminator region [21]. The latter method can also produce sufficient gene expression to elicit immune responses in mice without the need for constructing sets of plasmid vectors. This approach also allows different fragments of genes to be used by the use of random primers or primers selective for particular fragments in the PCR reaction used to generate the coding regions.

The next step after demonstrating protection in a challenge model is to deconvolute the pool and identify individual antigens that are responsible for protection. In theory, this can be done by immunizing with multiple distinct pools in a matrix array, where the intersection of each row and column contains a unique set of genes. However, mixtures of antigens may synergize to provide a protective immune response. Thus, when the mixtures are broken up into smaller sets, protective efficacy may be lost. Alternatively, some gene products may interfere with protection and could cancel out an active component in a pool. The concept of beginning with random expression libraries and carrying through to identification of single antigens has not yet been reported. This may be an indication of the complexity of the biological phenomena that underlie immunity to complex microorganisms.

A more conservative approach to the screening of antigens in complex pathogens has been attempted using individual genes cloned from expression libraries. Here, the limitations are experienced not at the macrolevel, as in complex mixtures of genes, but at the microlevel, as in the expression and trafficking of the single antigens themselves. For example, expression of individual genes may be limited by nucleic acid content (A-T-rich genes may form less stable mRNA), or may require cofactors for transport of message out of the nucleus. Amino acid content, for example, high hydrophobicity, may result in insolubility or in a requirement for a cofactor provided by the pathogen for optimal assembly, resulting in inefficient processing and presentation. Sequence elements that target proteins for degradation or add bulky posttranslational modifications may bias T-cell recognition or prevent processing altogether. Hepatitis C virus provides examples of many of these limitations: insoluble proteins, cotranslational assembly of subunits, and glycosylation sequences that are not used in the native virion but are glycosylated when the same proteins exit the cell through normal secretory pathways [87]. The end result in such a situation is that a quick experiment may determine whether an antigen performs well or poorly as a DNA vaccine, but detailed study and a comprehensive understanding of the structural biology of the pathogen may be required to create a functional DNA vaccine for an antigen of interest. Thus, the level of understanding of the antigen that is required to create a good DNA vaccine has served as a limitation to the use of this technology for screening approaches to vaccination. Nevertheless, it has been possible using directed screening of a cDNA expression library from the tick *Ixodes scapularis* to identify individual protective antigens in an animal challenge model [86].

5.4 The Development Path of a DNA Vaccine Candidate

An overview of a vaccine development process along with basic scientific, technical, and economic aspects of vaccine research and development is given in Chapter 4. Most of the contents of that chapter also apply to DNA vaccines and will not be repeated here. Rather, in the following section, we will concentrate more on regulatory quality and safety requirements for potential DNA vaccines.

Anticipating the need for an early regulatory guidance, specific points to consider have been issued by the Food and Drug Administration (FDA) in the United States and by the World Health Organization (WHO) [88, 89]. These specific guidelines should be viewed in the context of the entire regulatory framework. Therefore, a limited list of other relevant guidelines is also given in the reference section [90]. Partly, these guidelines are not directly applicable to vaccines in general (including DNA vaccines) or even expressively exclude these from their scope. Nevertheless, they describe the basic rationale and define a technical state of the art that must be met. An overview on how to apply regulatory requirements pragmatically to biologicals and vaccines is given in Ref. [91].

Several DNA vaccines have been tested in clinical trials. As such, DNA vaccines have progressed part of the way through the development process. However, these clinical trials were done at an early stage and were more or less extended studies of research candidates. Such early stage clinical trials differ greatly from those required for licensable products, as usually only an abbreviated preclinical development has been performed in order to make clinical trial products available at the earliest possible time. Many more detailed quality analyses and very long, complex, and expensive safety studies will have to be performed to arrive at a marketable DNA vaccine that is sufficiently safe and efficacious to justify its widespread application in healthy individuals.

5.4.1 Quality of Starting Materials

Any pharmaceutical product needs to be thoroughly and completely defined and characterized by adequate analytical methods. This includes all starting materials, the production process, purified bulk materials, the formulated vaccine, and any excipient, adjuvant, or other constituent of the vaccine and may well mean that in total a set of 100 or more analytical methods must be applied. Table 1 explains how essential starting materials of a DNA vaccine need to be tested and characterized, mainly in order to provide sufficient information for appropriate risk and safety evaluations and also for a proper and reproducible specification or definition of the product. For safety reasons, any unintended by-product that could be expressed by the

Tab. 1 Characterization and quality aspects of starting materials of a plasmid DNA vaccine

Plasmid
Construction of the entire plasmid
Detailed functional map
Source and function of plasmid components
Regions of eukaryotic origin
Antigen-coding sequence
Identification
Origin
Means and ways of isolation
Sequence
Bacterial cells to produce plasmids
Established Master Cell Bank and Working Cell Bank
Confirmed identity
Absence of microbial contaminants, bacteriophages
Stability upon passaging
Defined maximum passage number
Use only if absolutely unavoidable:
• Retroviral-like long terminal repeats
• Oncogene sequences
• Extended sequences with homologies to the human genome
• Sequences encoding for cell growth-regulating functions
• Alternative, unintended reading frames

plasmid should be avoided. This applies to oncogenes, alternative open reading frames, gene-regulating (long terminal repeat) sequences, and to resistance markers based on allergy-prone antibiotics, such as penicillins. If antibiotic-selection markers are used, one should preferably use kanamycin or neomycin, as these are less often used for critical clinical indications. The minimum levels of these antibiotics in the final vaccine must be specified and should be below levels that could cause unintentional effects. The same degree of characterization applies to any formulation component and to the bacterial cells used to produce the plasmids. The cells must come from established and pretested cell banks, and these are to be used within a defined number of passages, during which time the bacteria and plasmids remain stable.

5.4.2 Process Development and In-process Controls

Purification of plasmids at a small scale normally follows well-established methods, which can be applied to produce clinical trial product of adequate quality and are adaptable to a larger scale. But before a plasmid can be made at a final product scale, significant investments into facilities and equipment must be made to comply with current Good Manufacturing Practices (cGMP). Furthermore, there are normally various technical difficulties to be solved, as it is never trivial to scale up even a seemingly simple process. For example, the control of microbial contaminations becomes a much more demanding task during large-scale operations and the inevitable, longer holding times often adversely affect the product quality and yields. If complex formulations are envisaged, this will certainly compound these scale-related difficulties.

The purification process must be adequately controlled by a range of methods for monitoring and quantifying impurities, such as RNA and genomic DNA, endotoxins, bacterial proteins, carbohydrates, and impurities derived from starting materials or from substances introduced during the process, for example, by chromatography column matrix bleeding. Degraded plasmids (linear and relaxed, open circular forms, dimers), modified or inadequately complexed/formulated plasmids could be less efficient and should therefore be monitored and limited by suitable control methods (compare Table 2). Much experience has already been accumulated on how to produce and analyze DNA vaccine and gene therapy plasmids [92].

A stable manufacturing process should ideally yield a consistent quality at all intermediate steps, so that upper limits for impurities and lower limits for the product's quality can be adequately controlled by keeping the process constant. In these cases, it may be sufficient to prove and validate the consistency of the process. But for critical components and in those cases where limits are narrow, analyses will be necessary and advisable to monitor the quality.

But what are acceptable limits for impurities? The more formal answer to this question is that preclinical and clinical data should justify the acceptability of the chosen limits, that is, neither the efficacy nor the safety of the vaccine is adversely affected by the remaining impurities. Since this is not really a measure that can be applied during process development, the basic rationale should be that whatever is avoidable by current standards must be avoided. In practical terms, this means that state-of-the-art technologies should be applied and reasonable efforts should be made to achieve the highest levels of purity. If additional measures would adversely affect vaccine quality (stability, specific activity) or significantly reduce the yield without giving any benefit, these would appear inappropriate.

Tab. 2 Characterization and quality testing of a DNA vaccine plasmid purification process

Elimination of impurities
Bacterial RNA and genomic DNA
Endotoxins
Bacterial proteins, carbohydrates, and other impurities
Media components and substances used or added during purification
Product-related impurities
Linear plasmid DNA
Open circular plasmids
Dimeric or oligomeric plasmid components
Complexed (e.g. formulated, encapsulated, etc.) plasmids versus noncomplexed proportions
• **Acceptable amounts and upper/lower limits to be defined by process validation or alternatively by analyses of each batch**
• **Specifications to be justified by preclinical and clinical safety data**

5.4.3
Quality of the Final Product

As listed in Table 3, quality of the vaccine must again be confirmed for bulk plasmids, for the formulated, and for the filled final product. Apart from routine tests for plasmid identity, purity, and sterility, consistency of the product and process must be demonstrated in at least three consecutive runs of the entirc process. These runs must result in a product that meets all predefined specifications. Stability of DNA vaccines must be evaluated by long-term studies to demonstrate that the defined specifications are met until the end of the envisaged shelf live. For naked DNA, it is not anticipated that stability will be an issue.

The absolute requirement to measure the potency of each DNA vaccine lot represents particular challenges, as a potency assay should quantitatively measure the active ingredient for its ability to raise adequate immune responses. Most likely, this will require titrations of the vaccine in an animal model. If certain levels of antibodies can be correlated with protective effects, measurements of the induced immune responses may be sufficient. However, for most new vaccines such correlations do not exist, hence challenge infection studies (if available) with determinations of protective doses may be required. For naked DNA, it could be argued that potency assays based upon a quantitative measurement of expression after transfection of suitable cell lines could be sufficient, since this could correlate with antigen expression in vivo. However, formulated DNA vaccines may not behave similarly in vitro and in vivo.

5.5
Preclinical Safety and Efficacy

Before a new vaccine is tested in clinical trials, a reasonable set of data should be accumulated to provide evidence that

Tab. 3 Characterization and quality testing of bulk plasmids and formulated vaccines

Plasmid identity	Partial or complete sequence verification
Plasmid purity	Absence or degree of denaturation or degradation Specified minimum level of supercoiled DNA
Residual impurities	Unavoidable impurities within specified limits Sterility (absence of microbial contaminants)
Stability	Real-time studies in final containers (mainly for supercoiled plasmid DNA) • Specifications must be met for entire shelf life Accelerated stability studies at elevated temperature and other strenuous conditions useful and recommended
Consistency	3 (−5) consecutive batches in final facilities and with final equipment • Intermediates and products must meet specifications.
Potency assay	Quantitative measurement of the active component of the vaccine by, for example, – Titration in a suitable animal model – Quantitative measurement of expression after transfection of cell lines • Comparison with an internal reference standard

the vaccine can be safely administered to humans and is expected to be effective. For early clinical trials, a complete safety evaluation is normally not yet available. At a minimum, local reactions and toxicity upon single and repeated applications should be studied. It may be difficult to provide preclinical efficacy data for a new vaccine indication, if reliable models are not yet available. But even if permission to perform clinical testing were granted without supporting efficacy data from model studies, what would be an adequate dose range to be studied and how are the clinical studies to be evaluated for efficacy? Preclinical model studies and a good understanding of the protective function of the vaccine beyond its capacity to induce immune responses are critical before clinical trials are commenced. Studies in larger animal species can, in many cases, yield similar or even better information at lower risk and less cost than a premature Phase I clinical trial with its inherently limited possibilities to assess efficacy. There may be exceptions, where the vaccine's mode of action requires tests in humans, because the vaccine is targeted very specifically at certain cell types or is dependent on highly specific immune responses.

5.5.1
Pharmacodynamic and Pharmacokinetic Properties of the Vaccine

Pharmacodynamic and pharmacokinetic studies are normally applied only to chemical drugs; in the field of vaccines, these terms are uncommon. Pharmacodynamic effects of a medicinal product describe its precise mode of action, dose-effect relationships, and side effects that are related to the mode of action. Pharmacokinetic properties of a medicinal (drug) product include its in vivo adsorption, distribution, metabolism, and excretion. Whereas for drugs pharmacodynamics and pharmacokinetics are mandatory and provide essential information about basic safety and efficacy characteristics, these are not normally studied in depth for conventional vaccines, as there are technical and scientific limitations. However, for DNA vaccines, an in-depth knowledge of their mode of action, distribution, persistence, or elimination in vivo will be important to foresee and analyze potential risks and side effects.

Primary pharmacodynamic studies of a vaccine should provide an understanding about the immunological and protective mode of action along with dose-response relationships. (Secondary pharmacodynamics mainly reveal unintentional side effects and will be covered later in the context of safety studies). Regarding the humoral immune responses, the duration and titers of antibodies should be measured and specific functional responses should be studied, including, for example, virus-neutralizing, bactericidal, and complement-binding antibodies, or antibody-dependent cytotoxicity (ADCC). Classes and subclasses of antibodies can be evaluated to define the types of immune responses induced. For example, humoral immune responses predominantly induce IgG1, whereas cellular responses are more associated with IgG2a antibodies.

Since DNA vaccines have the potential to induce cytotoxic T-cells (CTL) more efficiently than conventional, inactivated vaccines, it may be desirable or necessary to study and characterize the specificity and functional effects of CTL responses. However, this represents a technical and practical challenge, because the assays to measure CTL are not strictly quantitative and immune cells from each vaccinated individual must be cultured and studied. Alternatively, cytokine profiles could be analyzed to identify the predominant immune regulatory pathway. In man, cellular (T-helper type 1) responses leading to CTL activation are mainly controlled and stimulated by IL-12 and Interferon-γ, whereas humoral responses are more associated with elevated levels of IL-4, 5, 10, and 13. If vaccine plasmids are coexpressing cytokines or costimulatory molecules of antigen-presenting cells, the function of these cytokines needs to be carefully assessed with particular attention to potential interactive dysfunctions. Other pharmacodynamic effects to be considered are immune complex formations and interactions with other vaccinations or therapies, which could be administered concurrently with the new vaccine.

Pharmacokinetic studies will be needed to understand how the vaccine plasmids are distributed in the body and internalized by certain cell types, whether they persist and express the encoded antigens or other sequences, and whether the plasmid DNA can be integrated into the cellular genome. Pharmacokinetic biodistribution studies in animal models should be evaluated by high-sensitivity PCR (polymerase chain reaction) methods using primers derived from the vaccine plasmid. When different tissues are analyzed, separation of genomic DNA from plasmid DNA will be needed for a distinction between integrated or nonintegrated plasmid DNA.

From studies in mice, it is to be expected that pure or encapsulated plasmids will mainly be distributed at the site of injection and in the lymphoid organs and will persist for at least a few months [93]. After intravenous injection, plasmids were disseminated to all examined tissues except to the brain and to the gonads [93, 94]. In contrast, after im injection, DNA was detected in most tissues within a few weeks, but thereafter only in injected muscle. A small number of plasmids may persist for long periods of time, as intramuscularly injected plasmid DNA was still found after more than a year and expressed a reporter molecule during the entire observation period of 19 months [95], which is almost the entire life span of a mouse. Plasmids delivered to the skin or propelled into upper cellular layers of the skin by the gene gun are mainly taken up by terminally differentiated keratinocytes, which will be lost after several days or weeks [96]. But, migrating dendritic or Langerhans' cells in the skin can also take up the plasmids and could possibly carry these to the draining lymph nodes [97, 98].

If a particular construct exhibits integration activity, that is, if plasmid sequences are found to be inserted into a host cell genome, one will need to assess risk benefits for the disease to be targeted. In other words, "... manufacturers should carefully evaluate whether to proceed with product development". This judgment reflects the views as expressed by the U.S. Food and Drug Administration in 1996 [88, see part V, F] and relates to the risks of insertional mutagenesis. Animal model studies led to the conclusion that the risks are low, as the probability of an integration event was assessed as being several orders of magnitude below the spontaneous mutation rate [99, 100]. However, evidence of genomic insertion must be considered only as a first step toward uncontrolled critical events, such as an activation of oncogenic sequences, deactivation of suppressor genes, or rearrangements. If integration of DNA vaccines is observed, very long and costly tumorigenicity studies may be needed. However, it must be kept in mind that such studies are meant to identify risks, but are rarely able to prove absence of a specific risk beyond any doubt.

5.5.2 Preclinical Safety and Toxicology Testing

Beyond standard toxicology and histopathology, the safety package of a DNA vaccine must be assembled by a variety of complementary methodological approaches, including biological, molecular biological, biochemical, immunological, and immune-histological methods. Not only the obvious risks but also all potential risks and unexpected or undesirable side effects of the vaccine must be assessed. Worst-case assumptions should be made wherever data is limited or inconclusive, for example, regarding the distribution in vivo or the longevity of persistence of plasmids and antigen expression.

Table 4 summarizes the main preclinical safety aspects to be considered and evaluated. Conventional toxicology studies followed by histopathological examination of all affected organs should be done in order to assess basic safety issues. These trials are normally conducted in two different species, for example, a rodent and a nonrodent species. Mice or rats and dogs are often chosen, but with regard to certain immunotoxicological risks, monkeys, primates, or even transgenic animals may be more appropriate to study DNA vaccines. Choosing adequate safety trials and models for vaccines (and in particular for DNA vaccines) may be very difficult, hence regulatory guidelines are also not very specific. Therefore, guidelines encourage developers to contact the regulatory authorities responsible for advice and consultation on the intended toxicology program.

Tab. 4 Preclinical safety testing of DNA vaccines

An evaluation of potential unexpected and undesirable effects ...	
must consider the vaccine's	*... in order to assess the*
• mode of application;	• toxicity of single doses/overdoses;
• intended application scheme;	• local tolerance
• formulation;	• toxicity upon repeated application with particular attention to potential risk factors, such as
• dose range;	– influences on specific organ systems (secondary pharmacodynamics)
• distribution in vivo;	– reproductive functions and fertility
• longevity of persistence and antigen expression;	– immunotoxicology
• acute and chronic effects of antigen expression and immune responses;	*If risk factors exist or have been identified, additional studies are required to address*
• incorporation of other active genes, such as costimulatory cytokines;	• embryonal, fetal, and perinatal toxicity;
• reactions of particularly sensitive individuals; and	• mutagenicity and tumorigenic potential
• potential of immunopathological reactions	

Local tolerability testing is an absolute requirement for any vaccine. One may reasonably expect that pure plasmids will be well tolerated, but technologies to improve the uptake of plasmids into cells or to render DNA vaccines more immunogenic may affect local tolerance. For example, coadministrations of hypo-osmotic solutions, local anesthetics, or cardiotoxin have been used in research, but induce cell necrosis at the injection site.

Animal toxicity studies using single doses of the vaccine and overdoses (e.g. 10-fold overdoses) will most likely not reveal systemic intolerances of a DNA vaccine. Trials applying repeated doses will be more informative and should thus be planned very carefully. The number of doses and the potency studied should correspond to the intended application scheme, which normally consists of a primer dose followed by one or two booster vaccinations. Safety margins could be extended by administering extra vaccinations with shorter time intervals and by increasing the dose. Histopathology must be done with all organs and tissues that show changes or are expected to contain vaccine plasmids. Particular attention must be paid to potential immunotoxicological side reactions: pathologic immune complex formation could cause obstructing protein deposits in the kidney tubuli and autoimmune reactions can be identified either histologically as inflammatory cell invasions in affected tissues or by elevated cell enzyme levels released as a consequence of a cell destruction.

Vaccines that are expected to affect physiological processes related to fertility, pregnancy, or fetal development would require various further studies regarding the vaccine's reproductive toxicology. The same applies if alterations to male or female germ-line cells would be detected during other toxicology studies. Suitable test animal species for these studies must be chosen on the basis of immunological and functional properties of the vaccine. Mice, rats, or hamsters are normally used, but may not always be the ideal choice for a DNA vaccine. Reproductive toxicology consists of three different elements, which can partly be combined in a common study protocol. Fertility studies would evaluate adverse effects on spermatogenesis, formation of ovarian follicles, conception, implantation, and organogenesis. Teratology studies would cover the organogenesis period and usually end near or at the term of delivery. Peri- and postnatal toxicity testing normally commences before mating or in early pregnancy, covers the entire pregnancy period, and extends over the entire lactation period until weaning. Recently, fertility studies have been requested more often in order to assess unexpected risks in female recipients of medicinal products. Thus, fertility studies may be needed for any new DNA vaccine that is intended also for women of childbearing ages. A full reproductive toxicology program, extending over two to three generations and combined with teratology testing may be required if vaccine plasmids have been found in gonadal tissues and could induce germ-line alterations. This will also raise the issue of whether the plasmids can be passed on to the next generation. Therefore, the theoretical and unforeseeable risk of plasmids persisting in the gonads would significantly complicate the safety assessment of a DNA vaccine.

Tumorigenicity studies may only be appropriate if a DNA vaccine shows a broad tissue distribution, persists for long periods of time, or is intended for frequent and chronic use. They would also be relevant if it is decided to continue the development of a DNA vaccine shown to induce genomic integration or known to contain oncogenic sequences or extensive sequence homologies to the human genome. Such long-term studies may, however, be dispensable in those rare cases in which the vaccine is targeted to a severely life-threatening, acute disease and is not intended for a prophylactic use.

DNA vaccines that coexpress cytokines or contain toxins or toxin conjugates may of course trigger a variety of unexpected immunological or even direct adverse effects. These will mainly be assessed during single and repeated dose toxicity trials, but further studies addressing specific questions may be needed. For vector and carrier microorganisms (e.g. salmonella or shigella species), specific studies to evaluate the likelihood and consequences of a distribution to unvaccinated individuals are mandatory, as these are also standard requirements for conventional live attenuated vaccines.

5.5.3 Immunotoxicology Aspects

DNA vaccines are derived from bacterial DNA plasmid and, thus, are able to stimulate anti-DNA antibodies. If these antibodies are cross-reactive with host chromosomal DNA, they could act like autoantibodies and induce autoimmune diseases (such as systemic lupus erythematosus) characterized by the accumulation of DNA, antinuclear antibodies, and complement in various organs along with local inflammatory responses. Specific "lupus-prone" mouse strains exist that develop a similar disease. Repeated application of DNA vaccines to these mice did not alter the onset or the course of the disease. In normal mice, anti-DNA antibodies were induced by DNA vaccines, but these remained far below those levels found in lupus-prone mice [101]. During clinical trials in humans, significant rises of anti-DNA antibodies have not been observed [102]. Thus, the risk of anti-DNA autoimmune responses seems low, but should be monitored during clinical trials.

Cells harboring vaccine plasmids and expressing foreign antigens normally present these antigens on their cell surface. They can not only stimulate an immune response but can as well become the target of an immune attack, resulting in inflammatory reactions and in cell-mediated cytolysis in tissues with high expression rates. These effects may be more prominent after a second or repeated application of the vaccine. Acute inflammation of muscle tissues associated with the destruction of myocytes has in fact been observed during some animal studies [103, 104], whereas other studies detected no anti-myocyte antibodies or muscle tissue reactions [101]. Therefore, monitoring of muscle cell enzyme levels during early clinical trials should be considered to assess the risks associated with tissue destruction and autoantibodies. So far, there have been no clinical reports about such problems. However, a higher efficiency of plasmid uptake and expression or plasmids targeted to specific organs may result in quite different findings. In principle, animal toxicology studies and histological evaluations should be able to detect such autoimmune effects, but due to animal species and human genotype differences, the matter needs to be studied mainly during clinical trials.

A classical way of actively inducing immune tolerance, for example, against allergens, would be to repeatedly administer low doses of the same antigen over a long period of time. In principle, this could also occur by a prolonged antigen expression after DNA vaccination. Neonates with their immature immune systems or individuals with reduced immune responses could be particularly prone to this kind of tolerance. In one study, tolerance was observed after DNA vaccination of neonatal mice. These mice did not respond to a second vaccination given several months later, even if the same antigen was injected directly as a conventional vaccine. If immunized at an age of two months and revaccinated four months later, mice responded well to the same DNA vaccine. Aged mice also showed reduced immune responses and reduced protection from infection [105, 106]. In contrast, several other studies did not reveal similar effects, thus immune tolerance appears to be associated only with certain antigens. As long as the underlying mechanisms remain unknown, preclinical studies could be useful to predict clinical results, particularly in those cases in which the vaccine is to be used in neonates, at advanced age, or in other immunocompromised individuals. As for any other vaccine intended for those target groups, separate clinical trials should also be done to assess safety and efficacy.

5.6 Clinical Safety and Efficacy Trials

A DNA vaccine candidate that has successfully passed all preclinical hurdles needs to be tested clinically by a three-phased scheme as outlined in Chapter 4. As defined by a European Council directive, a clinical trial means "...any investigation in human subjects intended to discover or verify the clinical, pharmacological and/or other pharmacodynamic effects of ... investigational medicinal product(s), and/or to identify any adverse reactions ...and/or to study absorption, distribution, metabolism, and excretion ... with the object of ascertaining ...safety and/or efficacy." [107]. This definition explains that the expectations for the conduct of clinical trials go far beyond efficacy and a basic safety assessment, but include pharmacodynamic and pharmacokinetic elements, which could be of particular relevance for DNA vaccines and even more so for vector organisms applied to DNA vaccines. Thus, clinical trials within that scope would be impossible without having established a relevant database from preclinical studies. Certain pharmacological and pharmacokinetic aspects cannot be studied directly in humans, so compromises must be made and the gaps must be filled by appropriate animal studies.

On the basis of the available preclinical data and on predefined clinical trial plans, ethics committees will evaluate whether the anticipated benefits and risks would justify clinical testing. In addition, local regulatory authorities must be informed and may deny consent. The clinical trial product must be made in compliance with Good Manufacturing Practices with increasing demands (e.g. regarding the qualification of equipment and facilities, product specifications, and validation of methods) as the trials proceed to later phases.

Phase I clinical trials mainly evaluate the vaccine's basic safety and also measure the immune responses induced. Safety aspects to be studied include, for example, general local and systemic reactions, potential immunological complications, and specific side effects on certain organs and tissues that might occur because of theoretical considerations or have been identified during toxicology studies. For live vector organisms, additional safety features, such as shedding and distribution of the organisms, will have to be studied in order to confirm preclinical safety evaluations and to rule out risks associated with the practical use of the vaccine.

Experience from published clinical Phase I studies will most likely lead to a dose-escalation study design, starting with doses that were shown to be of adequate immunogenicity during preclinical testing. According to existing – still rather limited – experience, increased DNA doses may then be needed to achieve immune responses in humans, similar to those seen in animal models. For example, antibody responses against a Plasmodium falciparum circumsporoite antigen were induced in mice at doses of a few micrograms of DNA per vaccination. The first trials of this vaccine in humans initially used doses from 20 µg to 100, 500, and finally up to 2500 µg, but even the highest doses did not stimulate any detectable antibodies. However, CTL responses in up to four out of five trial individuals in the highest dose group were noted [108, 109]. A similar experience has been seen with gene-gun delivery of plasmids, for example, for a DNA vaccine encoding hepatitis B surface antigen, where low antibody responses were observed in human trials [110]. Owing to the capacity of the gold particles to bind DNA, there have been limitations in the amount of DNA that can be delivered by the gene gun. So, repeated boosting and simultaneous applications at multiple sites were needed to deliver higher doses of DNA plasmids. In this case, the DNA vaccine finally induced antibody levels, which would have been considered as being protective, if induced by a conventional vaccine. The first example of a malaria DNA vaccine illustrate a dilemma, which is only, in part, specific to DNA vaccines. The vaccine induced CTL but no antibody response. The question now is whether CTL responses alone can be taken as a marker of protection, and if so, which level would be required? This question cannot be answered until the vaccine has been tested clinically under conditions that expose trial subjects to the parasites and monitor the number and severity of cases occurring in comparison to a control group. Until then, using CTL as an indicator of potency raises logistic and technical challenges, since measuring CTL responses requires viable peripheral blood mononuclear cells from each trial individual and highly specialized laboratories to perform the tests.

A Hepatitis B DNA vaccine may be a much more suitable candidate to establish a first DNA vaccine product and to successfully pass all clinical hurdles. There is a well-known, single antigen that confers protection, whose serological responses are easily measured and can be correlated with protective effects. Other DNA vaccine candidates include antitumor vaccines [111–113], which may pass clinical phases more rapidly than prophylactic vaccines, as their efficacy could be a directly measurable clinical effect, such as tumor regression or reduced or absent metastases. Furthermore, their use could be justified even with a partial effect and some unresolved safety concerns, whereas for a prophylactic vaccine for healthy people, higher standards must be met.

5.7
Registration and Licensing of DNA Vaccines

The formal process of licensing a new vaccine in different countries is described in Chapter 10 of this volume. In this chapter, only a few general aspects of licensing of an entirely new product and active principle will be briefly discussed.

Potential developers of a DNA vaccine product should not only consult specific guidelines and recommendations for vaccines and DNA vaccines but should also be advised to carefully monitor the regulations that evolve around gene therapy. There are some parallels and similarities between DNA vaccines and gene therapy approaches, thus insight may be gained about potential quality and safety aspects. For safety and toxicology studies and during the preparation of clinical trials, contact should be sought with regulatory authorities in the respective countries and preferably with the leading authorities, such as the European Medicines Evaluation Agency (EMEA) or the US FDA, who welcome such contacts and have implemented official channels to obtain a more binding opinion for specific questions. As the EMEA relies on rapporteurs from the individual country's authorities, certain national authorities (particularly those who are actively pursuing research in this field) may also be open to discuss development proposals on a more informal level. Through these consultations, developers will not only obtain useful advice on specific questions but may also determine where a specific application should be filed. There may occasionally be doubts regarding the regulatory pathways in the United States and whether a certain product (e.g. a DNA vaccine intended to induce antihormonal antibodies) will be considered a vaccine/biological product or a gene therapy/drug. These two categories are regulated quite differently and by different departments, but the definitions for these products partly overlap and are evaluated on a case-by-case basis.

DNA vaccines raise certain safety-related issues that may remain unsolved even after intensive studies. Hence, a scientifically grounded, quantitative risk evaluation should be made. On the basis of existing data from the developer's own studies, combined with published information, and complemented by reasoned assumptions where specific information is missing, such a risk assessment can be a valuable tool to aid the decision process. In the absence of quantitative data, worst-case assumptions should be made, but realistic case scenarios may be included to indicate a possible range. An example of how this done in a quantitative manner for the risk of insertional mutagenesis is given by Kurth [99].

As with other new technologies, it is possible that the first licensed DNA vaccines will receive conditional approval. In this case, postlicensing "Phase IV" clinical trials may be requested to address and monitor remaining unresolved issues by specific studies or on a larger scale. Furthermore, as for any medicinal product, effective and intensive pharmacovigilance procedures will be used to closely monitor the appearance of any side effects, while the vaccine is applied widely and routinely.

5.8 Conclusions

Studies of DNA vaccines in animal models have borne out many of the theoretical advantages of this approach over conventional methods of immunization. Fast and relatively simple to construct using standard molecular biologic techniques, they also have become easier to produce, owing to the development of various kits for producing plasmid in quantities needed for clinical trials. The antigens that are expressed from the plasmid DNA are in their native conformations and are processed for presentation to T-cells comparable to that of the native protein in viruses or intracellular bacteria. In laboratory animal studies, it is clear that this approach can provide protective immunity from infectious challenges in a variety of model systems. However, some important limitations remain. A good understanding of the structure of the antigen and its intracellular processing is required. Attempts to generate DNA vaccines of high quality without this understanding are often frustrating because of poor expression, antigenicity, or processing. The uptake of the DNA and the efficiency of its delivery can vary significantly among species. This is not only related to body mass but also to muscle size, volume of inoculum, and structure of muscle components. By far, the most drastic limitation is the lack of immunogenicity of these vaccines in humans. The basis for this difference between humans and animal models is still not well understood. Various method of targeting, particulate delivery, and formulation offer promise, but have yet to be fully tested in humans. The inability to compare an ineffective DNA vaccine with an effective one (since none yet exists) is a major handicap to understanding how to make DNA vaccines immunogenic in humans. A further consequence of the lack of effectiveness in early human trials is that the regulatory environment for DNA vaccines is not yet mature. Many Phase I studies have been done and regulatory agencies have formed a comprehensive set of expectations for which safety parameters are to be measured and how to measure them. However, since no DNA vaccine has advanced beyond Phase II, regulators have not yet confronted the need to create regulatory or safety requirements for registration studies. The future of DNA vaccines depends on one or more technologies, possibly not yet invented, that will give consistent high immune responses. Only when that problem has been solved will the manufacturing, clinical, and regulatory pieces of the development puzzle fall into place.

References

1. T. W. Dubensky, B. A. Campbell, L. P. Villarreal, *Proc. Natl. Acad. Sci. U. S. A.* **1984**, *81*, 7529–7533.
2. H. Will, R. Cattaneo, H. G. Koch et al., *Nature* **1982**, *299*, 740–742.
3. J. A. Wolff, R. W. Malone, P. Williams et al., *Science* **1990**, *247*, 1465–1468.
4. D. C. Tang, M. De Vit, S. A. Johnston, *Nature* **1992**, *356*, 152–154.
5. J. B. Ulmer, J. J. Donnelly, S. E. Parker et al., *Science* **1993**, *259*, 1745–1749.
6. E. F. Fynan, R. G. Webster, D. H. Fuller et al., *Proc. Natl. Acad. Sci. U. S. A.* **1993**, *90*, 11478–11482.
7. J. J. Donnelly, J. B. Ulmer, J. W. Shiver et al., *Annu. Rev. Immunol.* **1997**, *15*, 617–648.
8. S. Gurunathan, D. M. Klinman, R. A. Seder, *Annu. Rev. Immunol.* **2000**, *18*, 927–974.
9. J. R. Gebhard, J. Zhu, X. Cao et al., *Vaccine* **2000**, *18*, 1837–1846.
10. A. Bojak, D. Hammer, H. Wolf et al., *Vaccine* **2002**, *20*, 1975–1979.
11. E. Kurar, G. A. Splitter, *Vaccine* **1997**, *15*, 1851–1857.
12. M. Kwissa, V. K. Von Kampen, R. Zurbriggen et al., *Vaccine* **2000**, *18*, 2337–2344.
13. A. Kamei, S. Tamaki, H. Taniyama et al., *Virology* **2000**, *273*, 120–126.
14. Z. Q. Xiang, Z. He, Y. Wang et al., *Vaccine* **1997**, *15*, 896–898.
15. M. J. Hariharan, D. A. Driver, K. Townsend et al., *J. Virol.* **1998**, *72*, 950–958.
16. W. W. Leitner, H. Ying, D. A. Driver et al., *Cancer Res.* **2000**, *60*, 51–55.
17. O. Vidalin, A. Fournillier, N. Renard et al., *Virology* **2000**, *276*, 259–270.
18. M. M. Morris-Downes, K. V. Phenix, J. Smyth et al., *Vaccine* **2001**, *19*, 1978–1988.
19. C. Smerdou, P. Liljestrom, *Curr. Opin. Mol. Ther.* **1999**, *1*, 244–251.

20. C. M. Leutenegger, F. S. Boretti, C. N. Mislin et al., *J. Virol.* **2000**, *74*, 10447–10457.
21. K. F. Sykes, S. A. Johnston, *Nat. Biotechnol.* **1999**, *17*, 355–359.
22. I. F. Ciernik, J. A. Berzofsky, D. P. Carbone, *J. Immunol.* **1996**, *156*, 2369–2375.
23. A. Iwasaki, C. S. Dela Cruz, A. R. Young et al., *Vaccine* **1999**, *17*, 2081–2088.
24. G. Y. Ishioka, J. Fikes, G. Hermanson et al., *J. Immunol.* **1999**, *162*, 3915–3925.
25. T. Hanke, J. Schneider, S. C. Gilbert et al., *Vaccine* **1998**, *16*, 426–435.
26. J. Haas, E. C. Park, B. Seed, *Curr. Biol.* **1996**, *6*, 315–324.
27. J. Zur Megede, M. C. Chen, B. Doe et al., *J. Virol.* **2000**, *74*, 2628–2635.
28. S. L. Baldwin, C. D. D'souza, I. M. Orme et al., *Tuber. Lung Dis.* **1999**, *79*, 251–259.
29. C. Rush, T. G. Mitchell, P. Arside, *J. Immunol.* **2002**, *169*, 4951–4960.
30. M. P. Velders, S. Weijzen, G. L. Eiben et al., *J. Immunol.* **2001**, *166*, 5366–5373.
31. F. Rodriguez, J. Zhang, J. L. Whitton, *J. Virol.* **1997**, *71*, 8497–8503.
32. F. Rodriguez, S. Harkins, J. M. Redwine et al., *J. Virol.* **2001**, *75*, 10421–10430.
33. H. Ji, T. L. Wang, C. H. Chen et al., *Hum. Gene Ther.* **1999**, *10*, 2727–2740.
34. G. Deliyannis, J. S. Boyle, J. L. Brady et al., *Proc. Natl. Acad. Sci. U. S. A.* **2000**, *97*, 6676–6680.
35. A. Biragyn, K. Tani, M. C. Grimm et al., *Nat. Biotechnol.* **1999**, *17*, 253–258.
36. T. M. Ross, Y. Xu, R. A. Bright et al., *Nat. Immunol.* **2000**, *1*, 127–131.
37. J. S. Boyle, J. L. Brady, A. M. Lew, *Nature* **1998**, *392*, 408–411.
38. J. Wild, B. Gruner, K. Metzger et al., *Vaccine* **1998**, *16*, 353–360.
39. Y. H. Chow, W. L. Huang, W. K. Chi et al., *J. Virol.* **1997**, *71*, 169–178.
40. Y. Huang, W. P. Kong, G. J. Nabel, *J. Virol.* **2001**, *75*, 4947–4951.
41. R. R. Amara, F. Villinger, J. D. Altman et al., *Science* **2001**, *292*, 69–74.
42. E. J. Gordon, R. Bhat, Q. Liu et al., *J. Infect. Dis.* **2000**, *181*, 42–50.
43. J. A. Wolff, J. J. Ludtke, G. Acsadi et al., *Hum. Mol. Genet.* **1992**, *1*, 363–369.
44. T. Tuting, W. J. Storkus, L. D. Falo, Jr *J. Invest. Dermatol.* **1998**, *111*, 183–188.
45. A. Porgador, K. R. Irvine, A. Iwasaki et al., *J. Exp. Med.* **1998**, *188*, 1075–1082.
46. M. Corr, A. Von Damm, D. J. Lee et al., *J. Immunol.* **1999**, *163*, 4721–4727.
47. O. Akbari, N. Panjwani, S. Garcia et al., *J. Exp. Med.* **1999**, *189*, 169–178.
48. M. A. Chattergoon, T. M. Robinson, J. D. Boyer et al., *J. Immunol.* **1998**, *160*, 5707–5718.
49. S. Casares, K. Inaba, T. D. Brumeanu et al., *J. Exp. Med.* **1997**, *186*, 1481–1486.
50. A. Bot, A. C. Stan, K. Inaba et al., *Int. Immunol.* **2000**, *12*, 825–832.
51. E. Manickan, S. Kanangat, R. J. Rouse et al., *J. Leukoc. Biol.* **1997**, *61*, 125–132.
52. J. B. Ulmer, R. R. Deck, C. M. Dewitt et al., *Immunology* **1996**, *89*, 59–67.
53. T. M. Fu, J. B. Ulmer, M. J. Caulfield et al., *Mol. Med.* **1997**, *3*, 362–371.
54. A. Y. Huang, P. Golumbek, M. Ahmadzadeh et al., *Science* **1994**, *264*, 961–965.
55. M. A. Chattergoon, J. J. Kim, J. S. Yang et al., *Nat. Biotechnol.* **2000**, *18*, 974–979.
56. S. Sasaki, R. R. Amara, A. E. Oran et al., *Nat. Biotechnol.* **2001**, *19*, 543–547.
57. U. Kumaraguru, R. J. Rouse, S. K. Nair et al., *J. Immunol.* **2000**, *165*, 750–759.
58. J. A. Leifert, S. Harkins, J. L. Whitton, *Gene Ther.* **2002**, *9*, 1422–1428.
59. S. C. Oliveira, J. S. Harms, R. R. Afonso et al., *Hum. Gene Ther.* **2001**, *12*, 1353–1359.
60. G. Widera, M. Austin, D. Rabussay et al., *J. Immunol.* **2000**, *164*, 4635–4640.
61. S. Zucchelli, S. Capone, E. Fattori et al., *J. Virol.* **2000**, *74*, 11598–11607.
62. A. M. Krieg, *Annu. Rev. Immunol.* **2002**, *20*, 709–760.
63. Y. Sato, M. Roman, H. Tighe et al., *Science* **1996**, *273*, 352–354.
64. R. A. Pontarollo, L. A. Babiuk, R. Hecker et al., *J. Gen. Virol.* **2002**, *83*, 2973–2981.
65. R. Weeratna, C. L. Brazolot Millan, H. L. Krieg Amdavis, *Antisense Nucleic Acid Drug Dev.* **1998**, *8*, 351–356.
66. M. Dupuis, K. Denis-Mize, C. Woo et al., *J. Immunol.* **2000**, *165*, 2850–2858.
67. M. J. Roy, M. S. Wu, L. J. Barr et al., *Vaccine* **2000**, *19*, 764–778.
68. M. Kumar, A. K. Behera, R. F. Lockey et al., *Hum. Gene Ther.* **2002**, *13*, 1415–1425.
69. D. H. Jones, C. D. Partidos, M. W. Steward et al., *Behring Inst. Mitt.* **1997**, 220–228.
70. M. G. Cusi, R. Zurbriggen, M. Valassina et al., *Virology* **2000**, *277*, 111–118.

71. W. O. Rogers, J. K. Baird, A. Kumar et al., *Infect. Immun.* **2001**, *69*, 5565–5572.
72. M. Fernandez-Alonso, A. Rocha, J. M. Coll, *Vaccine* **2001**, *19*, 3067–3075.
73. G. Zhang, V. Budker, P. Williams et al., *Hum. Gene Ther.* **2001**, *12*, 427–438.
74. G. Gregoriadis, *Curr. Opin. Mol. Ther.* **1999**, *1*, 39–42.
75. D. O'hagan, M. Singh, M. Ugozzoli et al., *J. Virol.* **2001**, *75*, 9037–9043.
76. K. Roy, H. Q. Mao, S. K. Huang et al., *Nat. Med.* **1999**, *5*, 387–391.
77. K. S. Denis-Mize, M. Dupuis, M. L. Mackichan et al., *Gene Ther.* **2000**, *7*, 2105–2112.
78. H. Lee, J. H. Jeong, T. G. Park, *J. Control Release* **2001**, *76*, 183–192.
79. R. Schirmbeck, S. A. Konig-Merediz, P. Riedl et al., *J. Mol. Med.* **2001**, *79*, 343–350.
80. M. A. Barry, W. C. Lai, S. A. Johnston, *Nature* **1995**, *377*, 632–635.
81. P. M. Smooker, Y. Y. Setiady, T. W. Rainczuk Aspithill, *Vaccine* **2000**, *18*, 2533–2540.
82. K. Manoutcharian, L. I. Terrazas, G. Gevorkian et al., *Immunol. Lett.* **1998**, *62*, 131–136.
83. C. P. Locher, K. F. Sykes, D. J. Blackbourn et al., *J. Med. Primatol.* **2002**, *31*, 323–329.
84. K. F. Sykes, M. G. Lewis, B. Squires et al., *Vaccine* **2002**, *20*, 2382–2395.
85. P. C. Melby, G. B. Ogden, H. A. Flores et al., *Infect. Immun.* **2000**, *68*, 5595–5602.
86. C. Almazan, K. M. Kocan, D. K. Bergman et al., *Vaccine* **2003**, *21*, 1492–1501.
87. M. Houghton, *Curr. Top. Microbiol. Immunol.* **2000**, *242*, 327–39.
88. Food and Drug Administration, Center for Biologics Evaluation and Research: Points to consider on plasmid DNA vaccines from preventive infectious disease indications, Docket No. 96 N-0400, 1996, http://www.fda.gov/CBER/gdlns/plasmid.pdf
89. WHO Guidelines for assuring the quality of DNA vaccines. Annex 3 in: WHO Technical Report Series, No. 876, 1998.
90. European (CPMP) and internationally harmonized (ICH) guidance documents with basic applicability to DNA vaccines: Note for guidance on specifications: Test procedures and acceptance criteria for biotechnological/biological products. CPMP/ICH/356/96 Note for guidance on preclinical pharmacology and toxicology testing of vaccines. CPMP/SWP/465/95 for guidance on preclinical safety evaluation of biotechnology-derived Note pharmaceuticals. CPMP/ICH/302/95 Note for Guidance on the quality, preclinical and clinical aspects of gene transfer medicinal products. CPMP/BWP/3088/99 Safety studies for gene therapy products. CPMP/SWP/112/98 Retrievable via http://www.emea.eu.int/ ICH guidelines are also available under http://www.ich.org/ich5.html.
91. J. P. Gregersen, Requirements for the preclinical pharmacology and safety assessment and: Outline of major registration requirements (Annex A in *Research and Development of Vaccines and Pharmaceuticals from Biotechnology. A Guide to Effective Project Management, Patenting and Product Registration* (Ed.: J. P. Gregersen), VCH, Weinheim, New York, Basel, Cambridge, Tokyo, 1994, pp. 119–147.
92. M. Schleef, *Plasmids for gene therapy and vaccination*, Wiley-VCH, Weinheim Germany, 2001.
93. S. E. Parker, F. Borellini, M. L. Wenk et al., *Hum. Gene Ther.* **1999**, *10*, 741–758.
94. L. Lunsford, U. Mckeever, V. Eckstein et al., *J. Drug. Target* **2000**, *8*, 39–50.
95. J. A. Wolff, J. J. Ludtke, G. Acsadi et al., *Hum. Mol. Genet.* **1992**, *1*, 363–369.
96. E. Raz, D. A. Carson, S. E. Parker et al., *Proc. Natl. Acad. Sci. U. S. A.* **1994**, *91*, 9519–9523.
97. C. Condon, S. C. Watkins, C. M. Celluzzi et al., *Nat. Med.* **1996**, *2*, 1122–1128.
98. C. A. Torres, A. Iwasaki, B. H. Barber et al., *J. Immunol.* **1997**, *158*, 4529–4522.
99. R. Kurth, *Ann. N. Y. Acad. Sci.* **1995**, *772*, 140–151.
100. T. Martin, S. E. Parker, R. Hedstrom et al., *Hum. Gene Ther.* **1999**, *10*, 759–768.
101. G. Mor, M. Singla, A. D. Steinberg et al., *Hum. Gene Ther.* **1997**, *10*, 293–300.
102. R. R. Macgregor, J. D. Boyer, K. E. Ugen et al., *J. Infect. Dis.* **1998**, *178*, 92–100.
103. H. L. Davis, C. L. Brazolot Millan, S. C. Watkins, *Gene Ther.* **1997**, *4*, 181–188.
104. J. L. Whitton, F. Rodriguez, J. Zhang et al., *Vaccine* **1999**, *7*, 1612–1619.
105. G. Mor, M. Yamshchikov, M. Sedegah et al., *J. Clin. Invest.* **1996**, *98*, 2700–2705.

106. D. M. Klinman, M. Takeno, M. Ichino et al., *Springer Semin. Immunopathol.* **1997**, *19*, 245–256.
107. European Parliament and Council Directive 2001/20/EC. *Official J. Eur. Commun. L 121/34*, (1.5.2001).
108. R. Wang, D. L. Doolan, T. P. Le et al., *Science* **1998**, *282*, 476–480.
109. T. P. Le, K. M. Coonan, R. C. Hedstrom et al., *Vaccine* **2000**, *18*, 1893–1901.
110. M. J. Roy, M. S. Wu, L. J. Barr et al., *Vaccine* **2001**, *19*, 764–778.
111. F. K. Stevenson, D. Zhu, M. B. Spellerberg et al., *Rev. Clin. Exp. Hematol.* **1999**, *9*, 2–21.
112. M. Mincheff, S. Tchakarov, S. Zoubak et al., *Eur. Urol.* **2000**, *38*, 208–217.
113. P. Walsh, R. Gonzales, S. Dow et al., *Hum. Gene Ther.* **2000**, *11*, 1355–1368.

6
Characterization and Bioanalytical Aspects of Recombinant Proteins as Pharmaceutical Drugs

Jutta Haunschild and Titus Kretzschmar
MorphoSys AG Lena-Christ-Strasse 48 D-82152 Martinsried Germany

6.1
Introduction

The development of recombinant proteins as pharmaceutical drugs demands robust, sensitive, and specific analytical assays to characterize the purified drug with respect to its physicochemical as well as biological features, and bioanalytical assays to quantify proteins or their activities in biological matrices. Well-established analytical assays are applied to determine characteristics such as purity/impurities, identity, quantity, stability, specificity, and potency of the purified recombinant protein during drug development. The determination of the purity and identity of a protein drug is a particularly challenging task since recombinant proteins are produced from living systems that inherently lead to protein variants (e.g. posttranslationally modified and/or fragmented proteins) with altered characteristics, which may be hard to separate from the original protein drug. In stability studies, those protein features are evaluated that might be subject to change during storage/handling of the drug. Specificity measurements lead to a closer understanding of drug-target interaction(s), which might result in early hints about possible side effects in clinical trials. Finally, potency determinations are used to quantify the biological activity of the therapeutic protein.

On the other hand, bioanalytical assays are necessary to determine and quantify the protein drug in biological fluids. For example, validated bioanalytical assays are the key in the quantitation of the protein drug in the course of pharmacokinetic studies (see Chapter 8). Especially in the case of humanized/human monoclonal antibodies, bioanalytical assay development in human serum/plasma is challenging because the therapeutic concentration of antibodies can be very low (0.1 to 10 $\mu g\ mL^{-1}$ or even lower), and because these antibodies are so similar to the native human antibodies that circulate in the blood in very high concentrations of 10 000 $\mu g\ mL^{-1}$.

Another major topic of scientific and regulatory consideration in the development of therapeutic proteins is the assessment of undesired immune responses to the drug that may lead to a reduction in efficacy and to adverse reactions. This assessment also requires validated bioanalytical assays, which allow to precisely measure the immune response.

In future, there will be an even greater emphasis on the (bio)analytical description of biological substances because of an increase in numbers of biologicals in clinical development, and of the advent of generics of biological drugs and the task of evaluating these compounds for clinical use. Developing methods and protocols for assessing bioequivalence of an original drug and its generics is a high priority for the FDA according to the current FDA Commissioner [1].

The scope of this article is to summarize the analytical methods used to characterize the purified recombinant protein in the different stages of drug development, and to summarize bioanalytical methods with focus on their validation and standardization as well as in the determination of immunogenicity of the therapeutic protein.

6.2 Characterization of Purified Recombinant Proteins

6.2.1 Purity and Impurities

The absolute purity of a biological substance is hard – if at all possible – to determine. Regular and sometimes only subtle protein modifications such as glycosylation, alternative disulphide bond formation, deamidation, oxidation, phosphorylation, acetylation, sulfation, sulfoxidation, γ-carboxylation, and pyroglutamate formation lead to protein variants that may have more or less different characteristics. Also, truncated protein variants might be generated by the presence of cryptic or alternative start sites of transcription, by premature stop of the peptide chain elongation process, or by the action of host cell peptidases. Peptide mapping and mass spectrometry (MS) usually achieve detection of most of such protein variants. Aggregation is another modification of a protein, which can be the result of, for example, underglycosylation, oxidation, and/or deamidation, and can be detected by size-exclusion chromatography. The amount of aggregated protein usually should stay below 5%. It is highly recommended to investigate the nature and potential toxicity of such alterations. To analyze these variants is an essential yet challenging task, as their physicochemical features might not be very different from each other. Owing to the possible presence of highly related protein variants in the preparation, it is recommended to determine the purity of a protein drug by at least two independent methods, that is, methods that use different physicochemical principles such as SDS-polyacrylamide gel electrophoresis and reverse-phase high pressure liquid chromatography (HPLC).

Besides these protein variants, so-called process-related impurities have to be considered. Of major concern are residual antibiotics from fermentation, enzymes and antibodies from chromatography columns and other column leachates, endotoxin from bacterial hosts [2], (retro-) viruses [3], bacteria, fungi, mycoplasma, prions, various other media components such as solvents, antifoam agents, heavy metal ions, as well as preservatives and expression host components. As for DNA contaminations, less than 10 to 100 pg per dose are allowed in the final drug product [4]. To check for the presence of antigenic expression host-related impurities, a polyclonal antibody serum to the "empty" host, that is, host cells that are not harboring the product-encoding gene, is very helpful. Moreover, whenever possible, specific impurity standards should be used for impurity quantification, and the limit of detection/-quantification (LOD/-Q) for impurity assays should be indicated. The acceptance limits should not be set higher than safety data justify, and it should not be lower than what is historically achievable by the manufacturing process and by reasonable analytical efforts.

In some instances, the protein drug is conjugated to effector functions such as radioisotopes, toxins, or other proteins such as cytokines that mediate a biological effect. Besides considering all aspects mentioned above and below for the individual components of the conjugate, special care has to be taken to determine the average coupling ratio as well as the amount of free components, if any, in the preparations.

6.2.2 Identity

Identity assays aim at confirming the molecular composition and, if technically possible and commercially reasonable, structure of the drug substance, and thus should be suited to allow the detection of even minor alterations in the molecular composition of the drug.

Amino acid analysis and peptide mapping are standard methods in the course of protein identification processes [5, 6]. Molecular mass determination of whole molecules as well as peptide fragments with accuracies of about 0.01% by either matrix-assisted laser desorption/ionization (MALDI)-MS for surface immobilized samples, or electrospray ionization (ESI)-MS for liquid samples, is another highly efficient protein identification method. These methods additionally support the identification of posttranslational modifications such as glycosylation, glycation, phosphorylation, sulfation, etc. [7–11].

Besides amino acid analysis and elaborated mass spectroscopy techniques, many more analytical methods are applied to support the identity examinations of the protein drug, such as determination of the extinction coefficient, isoelectric point, and crystal structure, as well as recording the nuclear magnetic resonance (NMR) and circular dichroism (CD) spectra and determining the chromatographic profiles from HPLC-runs as well as from capillary and polyacrylamide gel electrophoresis (CE and PAGE, respectively).

Automated systems as well as microfluidic devices ("lab-on-the-chip") for chromatographic separation of proteins and their subsequent analysis have a huge potential to dominate the analytics field in the future [12].

6.2.3 Quantification of the Protein

Many physicochemical assays are established to quantify the protein mass. It is determined by exploiting the extinction coefficient in optical density measurements or by colorimetric assays such as the Bradford, Lowry, bicinchoninic (BCA), and biuret assay [13, 14]. Albeit easy to perform, these colorimetric assays suffer from inaccuracies that are due to the use of inappropriate standards like bovine serum albumin. If relevant standards are not available, quantitative amino acid analysis [6], the (micro-)Kjeldahl nitrogen method [14, 15] or gravimetry as very accurate but time-consuming alternatives can be applied.

Bioassays and immunoassays are also exploited to quantify the protein amount, which have to be validated for accurate measurements and definitively require a reference standard (see Sect. 6.2.6.1).

6.2.4 Stability, Storage, and Sterility

Stability studies include the evaluation of those protein features that are susceptible to change during storage and might influence the quality, safety, and efficacy. The testing should cover physicochemical, biological, and microbiological aspects, as well as the preservative content such as antioxidants or antimicrobials.

Stability-indicating assays are validated quantitative analytical procedures that can detect drug alterations over time. They should include tests for integrity of the drug, potency, sterility, and, if applicable, moisture, pH, and preservative stability measured at regular intervals throughout the dating period [16–19].

One key parameter for stability testing is temperature. Real-time stability studies may be confined to the proposed storage temperature. Accelerated stability tests can be conducted at elevated temperatures exceeding standard storage temperatures. Data from accelerated stability studies are supportive but do not substitute for real-time data. They may help to validate the respective analytical assays, to elucidate the degradation profile of the drug, and to assess the drug stability under storage conditions other than those proposed.

Evaluation of the effect of humidity on drug stability may be omitted if the drug container gives appropriate protection against variations in humidity. Other parameters that, if indicated, have to be considered for stability testing are the effect of light exposure, container/closure drug interactions, and stability after reconstitution of a freeze-dried product. Sterility testing should be performed initially and at the end of the proposed shelf life (for details see [20, 21]).

Besides stability and sterility, other characteristics of the sample such as visual appearance (color, opacity, particulates), dissolution time, and osmolality also have to be described for the drug product in its final container.

Although the biological drug may be subject to substantial losses of activity over storage time, the regulatory authorities have provided little guidance concerning release and end of shelf-life specifications. It thus remains a case-by-case decision whether the loss of activity is considered acceptable.

6.2.5 Specificity and Cross-reactivity

Assays have to be designed that allow the evaluation of the specificity of a given drug-target interaction, especially when considering antibodies as a very important class of protein drugs. One may distinguish three types of assays for specificity assessment:

1. Binding assays (see also Sect. 6.2.6.1), which include appropriate positive and negative controls as well as target molecule controls that ideally consist of the closest target-related variant(s). Preferably, quantitative inhibition assays are performed with soluble target preparations, which distinctly enhance the confidence in the specificity of the drug-target interaction.
2. Determination of the molecular nature of the drug binding site by epitope mapping or by measuring the impact of carbohydrates on the binding site and of other modifications that might modulate the binding event.
3. Immunohistochemistry (IHC) to scan for cross-reactivity with human (and animal) tissues [22]. Quick-frozen surgical

samples are preferred over post-mortem tissue samples. Tissues from at least three unrelated human donors should be evaluated.

If with these assays cross-reactivity in vitro to nontarget molecules or tissues is detected, then testing in vivo for cross-reactivity in animal models, if available, is indicated.

6.2.6 Potency Determination

In the first part of this chapter, general aspects of potency determinations are described, whereas in the second part, special features of potency assessment in biological matrices are summarized.

6.2.6.1 General Aspects and Assays

Protein therapeutics often exert an exquisite biological function at very low concentrations, that is, in the pico- to nanomolar range. Sensitive methods for the determination of the biological activity have to be established that reliably allow the quantification of this prominent feature of protein drugs. It is of utmost importance to define a reference standard in quantitative terms, which is a difficult undertaking for protein drugs [23]. Owing to microheterogeneities, denatured inactive material, various glycosylation patterns, and quite different set-ups of potency assays, it remains an elusive challenge to define appropriate standards. Applying the mildest purification schemes starting from the natural protein source (and not from recombinant expression systems), using appropriate storage conditions to keep the protein in its active conformation, as well as robust and sensitive potency assay conditions are the essentials for preparation of a reference standard. With such reference in hand, the potency is routinely expressed in activity units per milligram of pure protein. The WHO is providing calibrated potency standards that are available for commercial and academic organizations [24]. If there is no official source, a reference standard should be of the highest possible purity that can be obtained with reasonable efforts, and should be fully characterized as described above.

As for antibodies, the prominent biological function is binding to the antigen. Besides the manifold antibody binding assays, isothermal titration calorimetry (ITC) could become an elegant means for reference standard evaluation. The fraction of binding antibody in the sample is determined by measuring the stoichiometry of the antigen–antibody interaction [25, 26]. For example, with a monovalent Fab antibody fragment, a 1 : 1 stoichiometry for binding to a monomeric antigen is expected, if all Fab molecules are in their active binding conformation. For correct interpretation of data, antibody-ITC relies on the presence of homogenous, epitope-presenting antigen preparations, and on not too high affinities of the interactions.

In the case that a protein exerts more than one biological activity, the assay that most closely reflects the clinical situation should be chosen. Assays for measuring further biological function(s) of the drug should also be established to assess possible side effects in vivo.

However, two protein preparations with identical specific activity units or functionality in vitro may behave very differently in vivo. For example, different glycosylation patterns may lead to very similar or even identical molecular masses and potency in vitro, but the serum half-life and immunogenicity of both variants may significantly vary from each other, and thus may lead to distinctly different effects in vivo.

In general, potency determination methods can be grouped into (1) biochemical assays that use defined reagents in vitro, and into (2) cell-based as well as (3) animal-based assays that rely on living systems.

Biochemical Assays In most biochemical assays, the specific binding of the protein drug to its target molecule(s) is determined. The target can be either the native or recombinant protein or cells expressing the target molecules. Typically, the law of mass action governs these assays. Today's most often exploited binding assay is the immunoassay that relies on antibodies as detection reagent. Historically, the radioimmunoassay (RIA), which uses radioisotope labeling to track the binding event, paved the way to most of the nonradioactive immunoassays such as the outstanding enzyme-linked immunosorbent assay (ELISA) and its many variations as, for example, sandwich ELISA, capture ELISA, inverted ELISA, and competitive ELISA (for review, see [27–29]). Besides immunoassays, receptor-binding assays and substrate-binding assays are also commonly applied.

The quality of any immunoassay heavily depends on the specificity, affinity, and stability of the used antibody reagents. Binding of the antibody reagent to the protein drug yields a signal via some enzyme, fluorescence, luminescence, or radioisotope label on the detector antibody. This signal is then transformed into binding constants (e.g. IC_{50} values) and/or concentrations.

Another very elegant binding assay is based on surface plasmon resonance (SPR) measurements. It is a label-free method in which one reaction partner is immobilized on a chip, while the other reactant in solution is flowing over the chip. Upon binding, an optical signal is generated and recorded as a function of time [30, 31]. With this fast method, high- and low-affinity interactions are detected with only minute amounts of sample. Hence, surface plasmon resonance determination has become a very attractive technique not only for comprehensive evaluation of the binding event but also for the detection of anti-drug antibodies in the sera of patients [32–34] (see also Sect. 6.4).

In general, biochemical assays are cheap and easy to perform, and allow high-throughput binding measurements at high accuracy and precision. On the other hand, the biological activity of the protein drug cannot be appropriately assessed by such assays since the biologically inactive fraction of the drug is often detected as well. As for antibodies, biochemical assays are key for evaluation of their binding characteristics, but antibody potency assays should also consider the anticipated mode of action of the therapeutic antibody, beyond the mere antigen-binding event for example, induction of signal cascades that trigger cell proliferation or cell death.

Cell-based Assays The basis for setting up cell-based assays for potency determination is the availability of responsive cells that are either immortalized, freshly isolated from tissues, or generated by engineering to obtain heterologous expression of the target molecule of choice. The typical read outs are cell proliferation, differentiation, apoptosis, cytotoxicity (e.g. for antibodies: ADCC (antibody-dependent cell-mediated cytotoxicity) and CDC (complement-dependent cytotoxicity)), chemotaxis, signal transduction, secretion of biologically active substances, or reporter gene approaches relying on green fluorescent protein or luciferase constructs. Owing to the complex reactive responses of cells, the thorough validation of cell-based assays is highly recommended for a meaningful interpretation of data [23]. For example, the addition of drug-neutralizing compounds to the assay is recommended as a check for specificity.

Critical aspects of cell-based assays include the cell immortalization method, cell culture history, passage number, stability of cell line, mycoplasma infection, surface marker pattern, and the effect of fetal calf serum in the medium.

Animal-based Assays Animal-based assays are briefly described here as potency assays, and not as in vivo proof-of-principle studies in preclinical development.

Animals, in particular non-human primates, as well as tissues and organs from animals provide a metabolizing environment and thus offer the advantage of being closer to the "real-life" situation in patients. Especially, bioavailability and toxicity aspects are thus integrated in animal-based assays. However, the exploitation of animal-based potency studies is significantly hampered by the fact that they only provide a low throughput at high costs, take long time while yielding high variability in the results, and finally raise serious ethical questions. Hence, animal-based assays are only rarely used for potency determinations (see, for example, [35]).

6.2.6.2 Potency Determination in Biological Matrices

Besides potency assessment of the purified drug, potency assays are also applied for quantification of the protein activity in biological matrices from animals/humans. Since the concentration of the therapeutic protein in the biological sample is generally low (pico- to nanomolar), standard chromatographic methods to enrich the protein prior to analysis are often not applicable. Hence, the functionality of the protein has to be determined in a complex matrix of other accompanying unrelated (or sometimes even related) proteins, non-protein macromolecules, and low molecular weight compounds such as salts and colored ingredients. During the past decade, many assays have been devised for measuring the potency of proteins under these conditions [36]. By spiking in the purified drug as well as the reference standard into the matrix and by testing series of dilutions with the matrix as diluent, the matrix effect can be assessed and adequately considered in the calculations. Special focus has to be directed towards the interference with endogenous native, drug-related protein(s) in the matrix that have to be subtracted from the signal. In general, the storage of biological samples at $-80\,^{\circ}C$ in aliquots is indicated in order to avoid degradation events.

Consequently, the thorough validation of potency assays, as well as their refinement from the very beginning of drug development on, is of key importance.

6.3 Validation of Bioanalytical Assays

This section focuses on the bioanalytical methods of validation applied to preclinical studies (nonhuman pharmacology/toxicology) and human clinical pharmacological studies such as bioavailability and bioequivalence studies requiring pharmacokinetics (PK) evaluation [37, 38]. Inherently, differences exist between bioanalytical methods applied to preclinical and clinical development of small molecules versus recombinant proteins (macromolecules). Whereas small molecules are commonly analyzed by chromatography and mass spectrometry (e.g. HPLC and LC/MS/MS), recombinant proteins are mainly characterized by biochemical (e.g. immunoassays such as ELISA) and cell-based assays. As the currently existing FDA guideline "Bioanalytical method validation" mainly focuses on small molecules [38], workshops on "Bioanalytical methods validation for macromolecules" were and are being held, which will soon result in a new guideline for the method validation of macromolecules [39]. For this reason, only guiding principles can be given in this section.

Several different assay formats, as already mentioned earlier, can generally be used such as ELISA, fluoroimmunoassay (FIA), dissociation-enhanced lanthanide fluoroimmunoassay (DELFIA), RIA, and SPR. For the quantification of proteins in biological matrices such as blood, plasma, serum or urine, most often ELISAs are validated.

In the following part, an overview on different validation aspects mainly considering ELISAs is presented. The study validation comprises the method establishment, pre-study and in-study validation.

6.3.1 Method Establishment

For setting up a bioanalytical method, calibration standards and quality controls (QC) in which the reference standard is spiked into blank samples are needed. The quality of the reference standard is pivotal to the success of the assay for deriving accurate measurements. The documentation of the reference standard ideally includes the lot number as well as certificates of analysis, stability, identity, and purity.

6.3.2 Pre-study Validation

Pre-study validation defines that the method produces reliable results. During pre-study validation, fundamental parameters such as selectivity, assay calibration, accuracy, precision, linearity, and stability are evaluated.

6.3.2.1 Selectivity

The selectivity is the ability of an analytical method to differentiate and quantify the analyte (protein) in the presence of other components in the sample. Selectivity investigations focus on reliable quantitation of the analyte against a background of interferences from endogenous matrix components. Measurements are assessed by spiking the analyte into the biological matrix (e.g. serum) from a representative number of individual subjects (at least six) at concentrations near the lower limit of quantitation (LLOQ, see below), [37].

6.3.2.2 Assay Calibration

A calibration (standard) curve describes the concentration-response curve typically including more than eight calibrators and additional ones serving as anchor points thus facilitating curve fitting. All calibrators are prepared in duplicates in the matrix analyzed. The concentration-response relationship is most often fitted with a four- to five-parameter logistic model.

6.3.2.3 Accuracy and Precision

QC samples are spiked samples used to evaluate accuracy and precision. Accuracy describes the mean deviation of the QCs from the target (nominal, true) concentration and is mainly provided in percent deviation. Precision describes the closeness of QCs and is mainly expressed as coefficient of variation (CV) in percent.

At least three sets of QCs representing the entire range of the calibration (standard) curve are included: low, medium, and high QC. The low QC sample often serves also as the LLOQ, and the high QC as the upper limit of quantitation (ULOQ) both of which can be measured with acceptable accuracy and precision.

Accuracy and precision are determined using a minimum of five to six determinations per concentration and are assessed for "intra-assay" (intra-batch or intra-run) and "inter-assay" (inter-batch or inter-run) conditions. Accuracy should be within ±20 (30)% of the target concentration and precision should not exceed ±20 (30)% of CV [37, 38].

6.3.2.4 Linearity

As samples in immunoassays are generally diluted, the linearity has to be determined by keeping the matrix component constant and diluting the protein over the expected dilution range.

6.3.2.5 Stability

The stability of the protein in the biological matrix at intended storage temperatures (e.g., −20 °C or −80 °C) is assessed by determining freeze-thaw stability at a minimum of three cycles, short-term room temperature stability, and long-term storage stability for the time period of typical storage times (e.g. six months).

6.3.3 In-study Validation

Currently, the following in-study validation acceptance criteria are recommended: at least four of every six QC samples should be within about 20 (30)% of their respective nominal value. Two of the six QC samples may be outside the 20 (30)% of their respective nominal value, but not both at the same concentration. Thus, QC results cannot be reported from a truncated standard curve. For example, a typical 96well ELISA-plate should contain the following samples each in duplicates: 8 to 10 calibration samples, 1 blank, 3 QCs and 34 to 36 unknowns.

If the bioanalytical method is performed according to good laboratory practice (GLP), the method is described in a standard operating procedure (SOP) and the validation method is reported accordingly. In general, validated methods are used in preclinical development for toxicokinetic studies and in clinical development for all studies in which pharmacokinetics is evaluated.

6.4 Immunogenicity of Recombinant Proteins

Nowadays, many biotherapeutics in clinical trials are of human or humanized composition. In patients, one would expect these drugs not to be recognized and attacked by the immune system. But apparently all therapeutic proteins elicit anti-drug antibodies to a varying extent as summarized in recent overviews [40–42].

Hence, immunogenicity of recombinant proteins is a high-profile concern for industry and regulatory authorities.

6.4.1 Examples

The following section describes (1) examples for the incidence of anti-protein drug antibodies, (2) the potential impact of such antibodies, and (3) two cases of clinical consequences.

Incidences The following short list includes murine, chimeric, and humanized monoclonal antibodies as well as interferons and interleukins against which anti-protein antibodies have been raised in patients:

(a) *Murine antibodies*: OKT3 (anti-CD3): ~80% immune responses [42].

(b) *Chimeric antibodies*: Remicade (anti-CD20): ~10–57%; Simulect (anti-IL2 receptor): ≤2%; ReoPro (Fab, anti-GPIIb/IIIa): 7–19% [42].

(c) *Humanized antibodies*: Herceptin (anti-HER2): ≤0.1%; Zenapax (anti-IL2 receptor): 8% [42].

(d) *Interferons/interleukins*: Roferon (interferon-α2a): 20–50% [43]; Intron (interferon-α2b): 0–24% [43]; Betaferon (interferon-ß1b): ~44% [44]; Proleukin (interleukin-2): 47–74% [42].

Impact The induction of allergic anti-drug reactions such as the anaphylactic or delayed type is a relatively rare event. But anti-drug antibodies bind to the drug and might neutralize or modulate its bioactivity in in vitro assays as well as in vivo. Anti-protein drug antibodies might (1) have no impact at all in clinical settings (indeed, in many cases the presence of anti-drug antibodies yielded no detectable side effects or influence on the drug safety and efficacy), (2) affect the kidney clearance parameters and serum half-life due to antibody–drug complex formation, (3) neutralize and thus reduce the efficacy of the protein drug, or (4) in the worst case react with the endogenous, drug-related protein to deplete for a long duration or even forever the naturally occurring protein from the patient as seen with erythropoietin (EPO).

Cases Recently, serious red-cell aplasia reactions have been notified in France that are linked to anti-EPO antibodies [45]. The exact cause for the appearance of such antibodies is still unclear but changes in the production process might have played a role. Another example is the megakaryocyte growth and differentiation factor (MGDF) against which neutralizing anti-MGDF antibodies were raised in patients leading to severe thrombocytopenia [46, 47].

6.4.2 Factors Leading to Immune Responses

There is a plethora of reasons why therapeutic recombinant protein products can lead to an immune response [48]:

1. Obviously, the more the drug's primary sequence deviates from natural human sequences, the higher the likelihood of developing an immune response. Even a single amino acid deviation may elicit an anti-drug response [49]. As for antibodies, the variability within the sequences of the complementarity determining regions (CDR), which are mainly responsible for binding to the antigen, might lead to the induction of so-called anti-idiotypic antibodies. Antibodies are an excellent example of the steadily ongoing progress in the development of biotherapeutics. The first therapeutically relevant antibodies were of mouse origin provoking massive human anti-mouse antibody (HAMA) reactions in patients. During the past 10 to 15 years, more and more chimeric (human antibody Fc-part with grafted mouse variable region) and humanized (human antibody framework sequences with grafted mouse CDRs) antibodies have been developed

to minimize immune reactions. The first fully human antibody got market approval in 2003 (Humira™). With regard to antibodies one may now ask, how human is human? Antibodies often have acquired in vivo point mutations by somatic maturation processes to give a better fit to the antigen. Hence, there is nothing like one defined antibody species binding to a given antigen, but a plethora of antibodies, of which the sequences differ from individual to individual. Latest antibody libraries as source of therapeutic human antibodies such as HuCAL® (MorphoSys AG; Germany) thus rely on antibody sequences that are very close or even identical to the human germline sequences to further minimize the risk of eliciting immune reactions in a patient population [50–52]. The rationale behind this approach is that germline-based antibody sequences are more conserved in humans than somatically matured ones.

2. The glycosylation pattern of the recombinant protein drug might vary from that of the natural human protein owing to very different glycosylation capabilities of the expression host system used for production. For example, Gribben et al. observed the development of anti-recombinant human granulocyte macrophage-colony stimulating factor (rhGM-CSF) antibodies directed against the protein backbone, which is normally protected in the native protein by O-linked glycosylation but becomes exposed upon expression from yeast and *Escherichia coli* [53].
3. Impurities such as protein drug variants might lead to anti-impurity antibodies, although minor modifications such as protein oxidation or deamidation are assumed to be tolerated by the human immune system [54]. On the other hand, oxidation and deamidation often result in protein aggregation, which is known to substantially promote immunogenicity. Aggregation can also be promoted by the formulation and storage conditions as well as the production process per se [55–57].
4. The route of application has a substantial impact on eliciting anti-protein antibodies. Intramuscular and subcutaneous applications seem to be more prone to raise anti-protein antibodies as compared to intravenous application [58, 59].
5. The repeated administration of protein drugs as in chronic diseases results in a higher likelihood of developing anti-protein antibodies as compared to application in acute indications. In particular, nonphysiologically high doses of the drug may break the natural tolerance due to activation of otherwise clonally silenced B-cells. On the other hand, ultrahigh drug doses may have the opposite effect and induce tolerance by eradication of drug-reactive immune cells [60, 61]. Obviously, immunosuppressed cancer or transplantation patients most frequently have reduced levels of or even do not develop anti-drug antibodies [62].

Despite the progress in explaining the immunogenicity of protein drugs, unfortunately rather often immunogenicity cannot be assigned to the above listed factors, and the underlying mechanism(s) remain unknown.

In summary, it is of utmost interest to determine the level of anti-protein antibodies and their impact on the in vivo situation in order to allow the implementation of countermeasures to minimize immune reactions. One major challenge is to set up validated in vitro assays with appropriate sensitivity that include relevant positive controls for reliable detection of anti-drug antibodies (see below and [63]).

6.4.3 Methods to Determine Anti-protein Antibodies

Assays used for the detection of anti-protein antibodies generally fall into two classes: first, assays detecting antibodies that bind to a drug, and second, bioassays measuring antibodies that neutralize/modulate the biological effect of a drug.

In a first assay, the level of antibodies that simply bind to the protein is determined. As already described above, there are several different assay formats available (ELISA, FIA, DELFI, RIA, SPR), each having its advantages and disadvantages depending on the nature of the product. Calibration of anti-drug antibody assays is an area of considerable debate. As regulators have expressed concerns about bioanalytical data that are expressed in arbitrary units such as titers, it is emphasized that antibody reference standards (i.e. anti-drug immunoglobulins purified from antiserum) should be used for calibration so that results can be expressed in mass units (e.g. ng mL^{-1} immunoglobulin). According to Mire–Sluis (FDA), more than one positive control should be produced in nonhuman primates (or in immunoglobulin humanized mice), and the specific antibodies purified resulting in a polyclonal preparation [64]. A useful process may be affinity purification with three different stringencies providing high-, medium-, and low-affinity antibodies, which can be applied to qualify the assay. Sometimes also monoclonal anti-idiotypic antibodies are used for calibration purposes. The limit of quantitation in mass units should be measured by using, for example, a dilution series of the anti-drug antibody spiked into serum and by assessing the precision at higher dilutions. The nonspecific background (NSB) of the assay can best be determined by applying a significant number of negative control samples in order to provide a mean level of NSB. Mean NSB plus 2 or 3 standard deviations appears to be the most general practice for determining the nonspecific background value.

Binding anti-drug antibodies are ideally further examined for their capacity to neutralize/modulate a functionally relevant response of the drug. This often involves cell-based assays, similar in design to the ones used in potency assays.

Alternatively to direct measurement of anti-protein antibodies, there seems to be an increased interest in the use of T-cell responses as a marker of immunogenicity [64] (see also below).

In general, the selection of sampling times is crucial to appropriate anti-protein antibody detection. A predose sample is necessary to determine potentially preexisting antibodies. The timing of post-dose sampling depends on the frequency of the dosing regimen, but is best performed at time points when drug protein concentration in the systemic circulation is minimal in order to avoid interferences.

6.4.4 Prediction of Immunogenicity of Therapeutic Proteins in Humans

Besides the various causes for eliciting an immune response in patients, the protein drug sequence as such is subject to detailed investigations with the final goal of engineering out potentially immunogenic sequence stretches. Several approaches for the investigation of potential immunogenicity in humans are currently applied.

Animal Models Early experimental hints for protein immunogenicity may arise from animal models, especially from experiments with transgenic animals that are tolerant to the human protein [65], but might develop anti-protein variant antibodies. On the other hand, if the drug fails to stimulate an immune response in normal (nonengineered) animals, preferably in non-human primates, even when administered with some adjuvant, then there is a good probability of low immunogenicity in humans [66, 67].

Hydrophilicity There has been significant progress in the computational approach to predict immunogenic epitopes in protein drugs. Early experiments demonstrated that regions at the surface of proteins are more immunogenic than core regions. Surface sites such as N- and C-termini as well as interdomain loops contain rather hydrophilic amino acids that are accessible to the environment. The immune system tends to select epitopes from such sites to develop anti-protein antibodies [68, 69]. Thus, algorithms for hydrophilicity assessment have been applied to identify potentially immunogenic epitopes [70, 71].

Peptide-MHC Interaction Analysis The development of an anti-protein antibody response frequently requires activation of T-helper cells that induce B-cells to secrete specific antibodies. The key unit for regulation of this process is a trimolecular complex consisting of the T-cell receptor on T-cells, and the class II major histocompatibility complex (MHC) harboring an MHC-ligand on antigen presenting cells. The MHC-ligands are peptides derived from the degradation of either self or foreign proteins. During the past years, there has been substantial progress in the understanding of the interaction between the MHC and its peptide ligands. MHC molecules are peptide receptors with a certain degree of plasticity, being able to accommodate a great variety of different peptides provided they share some common features. For example, peptides naturally presented by the MHC are uniform in length and have a specific sequence motif, both defined by the respective MHC allele [72]. These allele-specific motifs allow prediction of T-cell epitopes as demonstrated by Rötzschke et al., and rendered possible algorithms to scan protein sequences for potential MHC-binding motifs [73, 74]. These algorithms were further developed and refined over the years to include latest structural data of the MHC as well as results from studies of antigen processing [75, 76]. With thrombopoietin, an example has been published that illustrates the power of these in silico analyses for antigenic epitopes [41]. Ideally, the results of an in silico analysis have to be experimentally confirmed by, for example, MHC-peptide binding assays or T-cell recognition assays. Alternatively, responses of specific, primed T-cells such as chromium-release, cytokine generation, or thymidine incorporation can be determined. Recently, an in vitro assay relying on naïve T-cells has been described for the assessment of immunogenicity [77]. The potential of the prediction of MHC-peptide interactions is nicely demonstrated with the J591 antibody that originated as a mouse monoclonal antibody against prostate-specific membrane antigen. This antibody has undergone a so-called DeImmunisation™ process (Biovation Ltd., UK), and Phase I clinical studies have been undertaken. To date, over 75 patients have received the modified J591 antibody without immune responses as verified by either standard ELISA or SPR assays (www.biovation.co.uk). One issue that is still open to prediction of immunogenicity by peptide-MHC interaction analysis is its restriction to T-cell epitopes. Ideally, algorithms have to be developed to cover B-cell epitopes as well.

Besides identification of antigenic sites of protein drugs by elaborated prediction tools and their removal by engineering approaches, a completely different approach is available to address immunogenicity problems. The modification of the protein drug by conjugating with polyethyleneglycol to mask antigenic sites is a major and successful alternative to "de-immunizing" protocols [78–80].

Despite all these efforts to predict protein drug immunogenicity in humans, the final proof for the presence of immunogenic epitopes are clinical trials in combination with sensitive and validated assays to reliably determine the level of anti-protein drug antibodies and their impact on drug safety and efficacy.

Since the number of recombinant protein drugs will substantially increase over the next decade, analytical as well as bioanalytical methods for the characterization of such macromolecules will definitively gain in importance. Especially the validation of bioanalytical methods is of key importance to allow an accurate description of protein drugs. This requires harmonized guidelines, which are not yet existing but seem to be on the way, to accelerate drug development.

Acknowledgment

We would like to thank Sabine Brettreich for preparing the manuscript.

References

1. M. Schuppenhauer, *BioCentury* Feb 3, **2003**, A5.
2. CDER/CBER/CDRH/CVR-FDA, Guidelines on validation of the limulus amebocyte lysate test as an end-product endotoxin test for human and animal parenteral drugs, biological products, and medical devices, December 1987.
3. ICH, Q5A Guidance for industry: viral safety evaluation of biotechnology products derived from cell lines of human or animal origin, September 1998.
4. W. Werz, R. G. Werner, *J. Biotechnol.* **1998**, *61*, 157–161.
5. C. Bayard, F. Lottspeich, *J. Chromatogr., B: Biomed. Sci. Appl.* **2001**, *25*, 113–122.
6. M. Fountoulakis, H. W. Lahm, *J. Chromatogr., A* **1998**, *826*, 109–134.
7. A. Dell, H. R. Morris, *Science* **2001**, *291*, 2351–2356.
8. J. J. Dalluge, *Curr. Protein Pept. Sci.* **2002**, *3*, 181–190.
9. J. Leushner, *Expert Rev. Mol. Diagn.* **2001**, *1*, 11–18.
10. F. K. Yeboah, V. A. Yaylayan, *Nahrung* **2001**, *45*, 164–171.
11. E. Mirgorodskaya, T. N. Krogh, P. Roepstorff, *Methods Mol. Biol.* **2000**, *146*, 273–292.
12. J. Khandurina, A. Guttman, *J. Chromatogr., A* **2002**, *943*, 159–183.
13. C. V. Sapan, R. L. Lundblad, N. C. Price, *Biotechnol. Appl. Biochem.* **1999**, *29*, 99–108.
14. R. P. Keller, M. C. Neville, *Clin. Chem.* **1986**, *32*, 120–123.
15. M. Thompson, L. Owen, K. Wilkinson et al., *Analyst* **2002**, *127*, 1666–1668.
16. FDA, Guideline for submitting documentation for the stability of human drugs and biologics, February 1987.
17. ICH, Q5C Guideline for industry: quality of biotechnological products: stability testing of biotechnological/biological products, July 1996.
18. ICH, Q1A (R1) Stability testing of new drug substances and products, August 2001.
19. ICH, Q1E Evaluation of stability data, February 2002.
20. U.S. Pharmacopeia, USP 26-NF 21, 2003.
21. *European Pharmacopoeia*, 4th ed., 2002, Sections 2.6.1, 5.1, European Directorate

for the Quality of Medicines (EDQM), Strasbourg.
22. FDA (CBER), Points to consider in the manufacture and testing of monoclonal antibody products for human use, February 1997.
23. A. R. Mire-Sluis, *Dev. Biol. Stand.* **1999**, *100*, 83–93.
24. International Biological Standards, http://www.nibsc.ac.uk/catalog/standards/preps/sub_cyto.html.
25. S. Leavitt, E. Freire, *Curr. Opin. Struct. Biol.* **2001**, *11*, 560–566.
26. W. H. Ward, G. A. Holdgate, *Prog. Med. Chem.* **2001**, *38*, 309–376.
27. S. Sivieri, A. M. Ferrarini, P. Gallo, *Mult. Scler.* **1998**, *4*, 7–11.
28. J. E. Butler, *J. Immunoassay* **2000**, *21*, 165–209.
29. R. S. Schrijver, J. A. Kramps, *Rev. Sci. Tech.* **1998**, *17*, 550–561.
30. M. A. Cooper, *Nat. Rev. Drug Discov.* **2002**, *1*, 515–528.
31. J. M. McDonnell, *Curr. Opin. Chem. Biol.* **2001**, *5*, 572–577.
32. S. J. Swanson, D. Mytych, J. Ferbas, *Dev. Biol. (Basel)* **2002**, *109*, 71–78.
33. N. R. Gonzales, P. Schuck, J. Schlom et al., *J. Immunol. Methods* **2002**, *268*, 197–210.
34. M. A. Takacs, S. J. Jacobs, R. M. Bordens et al., *J. Interferon Cytokine Res.* **1999**, *19*, 781–789.
35. J. Ashby, P. A. Lefevre, *J. Appl. Toxicol.* **2000**, *20*, 35–47.
36. A. R. Mire-Sluis, *Pharm. Res.* **2001**, *18*, 1239–1246.
37. K. J. Miller, R. R. Bowsher, A. Celniker et al., *Pharm. Res.* **2001**, *18*, 1373–1383.
38. FDA, Guidance for industry: bioanalytical method validation, May 2001.
39. AAPS, Workshop on bioanalytical method validation for macromolecules in support of pharmacokinetic studies, May 2003.
40. P. Chamberlain, A. R. Mire-Sluis, *Dev. Biol. (Basel)* **2003**, *112*, 3–11.
41. E. Koren, *Dev. Biol. (Basel)* **2002**, *109*, 87–95.
42. E. Koren, L. A. Zuckerman, A. R. Mire-Sluis, *Curr. Pharm. Biotechnol.* **2002**, *3*, 349–360.
43. P. Kontsek, H. Liptakova, E. Kotsekova, *Acta Virol.* **1999**, *43*, 63–70.
44. P. Kivisäkk, G. V. Alm, S. Fredrikson et al., *Eur. J. Neurol.* **2000**, *7*, 27–34.
45. N. Casadevall, J. Nataf, B. Viron et al., *N. Engl. J. Med.* **2002**, *346*, 469–475.
46. J. Li, C. Yang, Y. Xia et al., *Blood* **2001**,, *98*, 3241–3248.
47. I. Zipkin, *BioCentury* Sep 14, **1998**, 6(74), A8.
48. H. Schellekens, *Nat. Rev. Drug Discov.* **2002**, *1*, 457–462.
49. R. M. Maizels, J. A. Clarke, M. A. Harvey et al., *Eur. J. Immunol.* **1980**, *10*, 509–515.
50. A. Knappik, L. Ge, A. Honegger et al., *J. Mol. Biol.* **2000**, *296*, 57–86.
51. B. Krebs, R. Rauchenberger, S. Reiffert et al., *J. Immunol. Methods* **2001**, *254*, 67–84.
52. T. Kretzschmar, T. von Rüden, *Curr. Opin. Biotechnol.* **2002**, *13*, 598–602.
53. J. G. Gribben, S. Devereux, N. S. Thomas et al., *Lancet* **1990**, *335*, 434–437.
54. J. L. Cleland, M. F. Powell, S. J. Shire, *Crit. Rev. Ther. Drug Carrier Syst.* **1993**, *10*, 307–377.
55. E. Hochuli, *J. Interferon Cytokine Res.* **1997**, *17*, SS15–SS21.
56. J. C. Ryff, *J. Interferon Cytokine Res.* **1997**, *17*, SS29–SS33.
57. W. V. Moore, P. Leppert, *J. Clin. Endocrinol. Metab.* **1980**, *51*, 691–697.
58. W. E. Paul, (Ed.), *Fundamental Immunology*, 4th ed., Lippincott Williams & Wilkins Publishers, Philadelphia, 1999.
59. A. Braun, L. Kwee, M. A. Labow et al., *Pharm. Res.* **1997**, *14*, 1472–1478.
60. C. C. Goodnow, R. Brink, E. Adams, *Nature* **1991**, *352*, 532–536.
61. K. M. Shokat, C. C. Goodnow, *Nature* **1995**, *375*, 334–338.
62. F. Baert, M. Noman, S. Vermeire et al., *N. Engl. J. Med.* **2003**, *348*, 601–608.
63. A. R. Mire-Sluis, *Dev. Biol. (Basel)* **2002**, *109*, 59–69.
64. A. R. Mire-Sluis, Immunogenicity testing. Immunogenicity of therapeutic biological products, IABs international meeting, October 2001.
65. A. V. Palleroni, A. Aglione, M. Labow et al., *J. Interferon Cytokine Res.* **1997**, *17* (Suppl. 1), 23–27.
66. C. M. Zwickl, B. L. Hughes, K. S. Piroozi et al., *Fundam. Appl. Toxicol.* **1996**, *30*, 243–254.
67. D. Wierda, H. W. Smith, C. M. Zwickl, *Toxicology* **2001**, *158*, 71–74.

68. E. H. Kemp, E. A. Waterman, R. A. Ajjan et al., *Clin. Exp. Immunol.* **2001**, *124*, 377–385.
69. D. C. Benjamin, J. A. Berzofsky, I. J. East et al., *Annu. Rev. Immunol.* **1984**, *2*, 67–101.
70. J. Janin, *Nature* **1979**, *277*, 491–492.
71. M. Z. Atassi, *Eur. J. Biochem.* **1984**, *145*, 1–20.
72. K. Falk, O. Rötzschke, S. Stevanovic et al., *Nature* **1991**, *351*, 290–296.
73. O. Rötzschke, K. Falk, S. Stevanovic et al., *Eur. J. Immunol.* **1991**, *21*, 2891–2894.
74. K. C. Parker, M. A. Bednarek, J. E. Coligan, *J. Immunol.* **1994**, *152*, 163–175.
75. T. Sturniolo, E. Bono, J. Ding et al., *Nat. Biotechnol.* **1999**, *17*, 555–561.
76. M. Schirle, T. Weinschenk, S. Stevanovic, *J. Immunol. Methods* **2001**, *257*, 1–16.
77. M. M. Stickler, D. A. Estell, F. A. Harding, *J. Immunother.* **2000**, *23*, 654–660.
78. A. M. Chen, M. D. Scott, *BioDrugs* **2001**, *15*, 833–847.
79. N. Trakas, S. J. Tzartos, *J. Neuroimmunol.* **2001**, *120*, 42–49.
80. A. P. Chapman, *Adv. Drug Deliv. Rev.* **2002**, *54*, 531–545.

7
Biogeneric Drugs

Walter Hinderer
BioGeneriX AG, Mannheim, Germany

7.1
Introduction

Biopharmaceuticals, and especially therapeutic proteins, represent an exceptionally fast growing segment within the pharmaceutical market. In 2003, approximately 50 different recombinant proteins, which are applied as active ingredients in pharmaceuticals, are registered in Europe. These protein drugs gain more than 10% of the total market of pharmaceuticals, which was around US$400 billion in the year 2002. Moreover, considering new pharmaceuticals coming for approval, the percentage of recombinant proteins is expected to rise over 50% [1]. Among these biopharmaceuticals, many are blockbusters and most of them are high priced. In addition, the first wave of the therapeutic proteins will run off patent protection within the next five years and in principle this market segment will open up to competitors like the generic drug suppliers. Taking all these facts together, biopharmaceuticals might be highly attractive for the generic industry. Many financial analysts and market observers promise golden opportunities that lie ahead for biogenerics. It is believed that this cluster of products offers a multibillion Euro marketplace in the near future.

Nowadays, the field of biogenerics is frequently reviewed and controversially discussed. This chapter necessarily gives an incomplete overview of a very complex situation, for which neither a universal strategy nor a clear regulatory pathway exists today. It is written from the view of a German biogeneric company, BioGeneriX, a subsidiary of a big generic manufacturer, ratiopharm, and unavoidably mixed up with personal interpretations of the author. The decision of ratiopharm to spin-off a specialized company responsible for biogeneric pharmaceuticals was based on thorough analyses of opportunities, risks, and strategic possibilities for this kind of business. This article will narrow the subject to practical aspects, emerging from four years of experience with the biogeneric drug development. Furthermore, I will mainly focus on the first wave of biogeneric drugs, which are products that run off patent in the next five years. Finally, it must be emphasized that this chapter is directed exclusively to the European situation. Considering that patents will expire in most cases substantially earlier in Europe than in the United States, the first wave of biogeneric drugs is awaited in Europe first. The differences in patent expiry between Europe and the United States is readily explained by the different patent laws.

7.2 Recombinant Therapeutic Proteins

The term "biopharmaceuticals," although having a wider definition, is often used equivalently to "therapeutic proteins." The first generation of these potent drug substances were exclusively derived from human or animal sources and applied in a replacement therapy of hormone or blood-clotting factor deficiencies in chronic diseases. Examples for these first generation of proteins used as pharmaceutical drugs are human growth hormone (hGH), isolated from cadaveric human pituitary glands, insulin, extracted from pig or bovine pancreas, and several human blood proteins, derived out of pooled plasma fractions, for example, albumin, immunoglobulins, or coagulation factors. These natural sources, however, are linked to many safety problems in the past, and the resulting protein products have caused the transmission of infective agents, leading to Creutzfeld–Jacob disease in the case of hGH [2], and chronic hepatitis B and C, and AIDS in the case of coagulation factor VIII and IX [3]. Further problems emerged from the immunogenicity of nonhuman proteins, for instance, in the case of bovine or porcine insulin [4], which can lead to allergic or other immune reactions like neutralization of the drug activity by antibodies in some cases.

Along with the emergence of methods of genetic engineering in the 1970s and 1980s, such as recombinant DNA technology and hybridoma technology, these natural proteins were more and more replaced by recombinant versions. Only few exceptions exist, for example insulins and factor VIII and IX, where natural therapeutic proteins keep a significant role in the market. The first human health care product derived from recombinant DNA technology was Eli Lilly's insulin (Humulin®) approved in the United States and Europe more than two decades ago [5]. This was a milestone for the biotechnology industry. The continuous improvement of these new technologies built the basis for the successful development of the young biotechnology industry. Some of the originally small biotech start-ups, such as Genentech, Amgen, or Biogen, grew up to become big pharmaceutical companies during the last two decades and today they represent a significant portion of the pharmaceutical industry.

Besides the safety aspect, the recombinant DNA technology provides further advantages. Most important are (1) the large-scale production of high amounts of protein with defined and homogeneous quality to lower costs; (2) the development of novel drugs, directed to new targets, which could not be isolated in sufficient amounts and qualities from natural sources; and (3) creating protein variants, muteins, having even improved properties over the natural polypeptides. Today, it would be a very difficult approach to register a natural protein, derived from living organisms or organs, in view of existing recombinant versions, the biological sources of which are regarded as less inferior by the regulatory authorities.

Considering all these aspects, one can easily conclude that a product portfolio of a pure biogeneric company will exclusively consist of recombinant proteins. Therefore, this overview is restricted to recombinant proteins, including glycoproteins.

7.3 Definition of Biogenerics

The commonly used term "biogenerics" is not appropriate and not accurate in the eyes of regulatory authorities. At the moment, the industry faces a rather unweighable regulatory situation that makes the definition of biogenerics, better named "comparable biotech products" or "biosimilar products" according to recent guidelines, a difficult task. However, it is essential for the following considerations to find a common understanding of what biogenerics really is. At first, we have to look at chemically synthesized generics and ask, "how are they defined?" By no means will this lead to a straightforward, uniform answer. Depending on the point of view, such as trademark, patent, or regulation, or even on the country where the question is asked, the answer can vary considerably. The knowledge about, and the occurrence of, generics depends on the health policy, which is different among the European countries.

At least three aspects seem uniform and helpful for a general definition of generics: Generics are pharmaceuticals that (1) appear after patent expiry, (2) are sold with a price reduction compared to the original product, and (3) have an active substance identical to the originator's one. In Table 1, some other items contributing to a full definition of generics

Tab. 1 Definition criteria for generic pharmaceuticals and their transferability to biogenerics

Definitions	***Chemical generic ("generic")***	***Biotech generic ("biogeneric")***
Pharmaceutical, launched after patent expiry of the active substance	Yes	Yes
Will be offered with a low price	Yes (significant price reduction)	Yes (moderate price reduction)
Will be sold under a generic name (INN)	Yes (exceptions: branded generics)	Yes/No (brands are more likely)
Will be distributed without or almost little product-specific promotional efforts	Yes/No	No
The active ingredient is qualitatively and quantitatively essentially similar to an original product	Yes (proven by physical/chemical analysis and bioequivalence)	Yes/No ("biosimilarity" to be confirmed by clinical studies)
Approved without the proof of efficacy and safety in patients. Only a bioequivalence study was required.	Yes	No (efficacy and safety proven by phase I and II/III studies)
The kind of European approval is facultative	Yes (mutual recognition or centralized procedure)	No (centralized procedure is mandatory)
A simplified dossier is sufficient for filing a MAA	Yes (reference to the originator's dossier)	No (full dossier required, no reference possible)

are mentioned. Since chemical generics have a well established and a simplified route of approval, this regulatory feature contributes substantially to their definition. Classical generics are typically approved on the basis of an abridged filing dossier, which makes use of the possibility to refer to the originator's dossier of approval. The applicant proves essential similarity utilizing analytical data and shows bioequivalence in a small pharmacokinetic study, usually performed with healthy volunteers. It is also common that generics are sold under the international nonproprietary name (INN), often without any product-specific promotional activities. Branded generics, however, are exceptions and they still have a significant market share, especially in Europe.

Looking at the aforementioned characteristics, how can a definition of biogenerics be achieved? What is fulfilled for biotechnology-derived generics? Table 1 gives the answer. The aforementioned general three items, namely, patent expiry, lower price, and essential similarity, are fulfilled except for the latter, although we expect price differences to the proprietor product to be more moderate and more stable compared to chemical generics, where substantial price erosion can occur within a few months after the end of patent protection. The "essential similarity," "comparability," or "biosimilarity" is an ongoing dispute between regulatory authorities, European commissions, and the different lobbies competing for political influence. For the time being, generic applicants need to file a more or less complete dossier, including preclinical and clinical studies showing efficacy and safety. As the approval of biopharmaceuticals is not only linked to the product but also to the process and the site of production, the standard route for registration of generics via reference to the originator's file is not possible. This will be elucidated in more detail in the following section. For these kinds of products, own brands and specific promotions are more likely than for the classical generics.

In summary, biogenerics are best defined as copies of therapeutic proteins, launched after patent expiry of the active pharmaceutical ingredient, and sold with a moderate price reduction. They have to be approved via the route of the centralized procedure in Europe and currently require a complete stand-alone dossier including clinical studies proving efficacy and safety.

7.4 Regulatory Situation

At present, there is neither in Europe nor in the United States a regulatory pathway for obtaining a generic type of approval for a biotech drug. To understand the position of the regulatory authorities, it is necessary to consider the specific differences between low-molecular weight substances and complex macromolecules like polypeptides. In contrast to synthetic small molecules, large proteins and especially glycoproteins have a complex tertiary structure that is sensitive to modifications in solution. Slight conformational changes may reduce efficacy and/or lead to an increase in immunogenicity. Degradation or oxidation of amino acid residues are examples for such undesired alterations, which can occur during the process or later on during the shelf life of the product. Aggregation, often favored by oxidation, can be a critical parameter during the production process and for the storage of bulk substance or final product. Sophisticated formulation is necessary to keep the protein as a monomer, stable in solution. Aggregation is correlated with an increase in immunogenicity. It is known that immunogenicity can cause severe clinical consequences at worst leading to life-threatening complications. Unfortunately, the analytical methods established today cannot fully predict the biological and clinical properties of a protein and cannot establish whether the structures of two biopharmaceuticals are completely identical [6]. According to these scientific arguments, a substantially abbreviated clinical program for biogenerics is currently not realistic. Preclinical and clinical trials demonstrating efficacy and safety is mandatory.

Another panel of arguments against an essential similarity or a pure comparability approach results from the specific impurity profile related to process and biological sources. This is the reason for the linking of the market authorization application (MAA) to product, process, and site of production, which build an inseparable package for the approval. Changing the process or transfer to another production site would require a new registration. With the help of batch record data, a comparability approach is possible in such cases and the 2001 guideline on comparability of the Committee for Proprietary Medicinal Products (CPMP) is appropriate [7]. Undoubtedly, biogeneric players have to apply new manufacturing processes to existing products, without having access to the methods of the innovators or to the material from intermediate steps. Thus, the claim for a comparability is difficult. Consequently, the authorities will regard these biogenerics case by case. The actual term used for "biogeneric drug" by the European Medicines Evaluation Agency (EMEA) is "biosimilar product"; a clear policy or a decision, however, is still not released. Several of the biogeneric companies are aware of this situation and will follow a conservative approach that entails running significant trials and seeking regulatory approval as stand-alone products [8].

What changes can be expected in the near future? The political influence of the different lobbies will put pressure on the emergence of a clear regulatory outline for follow-on biotech drugs. Unlike in the United States, where a fundamental legal differentiation is made between drugs and biologics, in the European Union the legislation is based on a single definition framework, and so the introduction of a legal provision and regulatory pathway for comparable biological medicinal products would be easier [9]. A new perspective is provided by draft Annex 1 to the European Directive 2001/83 [10], which sets out the legal basis for biogenerics to reference the originator product. This guideline still requires transposition into national legislation. On the basis of this policy, a comparability approach cannot be excluded for the future; however, this must not be mistaken for a classical generic pathway. Substantial, case-dependent preclinical and clinical work still would be necessary to convince authorities about comparable efficacy and noninferiority to the reference product. Their major concern seems to be immunogenicity (see also Sect. 7.6.1).

On the other hand, the Food and Drug Administration (FDA), though already allowing a slim approval pathway for follow-on versions of insulin and hGH, has no legal basis for an "Abbreviated New Drug Application" (ANDA) of a biotech product. If the FDA will change its view on comparability aspects is a matter of political discussions too. Whether the recent massive drug review reorganization within the FDA will go further in this direction is a speculation. Most categories of therapeutic proteins, including cytokines, growth factors, interferons (IFN), and enzymes are transferred from the Center for Biologics Evaluation and Research (CBER) to the Center for Drug Evaluation and Research (CDER). Insulin and hGH already belonged to CDER. It remains to be seen, if this will lead to benefits for the regulation of biogenerics in the United States. For more information on the US regulatory situation and the changes in the FDA, refer to recent reviews [11–13].

Defining categories according to different risk profiles may be a solution for future regulation. It should be possible to use scientific data to create categories of biologics that conceivably could be approved on the basis of relatively less clinical data and those that would require much more extensive human testing prior to approval [11].

In conclusion, biogeneric players are currently faced with a complex and unsatisfactory regulatory situation. A regulatory basis for an EMEA approval based on substantially abridged clinical trials is not present today. The first wave of biogeneric drugs now enters clinical phases. The companies have to decide whether to follow a classical stand-alone approach with extended preclinical and clinical development, which is a low-risk, high-cost strategy, or to speculate for simplifications and follow a slim clinical program. The latter is certainly a high-risk strategy.

7.5 Patent Situation

Patents are the most relevant and the most effective means of intellectual property protection. Obviously, the knowledge about relevant patents is of crucial importance for the generic business. The leading companies in this field have established strategies to launch their products immediately after patent expiry. Thus, they have to be sure, country by country, when corresponding patents will expire. The search of relevant patents, the understanding of the scope, the analysis of the patent family, and the examination of the legal status, including supplementary protection certificates (SPCs), have to be performed thoroughly. For big generic companies like ratiopharm, which started with this kind of business in 1974, this is mainly routine.

Unfortunately, the patent situation is considerably more complicated in the field of biotechnology. Some of the specific problems should be mentioned here. The total number of patents dealing with a specific protein can be incredibly high. For instance, our in-house patent searches for potential biogeneric target products revealed between 1000 and 5000 hits for a given therapeutic protein. These huge numbers of documents have to be restricted to those few that are really relevant and which cannot be circumvented. This requires specific biotech knowledge along with patent know-how. Patent analysis for a generic development is mainly a survey of the past, a historical work-up. If a patent expires in Europe today, it must have been filed 20 years ago. Although the methods used then are free for use today, they are in most cases old fashioned and not suitable for a state-of-the-art production process. Moreover, since patents are rapidly filed in the earliest stage of development, the methods described, in general, are far away from the final biopharmaceutical manufacturing process. The pressure to be the first in filing a patent in order to secure the intellectual property status was high and of strategic importance for the previously small biotech companies. Additionally, at that time when gene technology was just emerging, attorneys and examiners had only little experience, if any, with this subject, and the corresponding patent claims are often not clear and are difficult to interpret.

In addition, the majority of the relevant biotechnology patents are related to specific methods for production, sometimes even restricted to a specific expression system, rather to appear as broad substance patents. This is related to the fact that recombinant versions of already known natural proteins, in principle, lack novelty. The recombinant protein itself was not regarded as new if the natural protein was state of the art. Therefore, patents have to claim a recombinant protein in combination with modes of production, including vector constructs, expression systems, and purification methods, or more importantly, together with an application. It is obvious that for nearly all examples of products, which are mentioned in Table 2, one will not find a monopolistic market situation. Typically, parallel developments using distinct solutions lead to two or more competing products.

Besides "primary patents" that are basic patents that cannot be circumvented and whose expiry has to be waited for, there are other categories of patents that may appear as severe hurdles. One example is the pharmaceutical formulation of therapeutic proteins, a dangerous minefield also for classical generics. Numerous patents claim distinct product-specific formulations. These inventions are related to a more advanced development stage, and they usually expire several years later. I believe that some of the biogeneric players are not fully aware of this difficult situation. Again, it requires specific experience to find a gap within a dense net of formulation patents. The biogeneric products will probably appear with modified formulations on the market, which differ from the original products. Otherwise, the generic competitors have to wait some more years for launch or they would undergo the risk of being sued for patent infringement. An altered formulation, however, is a step further from a regulatory comparability approach (see above).

Another category of such "secondary patents" is applications claiming the therapeutic use of the drug for a new medical indication. These kinds of patents have to follow special rules for wording the claims. Nevertheless, they can hinder generic developments for a long time. An example is given by interferon alfa (IFN-α), which was claimed in recent years for its therapeutic application in hepatitis C–infected patients in combination with ribavirin, a classical antiviral compound (see also Sect. 7.6.7). Although both the drugs run off patent, their combined use in the major indication seems to be blocked at least until 2018.

Once a disturbing patent or patent application has been identified, there are in principle four strategic possibilities to overcome infringement: (1) waiting for its expiration; (2) taking a license; (3) working out a circumventing strategy, if possible; or (4) performing legal actions, such as filing an opposition or an appeal, or, during an advanced step, filing a nullity suit. It is assumed that the biogeneric companies will presumably adopt strategies (1) and (3).

An indispensable legal aspect for the generic industry is the so-called "Roche–Bolar"-type exemptions or "Bolar" provisions. So far, I have only considered infringement of patents by the substance itself or by methods used. However, even if a substance or method infringes the claims of a patent, there are exceptions that

Tab. 2 Putative targets of biogeneric developments on the European market

Recombinant human protein	***Branded product (EU market)***	***Originator companies***	***Launch (EU)***	***Active pharmaceutical ingredient (API)***	***Major indications***	***Expression system (host organism)***	***Patent expiry (without SPC)***
Erythropoietin (EPO)	Erypo, Eprex	Kirin, Amgen, Johnson & Johnson	1988	Epoetin alfa	Renal anemia (dialysis patients) CT-induced anemia (onocology)	Mammalian (CHO)	2005
	NeoRecormon	Genetics Institute, Boehringer Mannheim, Roche	1992	Epoetin beta			
Granulocyte colony-stimulating fac-tor (G-CSF)	Granocyte	Chugai	1993	Lenograstim	CT-induced neutropenia (oncology), bone marrow transplantation	Mammalian (CHO)	2006
	Neupogen	Amgen	1991	Filgrastim		Bacterial (*E. coli*)	2006
Granulocyte-macrophage colony-stimulating factor (GM-CSF)	Leucomax	Genetics Institute, Schering–Plough, Sandoz, Novartis	1992	Molgramostim	CT-induced neutropenia (oncology), leukemia, lymphoma	Bacterial (*E. coli*)	2005
	Leukine	Immunex, Amgen, Schering AG	2003?	Sargramostim	CT-induced neutropenia (oncology), leukemia, lymphoma	Yeast (*S. cerevisiae*)	

Human growth hormone (hGH)	Genotropin	Genentech, Pharmacia–Upjohn	1988	Somatropin	Growth deficiency in children, growth hormone deficiency in adults (1994)	Bacterial (*E. coli*)	2000
	Humatrope	Eli Lilly	1988				
	Norditropin	Novo Nordisk	1988				
	Zomacton	Ferring	1994				
	Saizen	Serono	1989			Mammalian (C127)	
Insulin	Huminsulin	Eli Lilly	1982	Insulin human	Diabetes mellitus	Bacterial (*E. coli*)	2003
	Insuman	Aventis	1999				
	Insulin Actrapid	Novo Nordisk	1991			Yeast (*S. cerevisiae*)	
Hepatitis B virus surface antigen (HbsAg)	Engerix-B	GlaxoSmith-Kline	1987	HBV surface antigen	Prophylaxis of HBV infection (active immunization)	Yeast (*S. cerevisiae*)	2001
	Gen H-B-Vax	Aventis-Pasteur, Chiron, Behring	1986				
Factor VIII	Kogenate	Genentech, Bayer	1994	Octocog alfa	Hemophilia A	Mammalian (BHK)	2004
	Helixate	Genentech, Aventis-Behring	1994	Octocog alfa			
	Recombinate	Genetics Institute, Baxter	1992	Rurioctocog alfa		Mammalian (CHO)	
	ReFacto	Pharmacia–Upjohn, Wyeth–Ayerst	1999	Moroctocog alfa			2005

(*continued overleaf*)

Tab. 2 *(continued)*

Recombinant human protein	***Branded product (EU market)***	***Originator companies***	***Launch (EU)***	***Active pharmaceutical ingredient (•)***	***Major indications***	***Expression system (host organism)***	***Patent expiry (without SPC)***
Interferon alpha (IFN-α)	Intron A	Biogen, Schering–Plough	1986	IFN-α-2b	Chronic hepatitis B and C	Bacterial (*E. coli*)	2001
	Roferon A	Genentech, Roche	1989	IFN-α-2a			
	Infergen	Amgen, Yamanouchi	1999	IFN-αcon 1			
Interferon beta (IFN-β)	Betaferon	Chrion, Schering AG	1995	IFN-β-1b	Multiple sclerosis, relapsing-remitting and secondary progressive forms	Bacterial (*E. coli*)	2003
	Avonex	Biogen	1997	IFN-β-1a		Mammalian (CHO)	2001 (2012)
	Rebif	Serono	1998	IFN-β-1a			

certain actions using this protected matter are not regarded as infringement. Currently, nearly all member states belonging to the World Intellectual Property Organisation (WIPO) include experimental-use excemptions in their national patent laws. Typically, these exceptions are related to noncommercial research experiments. Several countries outside the European Union have installed the aforementioned Bolar provisions either in their patent law or in the rulings of pharmaceutical registration. What does Bolar provision mean? The term "Roche–Bolar" arose from a legal procedure held in 1984 in the United States, Roche Prods. Inc. versus Bolar Pharm Co. [14]. The court decided that testing of generic versions of pharmaceuticals during patent life was infringement and added no weight to the experimental-use excemption of the US patent law. Later, this decision was overruled by the Hatch–Waxman Act (section 505 of the FDA Act), which provided the legal basis for generic drug development and the filing of an ANDA (see section above) without infringing existing patents any longer. Briefly, a Bolar provision allows all developmental, testing, and experimental work required for the registration of a generic medicine to take place during the patent period of the original product. The purpose of such a provision is to ensure that generic medicines are on the market immediately after patent expiry. Bolar-type provision exists in the United States, as a part of the Hatch–Waxman Act, Canada, Japan, Australia, Israel, and some Eastern European countries such as Hungary, Poland, and Slovenia [15]. Among the European Union countries, with the exception of Portugal, no Bolar provision exists so far. There are enormous efforts taken by the generic lobby to influence a harmonized European Regulation following a Bolar-type exemption. It seems that the European Commission is now in favor of Bolar provisions in the European Union and has proposed an appropriate Bolar ruling under article 9 of the regulation for a European community patent. In addition, a recent proposal for an amending Directive 2001/83/EC included Bolar ruling under article 10. A valid Bolar provision would be a breakthrough for the generic industry in Europe.

The generic manufacturers traditionally make use of patent-free countries and countries with Bolar provisions to develop and manufacture their products and to perform there clinical bioequivalence studies. After patent expiry, they are free to transfer the production to their facilities in Europe. This is much more difficult for a biogeneric development, which requires a high technical standard according to the guidelines of the International Conference on Harmonisation (ICH). As already mentioned, biologics will be approved taking into consideration the production process and the production site. Thus, the European biogeneric companies are in a dilemma. They have to perform the complete development and production within patent-free territories or within the Bolar countries outside the European Union. They have to charge contract research organisations (CROs) and manufacturers (CMOs) or enter into strategic alliances with biopharmaceutical companies, for example, in Canada or Asia. On the other hand, they deal with multisource biotech of high complexity, which makes the transfer of the bioprocess to another site, from the technical and regulatory point of view, a time-consuming and expensive step, bearing always the risk of additional bridging studies. Furthermore, in the Bolar countries, the biotech production capacities available for the biogeneric companies, if any, are very limited. Strong competition between the biogeneric companies took place for these few resources and will continue further due to the increasing number of projects.

Considering everything, one can understand why the generic industry urgently craves for a Bolar-type ruling in the European Union. This regulation unequivocally should also cover biosimilar products. This is currently not clear for many Bolar provisions, including the Hatch–Waxman Act in the United States. Such a policy would allow the biogeneric industry to invest in dedicated multipurpose biotech facilities and would facilitate a market entry more close to the patent expiry.

7.6
Biogeneric Targets: First Wave

Table 2 summarizes potential target proteins for a first wave of biogeneric products. It must be emphasized that this is a personal selection and not a complete one. This table only includes those candidates that will have patent expirations up to the year 2008. Furthermore, the market volumes [16] have also been used for this selection. Owing to the high costs for the development of a biogeneric, only top-selling products are of interest to the companies. There are further exclusions for the selection of this panel. Recombinant proteins alone have been considered. Monoclonal antibodies are believed not to appear in a first wave. Although antibodies present a very important cluster of successful products, they are not discussed here. There are two reasons against a generic antibody development. Firstly, a copy of a monoclonal antibody seems possible on a functional rather than on a molecular level, leading to extended preclinical and clinical development. Secondly, antibodies require much higher volumes in production caused by relatively high dosages. Likewise, some other candidates are not considered because of difficult clinical development, for example the plasminogen activators.

7.6.1
Erythropoietin (EPO)

EPO exerts an exceptional attraction for the generic industry. EPO has the highest market volume among all therapeutic proteins and is a real blockbuster. It is not speculation to assume that nearly all of the biogeneric companies have an EPO in their pipeline. A recent review reported 16 companies developing this product and at least 7 of them have planned to enter the European market [8]. Three products are currently present in Europe: Eprex® (epoetin alfa) from Johnson & Johnson, NeoRecormon® (epoetin beta) from Roche, and Aranesp® (darbepoetin alfa) from Amgen. The former, Johnson & Johnson's Eprex®, is the best-selling genetically engineered drug ever, with $3.4 billion in sales in 2001 [17]. The latter, Amgen's Aranesp®, is a second-generation product, containing a mutein with prolonged bioavailability. This was achieved by the introduction of additional glycosylation sites. Aranesp® is certainly not in the current focus of biogeneric developments. A fourth product, Dynepo® (epoetin delta) from Aventis and Transkaryotic Therapies (TKT), is already approved. The method used by TKT was an activation of the human chromosomal *EPO* gene instead of cloning and expressing the gene in Chinese hamster ovary (CHO) cells. This looks like a typical patent strategy leading at the end to a biosimilar product. However, it will be very difficult for Aventis to maintain noninfringement before court. It is assumed that Aventis will wait as long as Kirin–Amgen's basic patent expires in December 2004. The question whether Dynepo® is a generic or not is justified. According to the definition given in Sect. 7.3, Dynepo® will likely not appear as a generic on the market because Aventis, a highly reputed innovator company, will sell the product with scientific arguments rather than by price reduction. A fifth EPO, produced with baby hamster kidney (BHK) cells, designated as epoetin omega and known for a long time [18], is again under clinical investigation. This product probably belongs now to Lek (Slovenia), which was the sponsor of the studies [19]. Lek was recently acquired by Novartis.

The analytical and biological differences between epoetin-alfa and -beta, two CHO-derived EPOs, do not seem to be of clinical relevance [20]. Nevertheless, a manufacturer of an EPO can apply for a new INN, because slight differences in the gylcosylation pattern cannot be avoided. EPO products applied today contain a very complex mixture of many different isoforms that just differ in their glycosylation structure. The capability for sufficient glycosylation is provided by the host cell, for example, CHO; however, upstream and downstream production procedures have a considerable influence on the final isoform composition. The necessary quality of a new EPO product is defined by a monograph of the European Pharmacopoeia [21], but the specifications mentioned are just minimum requirements. The true measure is set by the originator products. Nevertheless, the existence of a monograph is welcome for a generic approach.

Recently, there was growing concern in France about case reports of pure red-cell aplasia (PRCA) developing in 13 patients with chronic renal anemia treated with Eprex® (epoetin alfa). Neutralizing antibodies induced by EPO therapy cross-neutralize natural endogenous EPO, leading to severe transfusion-dependent anemia [22]. One year later, the number of reported PRCA cases, as of November 2002, had already increased to more than 175 patients. This reflects an incidence of 20 per 100 000 per year. The most likely explanation for this serious side effect is a subtle change in the EPO molecule, probably introduced by the manufacturing and/or formulation changes in 1998 [23]. There were two coincidences for the emergence of PRCA, (1) with the reformulation of Eprex®, mainly the withdrawal of albumin under pressure of the CPMP, and (2) with the shift from intravenous to subcutaneous administration [17]. Procrit®, the US version of Eprex®, is still formulated with human albumin and seems not to be correlated with PRCA. As a consequence, the EMEA recently restricted the use of Eprex® in renal anemia to intravenous administration.

It is likely that PRCA will have additional impact for generic versions of EPO. The CPMP is particularly concerned with the problem of immunogenicity and may require postmarketing monitoring for at least one year [17]. Furthermore, it cannot be excluded that this may also influence the requirements for approval of other protein drugs.

7.6.2 Colony-stimulating Factors (CSFs)

Two important endogenous growth factors, granulocyte-colony-stimulating factor (G-CSF) and granulocyte/macrophage-colony-stimulating factor (GM-CSF) regulate the proliferation and differentiation of progenitor cells within the bone marrow and the release of mature neutrophils into the peripheral blood. Cancer chemotherapy, which affects rapidly dividing cells, frequently leads to a side effect termed "neutropenia". Neutropenia is a decrease in counts of neutrophilic granulocytes in the peripheral blood and affects more than one in three patients receiving chemotherapy for cancer. Patients driven into neutropenia can develop fever and have an increased risk for infections. Life-threatening gastrointestinal and pulmonary infections occur, as does sepsis. A subsequent cycle of chemotherapy may have to be delayed until the patient has recovered from neutropenia. Recombinant human G-CSF and GM-CSF are effective pharmaceutical substances and have been successfully applied to treat chemotherapy-induced neutropenia. They restore the number of neutrophils in the blood and keep it above the critical level [24].

From a marketing point of view, G-CSF is more successful than GM-CSF. Scientific reasons for the preferential application of G-CSF in prophylaxis of neutropenia exist too. G-CSF is more specific for granulocytes. It takes a position within the regulation cascade of hemopoiesis, one step behind of GM-CSF and closer toward the differentiation of neutrophils. The dominant product on the market is Amgen's Neupogen® containing filgrastim, an *Escherichia coli*–expressed, recombinant human met-G-CSF. The second G-CSF product, Chugai's Granocyte®, is derived from recombinant CHO cells and glycosylated. With the exception of France, this product has only little market shares. This is explained by the respective marketing power of the pharmaceutical companies rather than by scientific reasons.

In 2002, Amgen additionally launched Neulasta®, a second-generation, PEGylated filgrastim. This product has a prolonged pharmacokinetic half-life profile and requires less frequent application. It is assumed that Neulasta® will expand the CSF market and will replace Neupogen® to some extent before the patents expire. The most attractive generic target among the CSF products is certainly Neupogen®. However, generic versions appearing after patent expiry then have to compete with Neulasta®. This will happen at the earliest in 2006. Nevertheless, the high cost of the therapy limits its widespread use [24], and a lower price for a generic G-CSF could justify the application of a first-generation drug and would even favor broader application.

Today, the market volume of GM-CSF seems too low to guarantee return of investment for a generic version within a reasonable time. Furthermore, GM-CSF and G-CSF compete in the indication of neutropenia. However, GM-CSF has some opportunities lying in new indications. One promising development is the successful treatment of active Crohn's disease [25]. The product of Immunex, Leukine® (sargramostim), which was not launched in Europe so far, was sold in 2002 to Schering AG, upon the acquisition of Immunex by Amgen. Sargramostim is a mutein and differs from the natural protein. It has a substitution of leucine in position 23, and carries heterogeneous yeast-type glycosylation, due to the *Saccharomyces* expression system applied. It can be assumed that Schering AG will launch this product in Europe and will further invest in the immunomodulatory potential of GM-CSF for therapy of morbus Crohn. In contrast, Leucomax (malgramostim; Novartis) is expressed in *E. coli* and not glycosylated. Likewise, this product was sold in 2002. Novartis transferred the product rights to its comarketing partner Schering–Plough. It remains to be seen if under new ownership the GM-CSF products will gain in market volumes in the near future.

7.6.3 Human Growth Hormone (hGH)

Recombinant hGHs are well-established products, launched in Europe in 1988. The market is highly competitive, as five different branded products are available for a limited number of patients (see Table 2). Some of the products, in certain patient groups, still have orphan drug status. The original indication of hGH was growth deficiency in children caused by the lack or insufficient endogenous production of the hormone. The goal of treatment in these cases is to stimulate linear growth, hopefully reaching a final height within the normal range. Nowadays, hGH has an expanded therapeutic spectrum in children [26] and is used additionally for adolescents and adult patients having secondary weight losses or muscular atrophy with or without hGH deficiencies [27]. All the five available recombinant growth hormones, four expressed in *E. coli* and one in murine cells, have identical amino acid sequence and are not glycosylated. These products are indeed "comparable" or "biosimilar"; they have the same INN, somatropin, and the same clinical efficacy. This is a good example where independent developments lead to very similar products. In this background, one can expect that the requirements for approval of a generic hGH are much less than, for example, for a glycoprotein like EPO. Considering also some further growth of the market, hGH is a promising candidate for a biogeneric product portfolio. With the exception of some specific patents, which block certain methods of production and formulation, hGH is already patent-free and additional products are awaited in the next years. There are several life cycle extension strategies by the originators moving in the direction of easier administration, such as needle-pen, needle-free injection, slow release formulations, or even oral and inhaled forms. This will drive generic companies to innovate too.

7.6.4 Insulins

Diabetes mellitus is a global epidemic affecting about 150 million people around the world. These numbers are believed to grow rapidly along with the increasing problems of age and obesity. It was estimated that the numbers will double within 25 years. Disease management of type 1 and type 2 diabetes is performed by insulin therapy. Three human-identical recombinant insulins are on the European market, including Eli Lilly's pioneer product Huminsulin® being the first genetic-engineered polypeptide drug launched [5]. This protein and Insuman® of Aventis are both synthesized in *E. coli*, whereas Novo Nordisk produces its Insulin Actrapid® in yeast. For the daily glycemic control in diabetics, there are usually three different types of insulin required: (1) a rapid-acting variant, (2) an intermediate-acting variant, and (3) a long-acting variant. These types differ in their pharmacokinetic profile accomplished by classical formulation strategies [28]. Briefly, regular insulin acts rapidly, the long-acting (retard) insulin is achieved by neutral-protamin-Hagedorn (NPH) formulation, and the intermediate variant is just a mixture of regular and NPH insulin.

Three novel insulin analogues, all muteins, are of importance. Regular insulin has the tendency to locally build dimers and hexamers upon injection, hence the release is not as rapid as desired. This prompted Eli Lilly and Novo Nordisk to develop muteins with an altered carboxy-terminus in the B-chain. Eli Lilly's solution for obtaining a more rapid insulin was a switch of the two amino acids in position 28 and 29, proline and lysine, thus preventing aggregation. The resulting protein was named "insulin lispro" and the corresponding product Humalog® was launched in 1996. Novo Nordisk reached the same effect by introducing aspartic acid instead of proline in position 28. The mutein was named "insulin aspart," and the corresponding product Novolog® was launched in 1999. In contrast, Aventis went the opposite direction toward a long-acting insulin obtained by exchange of the C-terminal asparagine with glycine in the A-chain, together with prolonging the C-terminus of the B-chain by attaching two arginine residues. The resulting product Lantus®, containing the mutein "insulin glargine," was approved in 2000. Taking together, these three novel insulins present expanding options in diabetes management [29].

The huge and continuously growing market of diabetes is in the focus of generic companies. The patents for regular human insulin expire in 2003. Therefore, generic versions will likely appear within the next years. Human insulin is a relatively small nonglycosylated protein, and the requirements of drug approval are believed to have a similar low extent as for hGH (see above). However, a disadvantage of insulin should be mentioned too. Because the dosages of insulin are very high, several tons of recombinant insulin have to be produced per year to supply all patients. Therefore, the manufacturers produce their insulins in largest scales, and the prices calculated per gram protein are relatively low. This could be a significant market entry barrier for generic insulins, which may have a limited potential for price reduction. It should be mentioned that ratiopharm and B. Braun have already launched a generic, semisynthetic human insulin in 2000.

7.6.5 Hepatitis B Vaccine

Vaccines are a special category of biological products, and only the recombinant vaccine for hepatitis B virus (HBV) is considered in this chapter. The prophylaxis of HBV infections is successfully performed using a subunit vaccine, based on a single recombinant protein, hepatitis B surface antigen (HBsAg). This antigen is produced efficiently in yeast and shows excellent immunogenicity. After seroconversion of anti-HBsAg, immunity is maintained for at least 10 years [30]. There are two products approved in Europe, Engerix-B® and Gen H-B-Vax® from GlaxoSmithKline and Aventis respectively. There are several reasons why these products are attractive generic targets. The patent protection for these products has already expired. In addition, the market volume is large and will further increase, especially in view of international HBV control programs already running or entering many countries in the next years. New trends for innovation should be mentioned too. Firstly, there are new generation vaccines combining HBsAg and Hepatitis A (HAV) immunogens, for example Twinrix® of GlaxoSmithKline, and, secondly, additional recombinant viral antigens of HBV are combined with HBsAg to provide against low- or nonresponder of HBsAg immunization.

7.6.6
Factor VIII (FVIII)

Hemophilia A is one of the most common inherited bleeding disorders and is caused by genetic deficiency of coagulation factor VIII (FVIII) [31]. The replacement therapy with plasma-derived FVIII and, later, recombinant FVIII has substantially improved the quality of life and the life expectancy of hemophilia A patients. In the early 1980s, significant numbers of hemophilia A patients have been infected with hepatitis viruses and HIV. The virus transmissions were caused by FVIII batches isolated from contaminated plasma pools [3]. Subsequently, this led to a dramatic change in the safety philosophy for plasma-derived products and for biologics in general. Likewise, there were demands for the development of safe recombinant FVIII, which was a scientific and a technical challenge. Activated natural FVIII mainly consists of a heterodimeric form of two different polypeptide chains linked via calcium ions. The heterodimer consists of 2351 amino acids and has a size of 127 kDa. Furthermore, there are 25 potential glycosylation sites, and the total molecular weight of the native glycoprotein is around 300 kDa. After all, in the early 1990s the first recombinant products appeared on the market. Today, four recombinant FVIII products are registered in Europe (Table 2), and indeed no transmission of hepatitis or HIV attributing to recombinant FVIII has been reported [32]. The latest product, Refacto® (moroctocog alfa), was launched in 1999 and is a mutein having a deletion of the B-domain, which renders the protein more stable. Refacto® (Wyeth) and Recombinate® (Baxter) contain CHO-expressed FVIII, whereas Bayer (Kogenate®) and Aventis (Helixate® use BHK cells for the expression of FVIII. The patents for FVIII will expire in 2004 and 2005 and further products could emerge. The market opportunities are promising. However, biogeneric developers would enter a highly difficult field. Recombinant expression, purification, and stabilization of FVIII requires considerable experience, more than that required for the aforementioned targets.

7.6.7
Interferons (IFN)

Interferons (IFN) are a class of related cytokines with multiple activities. They are grouped in three major categories, namely IFN-α, IFN-β, and IFN-γ, according to their different cellular origin. Besides their historically described antiviral activities, they are also known to possess immunomodulatory and cell-proliferative potential [32]. The nomenclature of the various IFNs in the past was chaotic and has changed over time. It is very difficult to relate descriptions in old papers or patents to the actual molecular definitions. Besides some antitumour applications, two widely used IFN therapies have been successfully established. These therapies are (1) IFN-α for the treatment of chronic hepatitis C virus (HCV) infections and (2) IFN-β for the treatment of multiple sclerosis (MS). These two applications are highly attractive for generic companies and the products should be discussed in more detail.

IFN-α, formerly "leukocyte IFN," is a mixture of many distinct protein homologues. At least 23 different *IFN-α* genes are known. For the recombinant production, Roche, respectively Genentech, decided to use IFN-α-2a, whereas Schering–Plough's product is related to IFN-α-2b. By contrast, Amgen has developed an artificial consensus IFN-α (IFN alfacon 1), probably following a patent strategy. All three recombinant proteins are produced in *E. coli*. The corresponding products, Roferon A®, Intron A®, and Infergen® are all approved for treatment of HBV and HCV infections. Emphasis must be added on two recent developments in HCV therapy, which strongly influenced generic strategies. At first, Schering–Plough introduced a combination therapy with IFN-α and ribavirin (Rebetol®), a classical chemical antiviral compound of ICN pharmaceuticals, until then only used for treatment of pulmonary infections of respiratory snycitium virus (RSV). This combination therapy was a breakthrough for treatment of chronic HCV infections and set a new standard within a short time. Secondly, Schering–Plough launched PEG-Intron® in 2000, a PEGylated IFN-α-2b based on Enzon's technology for PEGylation. This product offers an advantageous once-weekly administration, but more importantly, the clinical efficacy has been improved. Again, within a short time, the market switched nearly completely to the second-generation product. On the same way is Roche, which developed a PEGylated IFN-α-2a, Pegasys®, using the PEG technology of Shearwater. This product will be marketed together with its own ribavirin brand (Copegus®) for combination therapy. Because of synergistic intellectual property issues between Roche and Schering–Plough, it is assumed that these two companies will dominate the HCV market for a long time. This is a good example for successful life cycle management at the end of patent protection. Schering–Plough, for instance, has filed a panel of patent applications claiming the combination therapy with ribavirin and IFN-α. The introduction of PEGylated IFN-α will make it difficult, if not impossible, to sell nonPEGylated generic versions. Some generic companies, like Bioceuticals (Stada) and BioGeneriX (ratiopharm) have given up on plans for the development of IFN-α [33].

IFN-β, formerly "fibroblast IFN," has a long history of therapeutic applications. The latest and most important indication entered by this cytokine was multiple sclerosis (MS). MS is a chronic, noncurable, progressive, and neurological disease and is the prototype of an inflammatory autoimmune disorder of the central nervous system (CNS). The main pathological feature is demyelination of ganglions, which causes the various symptoms [34]. The use of IFN in MS has been studied for more than two decades; however, the mechanism of its action in MS is still not understood. Likewise, this holds true for the etiology of MS. Since 1995, IFN-β products are approved for treatment of relapsing-remitting and secondary progressive MS. Three products are on the market, which differ in several aspects. Betaferon® (Schering AG) contains a nonglycosylated mutein of IFN-β, expressed in *E. coli*. The active substance is termed "IFN-β-1b" and differs from the natural amino acid sequence by having serine instead of cysteine in position 17. In contrast, Avonex® (Biogen) and Rebif® (Serono) are both expressed in CHO cells. The proteins are *N*-glycosylated and have an identical sequence to the natural protein. The INN (IFN-β-1a) is the same for both glycoproteins, although slight differences in glycosylation exist. It must be emphasized that Betaferon® has a rather low specific activity of 32 million international units (IU) per milligram compared to Avonex, which has 200 IU/mg. Thus, the dosages for these two products, by amount of protein, differ by nearly one order of magnitude. Avonex® is indicated to be administered at 30 µg once a week, while Betaferon® is indicated at 250 µg to treat MS [35]. In contrast, the latest product Rebif® was administered with 44 μg three times a week. Moreover, whereas Avonex® is applied intramuscularly, Serono decided on a subcutaneous application of Rebif®. This high-dose strategy seems to be successful. Recent clinical trials showed that the therapy with Rebif® was significantly more effective than Avonex® [36]. This was of great importance for Serono, because they were able to overcome the existing orphan drug status in the United States. The question rose whether Rebif® is a generic or not. Basic patents have already expired, or have been rejected after oppositions. Undoubtedly, Rebif® is a copy of Avonex® and could claim for "biosimilarity." Also, the generic feature of a lower price might be fulfilled. The therapies with Rebif® and Avonex® have similar calculated costs based on time and patients. Thus, because of the higher dosage of Rebif®, the price for the substance IFN-β is indeed significantly lower for Rebif® than for Avonex®. However, in spite of these arguments, nobody, neither physicians, nor patients, nor Serono itself, would accept any relationship between Rebif® and a biogeneric product. Interestingly, Avonex® also has a biogeneric history. Paradoxically, the FDA approval of Avonex® was based along a comparability approach, just on biological, biochemical, and biophysical data, without clinical trials [35]. The background is as follows: Biogen and Rentschler Arzneimittel (Laupheim, Germany) founded a joint venture company, called Bioferon, for the development of IFN-β. The product received the name of the company, Bioferon®, and has gone into phase III trials. Later on, there was a breach between Rentschler and Biogen and the joint venture went into receivership during the trials. The rights for the clinical data stayed with Biogen and the rights for clone, process, and substance stayed with Rentschler. Hence, Biogen was obliged to develop a new cell bank and product. They used the clinical data of Bioferon® and showed analytical comparability to Avonex®. This was assumed to be a test case for biogeneric science [35]. Subsequently, on the basis of additional phase IV clinical data, Avonex® received the EMEA approval. Interestingly, Rentschler in cooperation with BioPartners is now redeveloping Bioferon®, which in principle is the reference product of the "biogeneric" Avonex®. They want to convince regulators once again about the comparability of these two products [37].

Generic developers of IFN-β should consider one drawback: although the patent situation allows free operation in Europe, there is one patent of Rentschler [38], expiring in 2012, claiming IFN-β preparations with improved glycosylation patterns. This patent, an inheritance of Bioferon, also covers the existing products. An opposition of Biogen was recently rejected by the European Patent Office. Therefore, a pure generic strategy would require license, if available, whereas a circumventing strategy would lead to a noncomparable product. Further risks should be mentioned too. In contrast to replacement therapies, like for EPO, hGH, insulin, or FVIII, the therapy with immunomodulators bears the risk of substitution. Especially in MS, it seems possible that in the near future IFN-β could be substituted by more efficient drugs. Several alternative substances are in clinical development or already approved. Alternatively and similar to the change in HCV standard therapy (combination with ribavirin), IFN-β could be used as adjuvant therapy in combination with chemical drugs. Furthermore, the benefits of IFN-β in MS, also in view of the high costs, are controversially discussed. At least the clinical effectiveness beyond the first year of treatment is called in question [39]. In conclusion, among the interferons, IFN-β-1a, is the most promising biogeneric candidate; however, the risk of substitution in the indication for MS is significantly higher than for other targets.

7.7 Biogeneric Developments and its Requirements

In principle, the requirements for the development of a therapeutic recombinant protein are independent of a generic strategy. The ICH guidelines provide a framework that is indispensable for an EMEA approval. The ICH tripartite documents define "good manufacturing practice" (GMP), which specifies the requirements and conditions for manufacturing the Active Pharmaceutical Ingredients API or final product. Furthermore, the ICH guidelines instruct, for example, the analysis of the expression constructs, cell hosts and substrates, viral safety evaluations, analytical procedures and their validation, and stability testing. In addition, the CPMP guidelines have been considered, too. As already outlined in Sect. 7.4, the regulatory prerequisites for approval of a biosimilar recombinant protein are far different from the abridged pathway applied for chemical generics. Biogenerics are more close to new drugs. The product development of a biotech product is characterized by three main sections: (1) process development, (2) development of analytical methods, and (3) preclinical and clinical development. Table 3 gives an overview of the various steps, which at the end have to be put together to obtain the market authorization under the legalities of drug approval. Noncompliance may have a severe impact on the time schedule and in the worst situation might be even irreversible. It is obvious that much less difficulties occur for biogeneric developers than for the originators. Many problems have been solved in advance and there is no need for a proof of principle. This results in abbreviated time lines. Figure 1 gives an example of the follow-up of a biogeneric project using CHO expression and shows the sequence and links of the various parts of development on a timely basis. This sequence is optimized for maximum overlapping. The product launch is realized seven to eight years after starting the project with molecular biology. By using *E. coli* as expression system, the project would be finished approximately one year earlier. In the following sections, emphasis is added to special features of biogeneric development.

7.7.1 Process Development

The required gene and its sequence are in the public domain. Typically, the DNA will be chemically synthesized. The selection of the host cell and the construction of useful expression systems is the standard technology of molecular biology. Biogeneric companies have to deal mainly with two expression systems, *E. coli* and CHO (see Table 2). Occasionally, *Saccharomyces* and BHK might be alternative expression hosts. The first milestone for the project is the finishing of the master cell bank (MCB) and the manufacturing working cell bank (WCB). Comprehensive analytical work is necessary, especially for the mammalian cells, to obtain all the safety data required for release of the cell banks into production. It should be mentioned that in the earliest step of development, for example for clone selection, it is indispensable to

Tab. 3 Development steps for a recombinant therapeutic protein

I. Process development	***II. Analytical development***	***III. Preclinical/clinical development***
Cloning of the gene	Definition of standards	Preclinical studies (two mammalian species)
Construction of the expression vector	Bioassays (cell-based in vitro assays or in vivo assays)	Toxicology in a rodent species (acute, chronic, subchronic)
Transfection of the host cell	SDS-polyacrylamid gel electrophoresis (SDS-PAGE)	Toxicology in a nonrodent species (acute, chronic, subchronic)
Selection of stable clones	Western blot (WB)	Safety pharmacology (cardiovascular, respiratory, renal, gastrointestinal, CNS, depending on drug)
Optimization of expression, culture media selection	Capillary zone electrophoresis (CZE)	Pharmacokinetic studies
Master cell bank (MCB)	Reversed phase HPLC (RP-HPLC)	
Working cell bank (WCB)	Size-exclusion HPLC (SEC-HPLC)	Phase I studies (healthy volunteers):
Characterization and safety of cell banks	Product-specific ELISA	Safety/Tolerance/Pharmacokinetic/Pharmacodynamic
Upstream procedures: Fermentation process (USP)	Host-cell-protein ELISA (HCP-ELISA)	
Downstream procedures. Purification scheme (DSP)	Residual DNA detection (picogreen, threshold)	Phase II studies (patients):
Optimization of individual process steps	N-terminal sequencing (Edman degradation)	Safety, proof of efficacy, dose finding
Stability and robustness of the process	C-terminal amino acid composition	
Introducing of GMP	Peptide mapping	Phase III studies (patients):
Consistency batches	Total amino acid content (upon hydrolysis)	Controlled safety and efficacy in specific indications
Validation	Carbohydrate analyses (total sugar, antennarity, sialic acids)	Multiple arms, vs. reference therapy or placebo controlled. Typically blinded studies and matched patient groups.
Pharmaceutical development (formulation)	MALDI-TOF spectroscopy (molecular weight)	
Development of fill & finish	Detection of free sulfhydryl moieties	Serological assays (ELISA, radioimmunoassay (RIA)) to quantify drug and antidrug. Neutralizing antibodies (inhibitory effect in bioassay)
Stability studies (holding steps, bulk material, final product)	Circular dichroism spectroscopy (CD-spectra)	
	Surface plasmon resonance (SPR, BIAcore)	

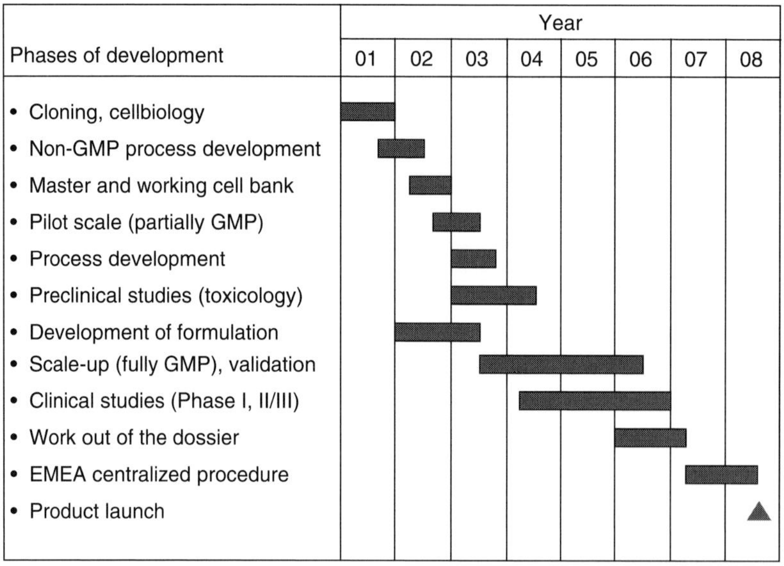

Fig. 1 Development of a biogeneric project (example for time schedule).

quantify and to analyze the product. This is easier to establish for biogeneric developments, because specific antibodies or complete ELISA products for clinical diagnosis are available for most products. Also, reference material can be obtained in high purity from the sales product.

In contrast, the development of the upstream procedures (USP) does not differ for a biogeneric. This has to be done independently. There is no access to any experience from the reference product. However, the fermentation techniques applied are standardized and routine in the hands of experienced manufacturers. Nevertheless, this is an important and time-consuming section, and difficulties may occur along with scale-up. The downstream process (DSP), which always contains a sequence of different chromatographic steps, accompanied by some filtration procedures, can be deduced from published literature, supported by experience in protein chemistry. It is very likely that biogeneric developments will end up with their own and unique DSP. There are many alternative methods available. The necessity of DSP is the removal of impurities, such as host cell proteins (HCP), DNA, endotoxins, pyrogens, and other process-related substances. The requirements on purity for a therapeutic grade protein are high. As discussed in Sect. 7.5, DSP is a field for secondary patent applications and the development often has to be performed along a circumventing strategy. The same is true for the development of the pharmaceutical formulation (see Sect. 7.5).

Much work and high costs are related to the stage of production of substance for use in clinical trials, especially for phase III. This requires introduction of GMP and huge amount of validation work later on. The phase III material has to be taken from the commercial scale. In contrast to originators, which have to walk more carefully step-by-step, considering feedbacks from preclinical or clinical phases, biogeneric companies have the possibility to move on and to produce all the materials from the final process and scale. This would substantially lower the risk for noncomparability of the various materials during different stages and thus avoids bridging studies. The development of the final product, which is a prefilled syringe in the majority of cases, shows no biogeneric-specific aspects. Because of patents, it is assumed that biogeneric products will appear with their own formulations, which differ more or less from the original product. The formulation of the reference product, in respect of stability, serves as the gold standard.

7.7.2
Development of Analytical Methods

Table 3 presents a selection of assays frequently taken for the characterization of a therapeutic protein. The field is complex and I can mention only few aspects. The assays are applied for in-process controls (IPC), batch release, extended characterization of the purified protein, process validation, or stability studies. Most of the assays are standard biochemical methods, which have to be adapted and validated for the specific protein. Biogeneric developers will relate their standards to the target product, especially if they have decided to go for a comparability strategy. In such a case, it seems wise to use the reference product throughout the development. In some cases such as EPO or FVIII, there are defined reference standards from the European Pharmacopoeia (Biological Reference Preparations, (BRP) available. Other sources for biological standards are the World Health Organisation (WHO) and the British National Institute for Biological Standards and Control (NIBSC). Of special importance is the potency of the protein, analyzed by specific bioassays. These assays determine the biologic activity, mostly in terms of international units (IU). Typical bioassays are (1) in vivo animal systems, for example mice (EPO), (2) cell-based proliferation assays (CSF) or antiviral assays (IFN). The potency to bind to the natural receptor can effectively be demonstrated with so-called "reporter gene activation assays", which transduce the receptor-binding signal via gene activation to an easily detectable marker.

For the characterization of the product, one has to consider purity, residual contaminants, biological activity, and physicochemical properties. Several sophisticated methods are required and have to provide structural evidence for a biosimilar product. The N-terminus and C-terminus have to be intact. Free sulfhydryl groups and disulfide bridges have to be in the right position. Altered versions of the proteins, which are termed as "product-related impurities", for instance caused by methionine oxidation or deamidation, truncated species, dimers, and aggregates, have to be characterized and quantified at very low levels. Additional analytical methods are utilized for glycoproteins. The analysis of the carbohydrate composition is mandatory. The antennarity structure and the specific number of sialic acids have a strong influence on potency and on the pharmacokinetic behavior in vivo. For the immunological detection of HCP impurities, a source- and process-specific test has to be developed on the basis of mock material used for immunization. Mock material is a kind of control protein preparation, without the product, just modeling the impurities. Although it is tremendous work to establish and validate all these assays, the biogeneric developers can refer to publications, European Pharmacopoeia in the best case, and charge specialized service laboratories that offer these methods. However, most of the methods required for the routine quality control, IPC and batch release, need to be installed at the site of manufacturing.

7.7.3 Preclinical and Clinical Development

This is the most significant part of the total development costs of a pharmaceutical. Biogeneric developers, however, underwent low risk for failure in clinical studies. This is the major difference in the development of a new protein drug. Depending on the kind of product and the degree of analytical comparability, the extent of clinical studies will vary considerably for biogenerics. This has been discussed in Sect. 7.4 in more detail. For the preclinical phase, the study program for a biogeneric drug can be reduced to a certain extent. Nevertheless, there are some irreplaceable parts showing toxicology, safety, and pharmacokinetics in two mammalian species. This is also mandatory for entering in human trials. The companies have to decide whether they should design their preclinical studies in direct comparison with the reference product or perform a stand-alone approach. Generally, published data are available for preclinical results of the original drug substance. This can be used as supporting material.

Before entering clinical studies, the formulation has to be fixed and stability data of at least three months are necessary. Most of the biogeneric products are liquid-formulated parenteralia presented as prefilled syringes. There are some important differences for the clinical trials of biogenerics. The studies are designed, at least partially, as comparator studies. Phase I will be performed with healthy volunteers or occasionally even with selected patient groups. The study has to show bioequivalence and comparable pharmacodynamics to the comparator product. To achieve bioequivalence, precise dosing per kilogram of body weight is essential. In addition, the phase I study also covers safety aspects. There is a residual risk for missing bioequivalence and thus showing non-comparability. This would have a severe impact on further clinical development. The classical separation between phase II and III studies (see Table 3) is not reasonable for biogenerics. Once proven an equivalent pharmacokinetic and pharmacodynamic response, a dose finding or a general proof of efficacy (phase II) does not seem necessary at all. A typical biogeneric phase III study is designed as a comparator study with two arms and crossing-over. This should show a comparable efficacy to the reference product and provide sufficient evidence that both products are interchangeable. In addition, a second part of the phase III study has to provide safety data based on a statistically calculated number of patients. The safety matter is the most important aspect throughout the different preclinical and clinical stages. The design of the clinical studies is one of the major questions to be addressed to members of the CPMP during a scientific advice procedure.

Finally, it should be mentioned that for both the animal studies and the clinical studies in humans, serological assays have to be developed in advance. These tests are required for the detection of the protein drug in serum or plasma (pharmacokinetic) as well as for the detection of specific antibodies evoked by the drug. If antibodies appeared, they have to be further analyzed for neutralizing activity. This could be demonstrated by an inhibitory effect in the bioassay.

7.8 Conclusions

Recombinant therapeutic proteins are in the sight of the generic industry since several years, and product developments have begun. The differences between chemical generics and biogenerics are dominated by the kind of approval, the long time lines, and in the overall high costs for a project. In these aspects, biogenerics resemble new drugs rather than classical generics. Significant investment and lack of biotech know-how are the two major barriers for the generic industry to enter this kind of business. The risks can be minimized by recruiting experienced staff and competent partners for development, production, and distribution. Key factors for success are excellence in biotech, management of alliances and partnering, legal competence, and marketing and distribution power. The first wave of recombinant proteins will run off patent from now until 2008. The target products have been presented in detail and many of them will have to compete with therapeutic equivalent biogenerics in the near future.

The common term "biogenerics" was used throughout the text, although it is disliked by the regulators. Other suggestions, such as "multisource products," "biosimilar products," or "comparable products" are used preferentially in relation to their approval. They all mean the same. It is of minor importance what kind of wording will be used at the end. To overcome any misunderstanding, it was necessary to define biogenerics unequivocally in direct comparison to chemical generics. Besides the regulatory conditions, market aspects are also included for a definition. Otherwise, biopharmaceutical products, which are accepted in the medical community as innovations, for example Rebif® (IFN-β-1a) or Dynepo® (epoetin delta), would be pure biogenerics. Indeed, it will be difficult in the future to classify products as biogenerics. This is due to additional innovations, own brands, multiple license strategies, and expected moderate price reductions.

The biogeneric business urgently requires the development of a more favorable legal framework. This situation, although evolving, has not yet become a reality. Two independent projections in the European Union are in progress: a more precise guidance for the approval of biogenerics and the ruling of a Bolar provision. The expected rulings would solve many conflicts the biogeneric developers actually deal with. Unfortunately, the recent concerns about the immunogenicity of Eprex® (epoetin alfa) leading to the life-threatening PRCA syndrome will not make it easier to approve generic erythropoietins.

The complex patent situation for biologics favors future litigation. Very likely, biogeneric companies will receive aggressive legal opposition from the originators. One of the strategies they can use, called "evergreening", means to file "secondary patents" some years after the basic patent, which claim essential parts of the old product, and thus prolonging the exclusive market period. In some cases, especially in the United States, this was discussed as a kind of patent law abuse. Besides patents, innovators have further methods to protect their products, which are not discussed in this overview. These are (1) data exclusivity periods (not directly relevant for biogenerics), (2) supplementary protection certificates (SPC), and (3) orphan drug status. Altogether, these legal instruments maintain a reasonable period of market exclusivity for the originator.

Another strategy of originators is life cycle management. The introduction of second-generation products at the end of patent expiry and possibly phasing out the predecessor product is a well-known strategy against generics. This was successfully demonstrated for IFN-α and indicates the risks for biogeneric companies.

Altogether, the competition between originators and generic companies put pressure on both sides to innovate. This results in benefits for patients (improved therapy) and the health system (lower costs).

References

1. K. Maleck, F. Pollano, *Eur. Biopharm. Rev.* **2001**, *3*, 19–21.
2. P. Brown, M. Preece, J. P. Brandel et al., *Neurology* **2000**, *55*(8), 1075–1081.
3. P. M. Mannucci, P. L. Giangrande, *Hematol. J.* **2000**, *1*(2), 72–76.
4. G. Schernthaner, *Diabetes Care* **1993**, *16* (Suppl. 3), 155S–165S.
5. R. E. Chance, B. H. Frank, *Diabetes Care* **1993**, *16* (Suppl. 3), 133S–142S.
6. H. Schellekens, *Nat. Rev. Drug Discov.* **2002**, *1*(6), 457–462.
7. Note for guidance on comparability of medicinal products containing biotechnology derived proteins as drug substance, CPMP/BWP/3207/00 (September 20, 2001).
8. S. Usdin, *BioCentury* **2003**, *11*(20), A1–A6.
9. *SCRIP World Pharm. News*, **2003**, 6, 2855.
10. Draft Annex I of Directive 2001/83/EC on the community code relating to medicinal products for human use.
11. K. Haan, S. Usdin, *BioCentury* **2002**, *10*(35), A1–A6.
12. T. L. Gerrard, *BioProcess Int.* **2003**, *1*(5), 38–42.
13. A. Dove, *Nat. Biotechnol.* **2003**, *21*, 495–498.
14. Roche Prods. Inc. v. Bolar Pharm Co, U.S. Court of Appeals for the Federal Circuit, Fed. Cir., 1984, 733F.2d 858.
15. National Economic Research Association (NERA), Policy Relating to Generic Medicines in the OECD. Study Carried Out on Behalf of the European Commission, 1998.
16. K. Robinson, *BioPharm Int.* **2003**, *14*(1), 24–26.
17. G. Kelley, *BioPharm Int.* **2002**, *15*(12), 44.
18. J. S. Powell, K. L. Berkner, R. V. Lebo, *Proc. Natl. Acad. Sci. U. S. A.* **1986**, *83*, 6465–6469.
19. A. Sikole, G. Spasovski, D. Zafirov et al., *Clin. Nephrol.* **2002**, *57*(3), 237–245.
20. P. L. Storring, R. J. Tiplady, R. E. Gaines Das et al., *Brit. J. Hematol.* **1998**, *100*, 79–89.
21. *Council of Europe, Pharmacop* **2002**, *1316*, 1123–1128.
22. N. Casadevall, J. Naaf, B. Viron et al., *N. Engl. J. Med.* **2002**, *346*(7), 469–475.
23. H. Schellekens, *Nephrol. Dial. Transplant.* **2003**, *18*, 1257–1259.
24. D. C. Dale, *Drugs* **2002**, *62* (Suppl. 1), 1S–15S.
25. B. K. Dieckgraefe, J. R. Korzenik, *Lancet* **2002**, *360*, 1478–1480.
26. M. J. Henwood, A. Grimberg, T., Moshang Jr, *Curr. Opin. Pediatr.* **2002**, *14*(4), 437–442.
27. P. Iglesias, J. J. Diez, *Expert Opin. Pharmacother.* **1999**, *1*(1), 97–107.
28. A. M. Gualandi-Signorini, G. Giorgi, *Eur. Rev. Med. Pharmacol. Sci.* **2001**, *5*(3), 73–83.
29. J. E. Gerich, *Am. J. Med.* **2002**, *113*(4), 308–316.
30. G. M. Keating, S. Noble, *Drugs* **2003**, *63*(10), 1021–1051.
31. J. Klinge, N. M. Ananyeva, C. A. Hauser et al., *Semin. Thromb. Hemost.* **2002**, *28*(3), 309–322.
32. M. De Andrea, R. Ravera, D. Gioia, et al., *Eur. J. Pediatr. Neurol.* **2002**, *6* (Suppl. A), A41S–A46S.
33. S. Usdin, *BioCentury* **2002**, *10*(17), A1–A13.
34. H. Wiendl, B. C. Kieseier, *Expert Opin. Investig. Drugs* **2003**, *12*(4), 689–712.
35. K. Haan, *BioCentury* **2002**, *10*(35), A7–A8.
36. F. Patti, A. Reggio, *Int. J. Clin. Pract. Suppl.* **2002** *(Sep)*; (131), S23–S32.
37. S. Usdin, *BioCentury* **2003**, *11*, A6.
38. Bioferon Biochemische Substanzen GmbH & Co, EP-529300-B1, 1992.
39. G. Filippini, L. Munari, B. Incorvaia et al., *Lancet* **2003**, *361*(9357), 545–552.

Part III
Therapeutic Proteins – Special Pharmaceutical Aspects

8 Pharmacokinetics and Pharmacodynamics of Biotech Drugs

Bernd Meibohm
University of Tennessee Health Science Center, Memphis, TN, USA

Hartmut Derendorf
University of Florida, Gainesville, FL, USA

8.1 Introduction

In the last two decades, an increasing fraction of pharmaceutical R&D has been devoted to biotechnology-derived drugs (biotech drugs) – large molecules such as soluble proteins, monoclonal antibodies, and antibody fragments, as well as smaller peptides, antisense oligonucleotides, and DNA preparations for gene therapy [1]. Biotech and genomic companies currently perform nearly one-fifth of all pharmaceutical R&D, a figure that is set to double within the next 10 years [2]. These biotech-related drug development efforts have so far been quite successful. Biotech products accounted for more than 35% of the 37 new active substances (NASs) that were launched in 2001, and it has been predicted that half of all NASs developed in the next 10 to 15 years will result from research into antibodies alone [1]. Numerous approved biotech drug products with blockbuster character underline this success – erythropoietin (Epogen®, Procrit®), abciximab (Rheopro®), and trastuzumab (Herceptin®) to name a few. Since the development of biotech drugs generally rests on a fundamental understanding of the related disease, their clinical development has also proven to be more successful than for conventional small molecules (new molecular entities: NCEs). Only 8% of the NCEs that entered clinical drug development between 1996 and 1998 reached the market compared to 34% of biotech drugs. On the basis of these facts, it can be predicted that biotech drugs will play a major, if not dominant, role in the drug development arena of the next decades. Thus, biotech-based medications might serve as the key for the aspired "personalized medicine" in the health care systems of the future [3].

The basis for the pharmacotherapeutic use of biotech drugs is similar to that of small molecules, a defined relationship between the intensity of the therapeutic effect and the amount of drug in the body, or, more specifically, the drug concentration at its site of action, that is, the exposure-response relationship. The relationship between the administered dose of a drug, the resulting concentrations in body fluids, and the intensity of produced outcome may be either simple or complex, and thus obvious or hidden. However, if no simple relationship is obvious, it would be misleading to conclude a priori that no relationship exists at all rather than that it is not readily apparent [4].

The dose-concentration-effect relationship is defined by the pharmacokinetic and pharmacodynamic characteristics of a drug. Pharmacokinetics comprises all processes that contribute to the time course of drug concentrations in various body fluids, generally blood or plasma, that is, all processes affecting drug absorption, distribution, metabolism, and excretion. In contrast, pharmacodynamics characterizes the effect intensity and/or toxicity resulting from certain drug concentrations at the assumed effect site. Simplified, pharmacokinetics characterizes "*what the body does to the drug*," whereas pharmacodynamics assesses "*what the drug does to the body*" [5]. Combination of both pharmacological disciplines by integrated pharmacokinetic/pharmacodynamic modeling (PK/PD modeling) allows a continuous description of the effect-time course directly resulting from the administration of a certain dose (Fig. 1) [6, 7].

Pharmacokinetic and pharmacodynamic principles are equally applicable to biotech drugs such as peptides, proteins, and oligonucleotides as they are to conventional small-molecule drugs. This also includes PK/PD-related recommendations for drug development such as the recently published exposure-response guidance document of the US Food and Drug

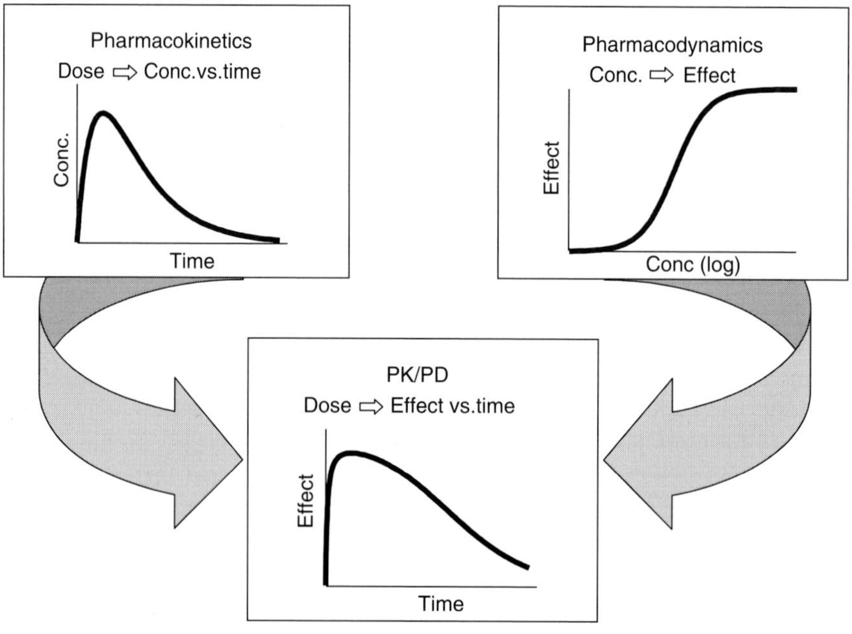

Fig. 1 Pharmacokinetic/pharmacodynamic (PK/PD) modeling as combination of the classic pharmacological disciplines pharmacokinetics and pharmacodynamics (from [7]).

Administration and the ICH E4 guideline of the International Conference on Harmonization of Technical Requirements for Registration of Pharmaceuticals for Human Use [8, 9]. Since biotech drugs are frequently identical or similar to endogenous substances, they oftentimes exhibit unique pharmacokinetic and pharmacodynamic properties. These may pose extra challenges and questions during their preclinical and clinical drug development that are different from small-molecule drug candidates and may require additional resources and unique expertise. Some of these problems and challenges will be discussed in the following.

8.2 Bioanalytical Challenges

The availability of an accurate, precise, and specific bioanalytical technique for the quantification of active drug moieties in plasma, blood, or other biological fluids is an essential prerequisite for the evaluation of the relationship between dose, concentration, and effect of biotech drugs. In analogy to small molecules, these analytical techniques have to be validated and have to meet prespecified criteria regarding accuracy, precision, selectivity, sensitivity, reproducibility, and stability, for example, those recommended by the US Food and Drug Administration [10–12]. Additional requirements for bioanalytical method validation for macromolecules have recently been published [11].

A further level of complexity is often added to the bioanalytics of biotech drugs by the fact that numerous of these compounds are endogenous substances that are already in the body before drug administration. Thus, the analytical technique will detect a so-called baseline level prior to drug exposure. This baseline level can either be constant or undergo complex changes, for example, circadian rhythms or irregular time courses. In order to characterize the clinical pharmacology of biotech drugs naturally present in the body, baseline values have to either be accounted for in the pharmacokinetic and pharmacodynamic analysis or be suppressed before exogenous drug administration [13–15]. The suppression of endogenous baseline levels is oftentimes facilitated via physiological regulation or feedback mechanisms. This approach was, for example, used to suppress the endogenous release of insulin and somatotropin (growth hormone, GH) via infusions of glucose and somatostatin respectively, prior to their exogenous administration for evaluation of their pharmacokinetics [16].

In contrast to the bioanalytics of small-molecule drugs, immunoassays and bioassays are frequently applied to quantify peptides and proteins in biological samples. Immunoassays are considered the analytical method of choice for concentration determinations of peptides and proteins as they can relatively rapidly and easily be performed [17]. These generally comprise enzyme immunoassays as well as radioimmunoassays with poly- and monoclonal antibodies. While both techniques have a sufficient sensitivity and reproducibility, their specificity for the active drug compound to be quantified has to be carefully evaluated during the assay validation process. Immunoassays may be insensitive to relatively minor changes in the primary or secondary structure of proteins. For recombinant interferon-γ, for example, bioavailability was reported to be >100% for subcutaneous compared to intravenous administration, which was produced by assay artifacts due to slightly modified degradation products [16]. Other potential interferences with immunoassays include matrix effects, specific binding proteins, proteases, and cross-reactivity toward endogenous proteins [17].

Bioassays are frequently used as an alternative or in addition to immunoassay techniques. Bioassays, in contrast to immunoassays, quantify not the pharmacologically active substance, but its biological activity, for example, in cell culture models based on cell differentiation, cell proliferation, or cytotoxicity as well as gene expression assays or whole animal models. Frequent major problems with bioassays comprise a high variability in the measured parameters, lack of precision, and their time- and labor-intensive performance. Furthermore, bioassays oftentimes also lack specificity for the measured compound, as they may also detect the response to bioactive metabolites [16, 17].

Because of some of the problems with bioassays and immunoassays, liquid chromatography (LC)-based techniques are increasingly applied as an alternative. While modern LC-based assays have a comparable sensitivity to immunoassays, they oftentimes are characterized by a higher selectivity [18, 19]. Müller et al., for example, used LC/mass spectrometry with matrix-assisted laser desorption ionization in ex vivo pharmacokinetic studies in combination with enzyme inhibition experiments to investigate the complex metabolism of dynorphin A1-13, a peptide with opioid activity, up to the fifth metabolite generation [20, 21].

Biodistribution studies for peptides are frequently performed with radioactively labeled compound. The radioactivity can either be introduced by external labeling with ^{125}I or by internal labeling of already present atoms, for example, via addition of ^{3}H, ^{14}C, or ^{35}S radioactively labeled amino acids to cell cultures producing recombinant proteins. External labeling chemically modifies the protein, which may affect its activity and pharmacokinetics. Internal labeling circumvents these potential problems, but has the disadvantage that radioactively labeled amino acids might be reused in the endogenous amino acid pool. Thus, when using radioactive labeling, it is generally necessary to investigate whether the physicochemical and biological properties of the proteins are unchanged. In addition, it is crucial to differentiate whether the measured radioactivity represents intact protein, labeled metabolites, or the released label itself. Radioactivity that can be precipitated with trichloroacetic acid, for example, can be used to delineate active protein from released label and metabolites of small molecular weight [16].

8.3 Pharmacokinetics of Peptides and Proteins

Although traditional pharmacokinetic principles are also applicable for peptides and proteins, their in vivo disposition is to a large degree affected by their physiological function. Peptides, for example, which frequently have hormone activity, usually have short elimination half-lives, which is desirable for a close regulation of their endogenous levels and thus function. Contrary to that, transport proteins like albumin or antibodies have elimination half-lives of several days, which enables and ensures the continuous maintenance of necessary concentrations in the blood stream [16]. The reported terminal half-life for SB209763, a humanized respiratory syncytial virus monoclonal antibody, for example, was reported as 22 to 50 days [22].

8.3.1 Absorption

Traditionally, the largest obstacle for a successful pharmacotherapy with peptide and protein drugs is their delivery to the desired site of action. A clinically usable absorption of exogenously applied peptides and protein after oral application with conventional dosage forms is usually not present [19, 23]. This lack of systemic bioavailability is mainly caused by two factors, high gastrointestinal enzyme activity and the function of the gastrointestinal mucosa as absorption barrier. There is substantial peptidase and protease activity in the gastrointestinal tract, making it to the most efficient body compartment for peptide and protein metabolism. Furthermore, the gastrointestinal mucosa presents a major absorption barrier for water-soluble macromolecules like peptides and protein [13, 19]. This is at least for peptides complemented by the functional system of cytochrome P450 3A and *p*-glycoprotein activity [24–26].

The lack of activity after oral administration for most peptides and proteins resulted in the past besides parenteral application into the utilization of nonoral administration pathways, for example, nasal, buccal, rectal, vaginal, percutaneous, ocular, or pulmonary drug delivery [27]. Drug delivery via these administration routes, however, is also frequently accompanied by presystemic degradation processes. Bioavailability of numerous peptides and proteins is, for example, markedly reduced after subcutaneous or intramuscular administration compared to their intravenous administration. The pharmacokinetically derived apparent absorption rate constant is thus the combination of absorption into the systemic circulation and presystemic degradation at the absorption site. The true absorption rate constant k_a can then be calculated as

$$k_{\mathrm{a}} = F \cdot k_{\mathrm{app}}$$

where F is the bioavailability compared to intravenous administration. A rapid apparent absorption rate constant k_{app} can thus be the result of a slow absorption and a fast presystemic degradation, that is, a low systemic bioavailability [13].

8.3.2 Distribution

Whole body distribution studies are essential for classical small-molecule drugs in order to exclude tissue accumulation of potentially toxic metabolites. This problem does not exist for protein drugs in which catabolic degradation products are amino acids recycled in the endogenous amino acid pool. Therefore, biodistribution studies for peptides and proteins are primarily performed to assess targeting to specific tissues as well as to identify the major elimination organs [28].

The volume of distribution of proteins is usually small and limited to the volume of the extracellular space because of their high molecular weight and the related limited mobility because of impaired passage through biomembranes [29, 30]. After intravenous application, peptides and proteins usually follow a biexponential plasma concentration–time profile that can best be described by a two-compartment pharmacokinetic model [13]. This has, for example, been described for leuprorelin, a synthetic agonist analog of luteinizing hormone-releasing hormone (LH-RH) [31], for clenoliximab, a macaque-human chimeric monoclonal antibody specific to the CD4 molecule on the surface of T-lymphocytes [32], and for AJW200, a humanized monoclonal antibody to von

Willebrand factor [33]. The central compartment in this model represents primarily the vascular space and the interstitial space of well-perfused organs with permeable capillary walls, especially liver and kidneys, while the peripheral compartment comprises the interstitial space of poorly perfused tissues like skin and (inactive) muscle [19].

Thus, the volume of distribution of the central compartment in which peptides and proteins initially distribute after intravenous administration is typically 3 to 8 L, approximately equal to slightly higher than the plasma volume [19] (approximate body water volumes for a 70-kg person: interstitial 12 L, intracellular 27 L, intravascular 3 L) [34]. The total volume of distribution (V_d) is frequently 14 to 20 L, not more than twice the initial volume of distribution (V_c) [13, 28]. This distribution pattern has, for example, been described for the somatostatin analog octreotid (V_c 5.2–10.2 L, V_d 18–30 L), the t-PA analog tenecteplase (V_c 4.2–6.3 L, V_d 6.1–9.9 L), and the glycoprotein IIb//IIIa inhibitor eptifibatide (V_c 9.2 L) [35–37]. Active tissue uptake and binding to intra- and extravascular proteins, however, can substantially increase the volume of distribution of peptide and protein drugs, as, for example, observed with atrial natriuretic peptide (ANP) [38].

There is a tendency for V_d and V_c to correlate with each other, which implies that the volume of distribution is predominantly determined by distribution in the vascular and interstitial space as well as unspecific protein binding in these distribution spaces. The distribution rate is inversely correlated with molecular size and is similar to that of inert polysaccharides, suggesting that passive diffusion through aqueous channels is the primary distribution mechanism [19].

Distribution, elimination, and pharmacodynamics are, in contrast to conventional drugs, frequently interrelated for peptides and proteins. The generally low volume of distribution should not necessarily be interpreted as low tissue penetration. Receptor-mediated specific uptake into the target organ, as one mechanism, can result in therapeutically effective tissue concentrations despite a relatively small volume of distribution [28]. Nartogastrim, a recombinant derivative of the granulocyte-colony stimulating factor (G-CSF), for example, is characterized by a specific, dose-dependent, and saturable tissue uptake into the target organ bone marrow, presumably via receptor-mediated endocytosis [39].

8.3.3 Protein Binding

It is a general pharmacokinetic principle, which is also applicable to peptides and proteins, that only the free, unbound fraction of a drug substance is accessible to distribution and elimination processes as well as interactions with its target structure (e.g. receptor) at the site of action. Hence, the activity of a drug is better characterized by its free rather than total concentration if there is no constant relationship between free and total drug concentration.

Physiologically active endogenous peptides and proteins are frequently interacting with specific binding proteins that are involved in their transport and regulation. Furthermore, interaction with binding proteins may enable or facilitate cellular uptake processes and thus affect the drug's pharmacodynamics. Specific binding proteins were, for example, described for IGF-1 (insulin-like growth factor), t-PA, interleukin-2, and somatotropin [16].

Six specific binding proteins were identified for IGF-1, with one binding at least 95% of IGF-1 in plasma. Since the binding affinity of IGF-1 to this binding protein is substantially higher than IGF receptors, the binding protein is assumed to have a reservoir function that protects the body from insulin-like hypoglycemia. Furthermore, the elimination half-life for bound IGF-1 is significantly longer than for free IGF-1, since only the unbound IGF-1 is accessible to elimination via glomerular filtration or peritubular extraction [28, 40].

Somatotropin, another example, has at least two binding proteins in plasma [16]. This protein binding substantially reduces somatotropin elimination with a tenfold smaller clearance of total compared to free somatotropin, and also decreases its activity via reduction of receptor interactions.

Apart from these specific bindings, peptides and proteins may also be nonspecifically bound to plasma proteins. For example, metkephamid, a met-enkephalin analog, was described to be 44 to 49% bound to albumin [41], and octreotid is up to 65% bound to lipoproteins [35].

8.3.4 Elimination

Peptide and protein drugs are nearly exclusively metabolized via the same catabolic pathways as endogenous or dietetic proteins, leading to amino acids that are reutilized in the endogenous amino acid pool for the de novo biosynthesis of structural or functional body proteins.

Nonmetabolic elimination pathways such as renal or biliary excretion are negligible for most peptides and proteins. Amino acids as well as some peptides and proteins such as immuneglobuline A, however, are excreted into the bile [19]. For octreotid, biliary excretion is, at least in rat and dog, an important elimination pathway [35]. If biliary excretion of peptides and proteins occurs, it generally results in subsequent metabolism of these compounds in the gastrointestinal tract (see Sect. 8.3.4.2) [13].

The elimination of peptides and proteins can occur unspecifically nearly everywhere in the body or can be limited to a specific organ or tissue. Locations of intensive peptide and protein metabolism are liver, kidneys, gastrointestinal tissue, and also blood and other body tissues. Molecular weight determines the major metabolism site as well as the predominant degradation process [13, 42] (Table 1).

The metabolism rate generally increases with decreasing molecular weight from large to small proteins to peptides, but is also dependent on other factors like secondary and tertiary structure as well as glycosylation. The clearance of a peptide or protein describes the irreversible removal of active substance from the extracellular space, which also includes cellular uptake besides metabolism. Because of the unspecific degradation of numerous peptides and proteins in blood, clearance can exceed cardiac output, that is, >5 L/min for blood clearance and >3 L/min for plasma clearance [13]. Investigations on the detailed metabolism of peptides and proteins are relatively difficult because of the myriad of molecule fragments that may be formed.

8.3.4.1 Proteolysis

Proteolytic enzymes such as proteases and peptidases are ubiquitously available throughout the body, but are especially localized in blood, in the vascular endothelium, and also on cell membranes and within cells. Thus, intracellular uptake is per se more an elimination rather than a distribution process [13]. While peptidases and proteases in the gastrointestinal tract

Tab. 1 Molecular weight as major determinant of the elimination mechanisms of peptides and proteins. As indicated, mechanisms may overlap. Endocytosis may occur at any molecular weight range (modified from [19, 28])

Molecular weight	*Elimination site*	*Predominant elimination mechanisms*	*Major determinant*
<500	Blood, liver	Extracellular hydrolysis Passive lipoid diffusion	Structure, lipophilicity
500–1000	Liver	Carrier-mediated uptake Passive lipoid diffusion	Structure, lipophilicity
1000–50 000	Kidney	Glomerular filtration and subsequent degradation processes (see Fig. 2)	Molecular weight
50 000–200 000	Kidney, liver	Receptor-mediated endocytosis	Sugar, charge
200 000–400 000		Opsonization	α_2-macroglobulin, IgG
>400 000		Phagocytosis	Particle aggregation

and in lysosomes are relatively unspecific, soluble peptidases in the interstitial space and exopeptidases on the cell surface have a higher selectivity and determine the specific metabolism pattern of an organ [19]. The proteolytic activity of subcutaneous tissue, for example, results in a partial loss of activity of subcutaneously compared to intravenously administrated interferon-γ.

8.3.4.2 Gastrointestinal Elimination

For orally administered peptides and proteins, the gastrointestinal tract is the major site of metabolism. Presystemic metabolism is the primary reason for their lack of oral bioavailability. Parenterally administered peptides and proteins, however, may also be metabolized in the intestinal mucosa following intestinal secretion. At least 20% of the degradation of endogenous albumin takes place in the gastrointestinal tract [13].

8.3.4.3 Renal Elimination

For parenterally administered and endogenous peptides and proteins, the kidneys are the major elimination organ if they are smaller than the glomerular filtration limit of ~60 kD, although the effective molecule radius is probably the limiting factor. The importance of the kidneys as elimination organ could, for example, be shown for interleukin-2, M-CSF and interferon-α [16, 19].

Various renal processes are contributing to the elimination of peptides and proteins (Fig. 2). For most substances, glomerular filtration is the dominant, rate-limiting step as subsequent degradation processes are not saturable under physiologic conditions [13, 43]. Hence, the renal contribution to the overall elimination of peptides and proteins is reduced if the metabolic activity for these proteins is high in other body regions, and it becomes negligible in the presence of unspecific degradation throughout the body. In contrast to that, the contribution to total clearance approaches 100% if the metabolic activity is low in other tissues or if distribution is limited. For recombinant IL-10, for instance, elimination correlates closely with glomerular filtration rate, making dosage adjustments necessary in patients with impaired renal function [44].

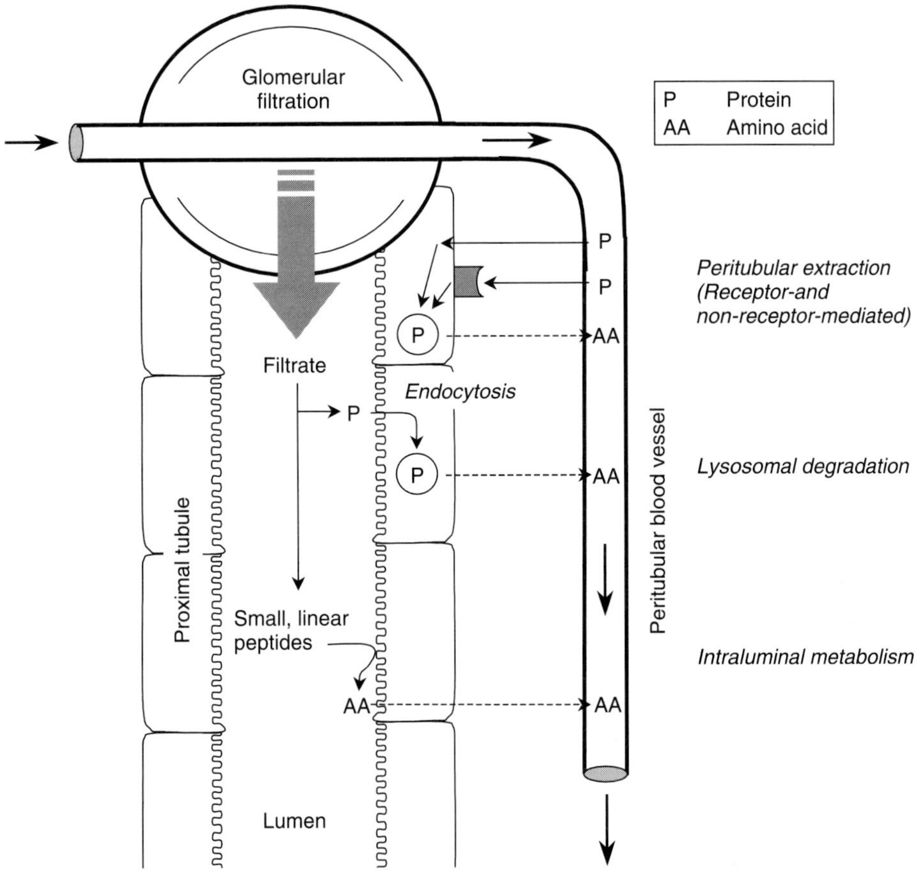

Fig. 2 Renal elimination processes of peptides and proteins: glomerular filtration, intraluminal metabolism, tubular reabsorption with intracellular lysosomal metabolism, and peritubular extraction with intracellular lysosomal metabolism (modified from [43]).

After glomerular filtration, small linear peptides undergo intraluminal metabolism, predominantly by exopeptidases in the luminal brush border membrane of the proximal tubules. The resulting amino acids are transcellularly transported back into the systemic circulation [28, 43]. Larger peptides and proteins are actively reabsorbed in the proximal tubules via endocytosis. This cellular uptake is followed by addition of lysosomes and hydrolysis to peptide fragments and amino acids, which are returned to the systemic circulation [16, 43]. Therefore, only minuscule amounts of intact protein are detectable in urine. An additional renal elimination mechanism is peritubular extraction from post glomerular capillaries with subsequent intracellular metabolism, which has, for example, been described for vasopressin and calcitonin [19, 43].

8.3.4.4 Hepatic Elimination

Apart from proteolysis and the kidneys, the liver substantially contributes to the metabolism of peptide and protein drugs. Proteolysis usually starts with endopeptidases that attack in the middle part of the protein, and the resulting oligopeptides are then further degraded by exopeptidases. The ultimate metabolites of proteins, amino acids, and dipeptides are finally reutilized in the endogenous amino acid pool. The rate of hepatic metabolism is largely dependent on specific amino acid sequences in the protein [28].

An important first step in the hepatic metabolism of proteins and peptides is the uptake into hepatocytes. Small peptides may cross the hepatocyte membrane via passive diffusion if they have sufficient hydrophobicity. Uptake of larger peptides and proteins is facilitated via various carrier-mediated, energy-dependent transport processes. Receptor-mediated endocytosis is an additional mechanism for uptake into hepatocytes (see Sect. 8.3.4.5) [28]. In addition, peptides such as metkephamid can already be metabolized on the surface of hepatocytes or endothelial cells [41].

8.3.4.5 Receptor-mediated Elimination

Receptor binding is usually negligible compared to total amount of drug in the body for conventional small-molecule drugs and rarely affects their pharmacokinetic profile. In contrast to that, a substantial fraction of a peptide and protein dose can be bound to receptors. This binding can lead to elimination through receptor-mediated uptake and subsequent intracellular metabolism. The endocytosis process is not limited to hepatocytes, but can occur in other cells as well, including the therapeutic target cells. Since the number of receptors is limited, their binding and the related drug uptake can usually be saturated within therapeutic concentrations. Thus, receptor–mediated elimination constitutes a major source for nonlinear pharmacokinetic behavior of numerous peptide and protein drugs, that is, a lack of dose proportionality.

M-CSF, for example, undergoes besides linear renal elimination a nonlinear elimination pathway that follows Michaelis-Menten kinetics and is linked to a receptor-mediated uptake into macrophages. At low concentrations, M-CSF follows linear pharmacokinetics, while at high concentrations, nonrenal elimination pathways are saturated, resulting in nonlinear pharmacokinetic behavior (Fig. 3) [45, 46]. Other examples for receptor-mediated elimination are insulin, t-PA, epidermal growth factor (EGF), ANP, and interleukin-10 [19, 28, 38, 44, 47].

Eppler et al., for example, had to develop a mechanism-based, target-mediated drug distribution model in order to accurately describe the nonlinear pharmacokinetics of a recombinant human vascular endothelial growth factor ($rhVEGF_{165}$) in patients with coronary artery disease [48]. Nonlinearity was caused by elimination of $rhVEGF_{165}$ by binding to specific and saturable high-affinity receptors followed by internalization and degradation.

8.3.5 Species Specificity and Allometry

Peptides and proteins exhibit distinct species specificity with regard to structure and activity. Peptides and proteins with identical physiological function may have different amino acid sequences in different species and may have no activity or be even immunogenic if used in a different species. The extent of glycosylation is another factor of species differences, for

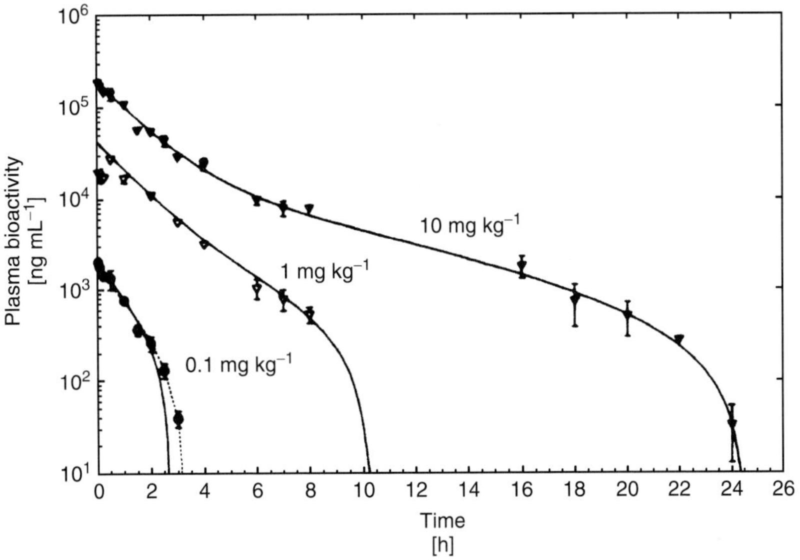

Fig. 3 Nonlinear pharmacokinetics of M-CSF, presented as measured and modeled plasma concentration–time curves (mean ± SE) after intravenous injection of 0.1 mg kg^{-1} ($n = 5$), 1.0 mg kg^{-1} ($n = 3$) and 10 mg kg^{-1} ($n = 8$) in rats (from [46]).

example, for interferon-α or erythropoietin, which may alter the drug's clearance. This is of particular importance if the production of human proteins is performed using bacterial cells [16].

Extrapolation of animal data to predict pharmacokinetic parameters by allometric scaling is an often-used tool in drug development with multiple approaches available at variable success rates [49–52]. In the most frequently used approach, pharmacokinetic parameters between different species are related via body weight using a power function:

$$P = a \cdot W^b$$

where P is the pharmacokinetic parameter scaled, W is the body weight in kilograms, a is the allometric coefficient, and b is the allometric exponent. a and b are specific constants for each parameter of a compound. General tendencies for the allometric exponent are 0.75 for rate constants (i.e. clearance, elimination rate constant), 1 for volumes of distribution, and 0.25 for half-lives.

For most traditional, small-molecule drugs, allometric scaling is often imprecise, especially if hepatic metabolism is a major elimination pathway and/or if there are interspecies differences in metabolism. For peptides and proteins, however, allometric scaling has frequently proven to be much more precise and reliable, probably because of the similarity in handling peptides and proteins between different mammalian species [16, 28]. Clearance and volume of distribution of numerous therapeutically used proteins like somatotropin or t-PA follow a well-defined, weight-dependent physiologic relationship between lab animals and humans. This allows relatively precise quantitative predictions for toxicology and dose-ranging studies based on preclinical findings [53].

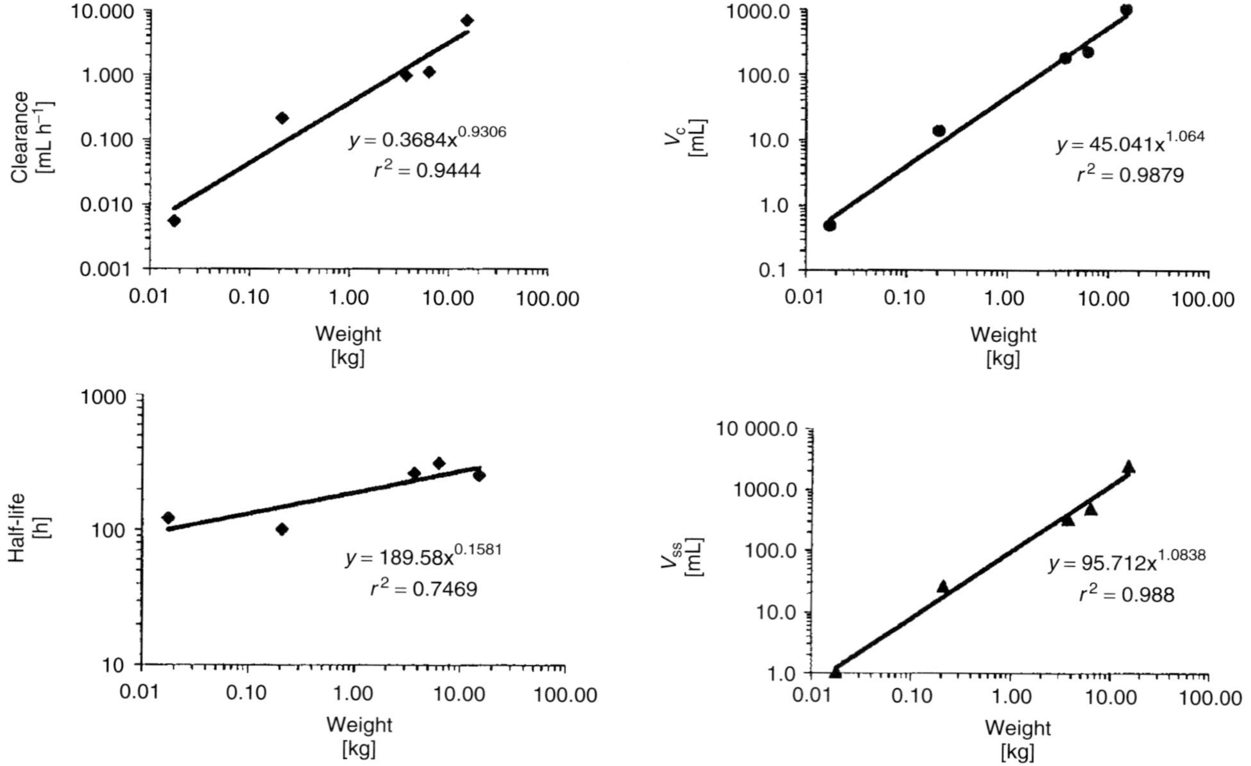

Fig. 4 Allometric plots of the pharmacokinetic parameters clearance, volume of the central compartment (V_c), volume of distribution at steady state (V_{ss}), and elimination half-life of rPSGL-Ig. Each data point within the plot represents an averaged value of the pharmacokinetic parameter with increasing weight from mouse, rat, monkey (3.74 kg), monkey (6.3 kg), and pig, respectively. The solid line is the best fit with a power function to relate pharmacokinetic parameters to body weight (from [54]).

Figure 4 shows the allometric plots of pharmacokinetic parameters for the recombinant, soluble, and chimeric form of P-selectin glycoprotein ligand-1 (rPSGL-Ig), an antagonist to P-selectin for the treatment of P-selectin–mediated diseases like thrombosis, reperfusion injury, and deep vein thrombosis. Human rPSGL-Ig pharmacokinetic parameters could accurately be predicted on the basis of data from mouse, rat, monkey, and pig using allometric power functions [54].

8.3.6
Immunogenicity

Because of the antigenic potential of proteins, formation of antibodies is a frequently observed phenomenon during chronic therapy with protein drugs, especially if human proteins are used in animal studies or if animal-derived proteins are applied in human clinical studies [16].

Most monoclonal antibodies are murine in nature and their systemic administration can lead to the development of human antimouse immunoglobulin antibody (HAMA) response, which is in most cases directed against the constant regions of the immunoglobulin. Genetically engineered mouse–human chimeric antibodies try to minimize this immunogenicity in man by joining variable domains of the mouse to the constant regions of human immunoglobulins [55]. The anti-EGFR (epidermal growth factor receptor) IgG monoclonal antibody cetuximab is an example of a murine–human chimeric antibody currently under clinical investigation for various cancer indications [56].

Extravascular injection is known to stimulate antibody formation more than intravenous application, most likely due to the increased immunogenicity of protein aggregates and precipitates formed at the injection site [57]. The presence of antibodies can obliterate the biological activity of a protein drug. In addition, protein–antibody complexation can also modify the distribution, metabolism, and excretion, that is, the pharmacokinetic profile, of the protein drug. Elimination can either be increased or decreased. Faster elimination of the complex occurs if the reticuloendothelial system is stimulated. Elimination is slowed down if the antibody–drug complex forms a depot for the protein drug. This effect would prolong the drug's therapeutic activity that might be beneficial if the complex formation does not decrease therapeutic activity [19, 28, 57]. Furthermore, antibody binding may also interfere with bioanalytical methods like immunoassays.

8.4
Pharmacokinetics of Oligonucleotides

Antisense oligonucleotides hold great promise as novel therapeutic agents designed to specifically and selectively inhibit the production of disease-related products, with fomivirsen being the first approved antisense oligonucleotide drug product [58]. So far, a significant body of preclinical and human pharmacokinetic data is only available for phosphothioate oligonucleotides (PONs).

Oral bioavailability is generally very low, ranging from 1 to 3%. Ongoing studies, however, indicate that oral bioavailability can be increased by the appropriate release of drug and permeability-enhancing excipients [59]. PONs have also been administered via subcutaneous, intradermal, and pulmonary application routes.

After intravenous administration, PONs follow generally two-compartment characteristics and are rapidly cleared from plasma, predominantly via distribution processes with a half-life of 0.5 to 1 h depending on the dose [60]. The ICAM-1 inhibitor alicaforsen, for example, has a distribution half-life of 1.0 to 1.2 h in humans [61, 62]. Plasma pharmacokinetics are nonlinear, with a more than proportional increase in area-under-the-curve (AUC) with dose that is most likely due to saturation of tissue uptake [58]. Figure 5, for example, shows the disproportional increase in systemic exposure after escalating doses of the HIV-inhibitor trecovirsen [63]. The plasma pharmacokinetics of various ONs are generally independent of their sequence and chemistry as plasma clearance is primarily determined by distribution processes.

After intravenous administration, PONs are detected in nearly all tissues and organs except for the brain and testes, suggesting significant transport barriers in these tissues. The extent of tissue uptake is dependent on the dose amount as well as dose rate. Major accumulation of PON occurs in liver and kidneys, and to a lesser extent in spleen, bone marrow, and lymph nodes, which seems to be independent of PON sequence. Chemical modification of the phosphothioate backbone structure, however, may alter protein binding and organ distribution. The mechanisms for uptake into target cells have not been fully elucidated yet, but these processes are energy-, temperature-, and time-dependent, and include most likely pinocytosis and podocytosis [64].

PONs are cleared from tissues by nuclease-mediated metabolism, with half-lives that vary between 20 and 120 h, depending on the organ or tissue. Successive removal of bases from the 3′-end is the major metabolic pathway in plasma, while both 3′ and 5′ exonuclease excision may occur in tissues. Exonuclease metabolism in plasma and tissues is

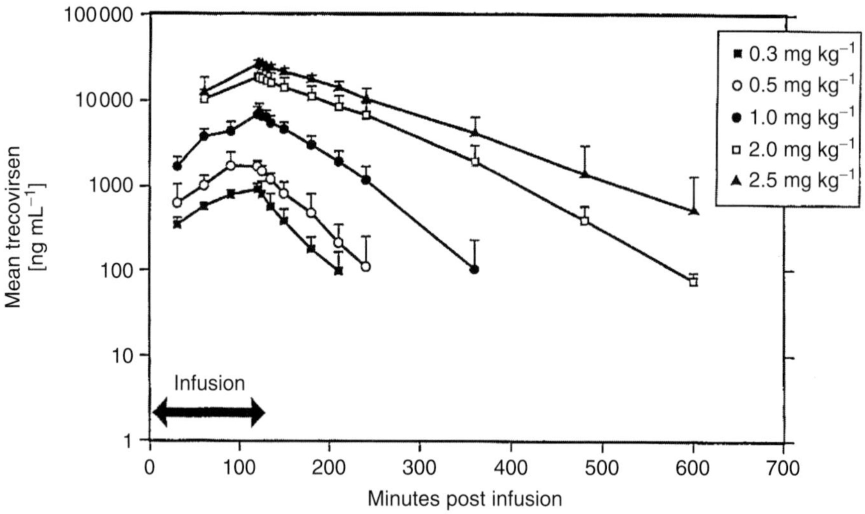

Fig. 5 Plasma concentration-time course of the antisense oligonucleotide trecovirsen in HIV-positive subjects, indicating a disproportional increase in systemic exposure after escalating doses. Trecovirsen was administered by 2-h intravenous administration to groups of 6 subjects (12 subjects for the 1.0 mg kg^{-1} dose group). (Reproduced from [63] with permission of Sage Publications, Inc.)

rapid, with 30 to 40% of PON having at least one nucleotide removed after 5 min in plasma. Endonuclease-mediated degradation of PONs is generally not observed [64].

PONs are highly bound to plasma proteins that protect them from renal filtration [65]. Plasma protein binding of ISIS2503, for example, ranged from 95 to 97% in rats and monkeys, but was saturable at high concentrations [66]. Urinary excretion is a major route of excretion for PONs, regardless of sequence or chemical structure, with the majority being shorter-length metabolites rather than unchanged parent drug [60]. Urinary excretion is nonlinear, with a greater fraction excreted at higher doses. Potential mechanisms include saturation of plasma protein binding as well as tubular reuptake mechanisms. Only a minor fraction of the dose is excreted into feces although enterohepatic recirculation has been suggested [58].

ISIS 104838 is a tumor necrosis factor-α (TNF-α) inhibiting second-generation ON containing five 2′-O-(2-methoxyethyl) modified (2′-MOE) nucleosides at the 3′- and 5′-terminus, respectively. The pharmacokinetic pattern for this second-generation PON was similar to first-generation PONs, except for a less pronounced nonlinearity in systemic exposure and a substantially prolonged terminal half-life of 27 ± 3.8 days, most likely due to the complete blockade of exonuclease digestion by the MOE modification [67].

8.5 Pharmacokinetics of DNA

In comparison to peptides, proteins, and oligonucleotides, much less is known about the pharmacokinetics of recombinant plasmid DNA used like a "drug" in the novel treatment approach of gene therapy.

The in vivo disposition of plasmid DNA and its complexes depends largely on its physicochemical characteristics, a strong negative charge and high molecular weight [68]. After intravenous administration in rats, pDNA is detected in all major organs including lungs, liver, kidney, and spleen. Low-level detection in the brain is most likely an artifact from residual blood, given that pDNA is unlikely to cross the blood-brain barrier [69].

After intravenous administration in mice, pDNA is rapidly eliminated from the plasma due to extensive uptake into the liver as well as rapid degradation by nucleases, with hepatic uptake clearance approaching liver plasma flow. pDNA is preferentially taken up by liver nonparenchymal cells, such as Kupffer and endothelial cells via receptor-mediated processes [70].

Analysis of the functional forms of pDNA in rats revealed that supercoiled pDNA rapidly disappears from plasma with a half-life of 0.15 min. Approximately 60% of supercoiled pDNA is degraded to open circular pDNA, which is subsequently nearly completely converted to linear pDNA. Conversion of open circular to linear pDNA followed Michaelis–Menten kinetics, while linear pDNA was removed with a half-life of 2.1 min. The slower elimination of open circular and linear pDNA compared to supercoiled pDNA was suggested to be related to a stronger interaction with plasma macromolecules that might offer some protection from plasma nuclease degradation [69].

8.6 Exposure/Response Correlations for Biotech Drugs

Since biotech drugs are usually highly potent compounds with steep dose-effect curve, a careful characterization of the dose-concentration-effect relationship should receive particular emphasis during the preclinical and clinical drug development process. Integrated pharmacokinetic/pharmacodynamic (PK/PD) modeling approaches have widely been applied for the characterization of biotech drugs. PK/PD modeling does not only allow for a continuous description of the time course of effect as a function of the dosing regime and comprehensive summary of available data but also enables testing of competing hypotheses regarding processes altered by the drug, allows to make predictions of drug effects under new conditions, and facilitates the estimation of inaccessible system variables [6, 71].

The application of PK/PD modeling is beneficial in all phases of preclinical and clinical drug development, with a focus on dosage optimization and identification of covariates that are causal for intra- and interindividual differences in drug response and/or toxicity [72]. It has recently further been endorsed by the publication of the Exposure-Response Guidance document by the U.S. Food and Drug Administration [8]. Mechanism-based PK/PD modeling appreciating the physiological events involved in the elaboration of the observed effect has been promoted as superior modeling approach as compared to empirical modeling, especially because it does not only describe observations but also offers some insight into the underlying biological processes involved and thus provides flexibility in extrapolating the model to other clinical situations [7, 73]. Since the molecular mechanism of action of biotech drugs is generally well understood, it is often straightforward to transform this available knowledge into a mechanism-based PK/PD modeling approach that appropriately characterizes the real physiological process leading to the drug's therapeutic effect.

In the following, the application of the three most common PK/PD modeling classes, direct link models, indirect link models, and indirect response models, will be discussed in more detail. In addition, extensions of these concepts and more complex approaches will be introduced in illustrative examples. However, it should be mentioned that PK/PD models for biotech drugs are not only limited to continuous responses as shown in the following but are also used for binary or graded responses. Lee et al., for example, used a logistic PK/PD modeling approach to link cumulative AUC of the anti-TNF-α protein etanercept with the American College of Rheumatology response criterion of 20% improvement (ARC20) in patients with rheumatoid arthritis [74].

8.6.1 Direct Link PK/PD Models

While drug concentrations are usually analytically quantified in plasma, serum, or blood, the magnitude of the observed response is determined by the concentration of the drug at its effect site, the site of action in the target tissue [6]. The relationship between the drug concentration in plasma and at the effect site may either be constant or undergo time-dependent changes. If equilibrium between both concentrations is rapidly achieved or the site of action is within plasma, serum or blood, there is practically a constant relationship between both concentrations with no temporal delay between plasma and effect site. In this case, measured concentrations can directly serve as input for a pharmacodynamic model. The most frequently used direct link pharmacodynamic model is the sigmoid E_{max}-Model:

$$E = \frac{E_{max} \cdot C^n}{EC_{50}^n + C^n}$$

with E_{max} as maximum achievable effect, C as drug concentration at the effect site, and EC_{50} the concentration of the drug that produces half of the maximum effect. The Hill-coefficient n is a shape factor that allows for an improved fit of the relationship to the observed data. Thus, a direct link model directly connects measured concentration to the observed effect without any temporal delay [5, 6].

Racine-Poon et al. provided an example for a direct link model by relating the serum concentration of the antihuman immunoglobulin E (IgE) antibody CGP 51901 for the treatment of seasonal allergic rhinitis to the reduction of free IgE via an inhibitory E_{max}-model [75]. Radwanski et al. used a similar approach to assess the effect of recombinant interleukin-10 on the ex vivo release of the proinflammatory cytokines TNF-α and interleukin-1β (IL-1β) in LPS-stimulated leukocytes [76].

8.6.2 Indirect Link PK/PD Models

The concentration-effect relationship of many biotech drugs, however, cannot be described by direct link PK/PD models, but is characterized by a temporal dissociation between the time courses of plasma concentration and effect. In this case, concentration maxima would occur before effect maxima, effect intensity would increase despite decreasing plasma concentrations and would persist beyond the time drug concentrations in plasma are no longer determinable. The relationship between measured concentration and observed effect follows a counterclockwise hysteresis loop. This phenomenon can either be caused by an indirect response mechanism (see Sect. 8.6.3) or by a distributional delay between the concentrations in plasma and at the effect site. The latter can conceptually be described by an effect-compartment model, which attaches a hypothetical effect-compartment to a pharmacokinetic compartment model. The effect compartment does not account for mass balance and only defines the changes in concentration at the effect site via the time course of the effect itself [5, 77].

An effect-compartment approach was, for example, applied by Gibbons et al. to quantify the reduction in mean arterial blood pressure by the antiadrenergic peptoid CHIR 2279 [78]. The same concept was used by Pihoker et al. to characterize the relationship between the serum concentration of the somatotropin-releasing peptide GHRP-2 and somatotropin (Fig. 6) [79].

8.6.3 Indirect Response PK/PD Models

The effect of most biotech drugs, however, is not mediated via a direct interaction between drug concentration and response systems, but frequently involves several transduction processes that include at their rate-limiting step the stimulation or inhibition of a physiologic process, for example, the synthesis or degradation of a molecular response mediator like a hormone or cytokine. In these cases, mechanism-based indirect response models should be

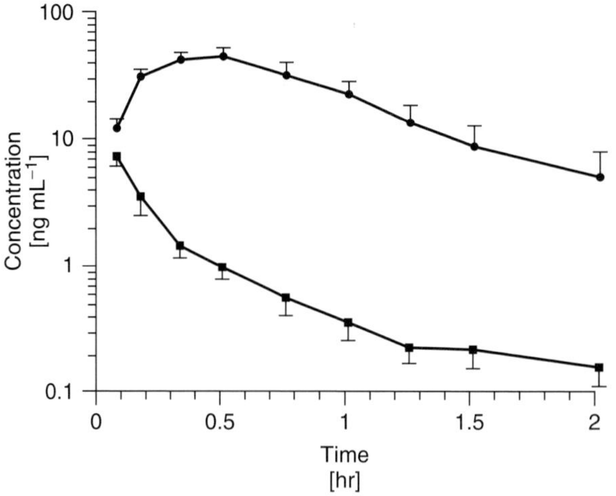

Fig. 6 Serum concentration profiles (mean ± SD) of GHRP-2 (growth hormone–releasing peptid-2; ■) and endogenous somatotropin (●) after intravenous administration of 1 μg kg^{-1} GHRP-2 in prepubertal children ($n = 10$). (from [79]. Copyright 1998, The endocrine Society.)

used that appreciate the underlying physiological process involved in mediating the observed effect. Indirect response models generally describe the effect on a representative response parameter via the dynamic equilibrium between increase or synthesis and decrease or degradation of the response, with the former being a zero-order and the latter a first-order process. Each of these processes can the stimulated or inhibited in four derived basic model variants [80–82].

Bressolle et al. used two variants of the indirect response models, stimulation of synthesis and of degradation processes, for modeling the effect of recombinant erythropoietin on the two response parameters free ferritin concentration (Fr) and soluble transferrin receptor concentration [83]. While erythropoietin reduces Fr, it increases Tfr (Fig. 7). The temporal change in both response variables can be described by the following equations:

$$\frac{\mathrm{d}Fr}{\mathrm{d}t} = k_{\mathrm{in},F} - k_{\mathrm{out},F} \cdot \left(1 + \frac{E_{\max} \cdot C_m^n}{EC_{50}^n + C_m^n}\right) \cdot Fr$$

$$\frac{\mathrm{d}Tfr}{\mathrm{d}t} = k_{\mathrm{in},T} \cdot \left(1 + \frac{E_{\max} \cdot C_m^n}{EC_{50}^n + C_m^n}\right) - k_{\mathrm{out},T} \cdot Tfr$$

with k_{in} as endogenous formation rate of Fr and Tfr, and k_{out} as first-order degradation rate constant, respectively. C_m is the erythropoietin concentration that was additionally modulated via a transduction process with 50-h delay.

Similarly, a modified indirect response model was used to relate the concentration of the humanized antifactor IX antibody SB249417 to factor IX activity in Cynomolgus monkeys as well as humans [84, 85]. The drug effect in this model was introduced by interrupting the natural degradation of Factor IX by sequestration of factor IX by the antibody.

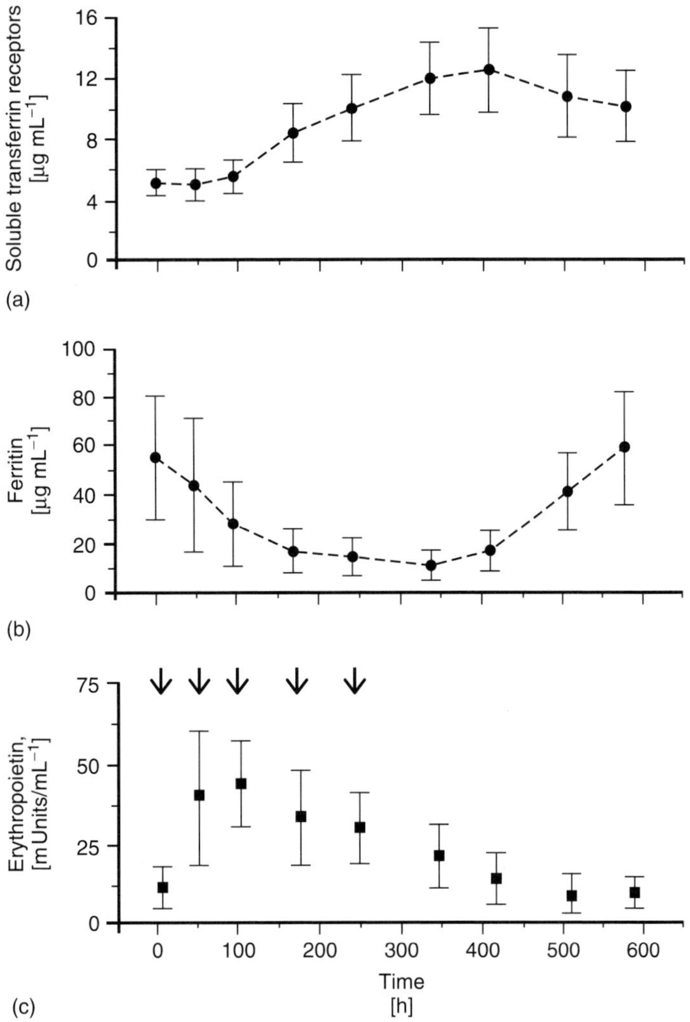

Fig. 7 Serum concentration-time course (mean ± SD) of erythropoietin (c) and its effect on the concentrations of free ferritin (b) and soluble transferrin receptor (a) after repeated subcutaneous administration of 200 U/kg recombinant erythropoietin in athletes ($n = 18$) (from [83]).

Indirect response models were also used for the effect of somatotropin on endogenous IGF-1 concentration, as well as the immune suppressive activity of the monoclonal antibody mAb 5c8 [86, 87].

Although physiologically related mechanism-based modeling should be preferred, an indirect response–based temporal dissociation between time course of concentration and effect can also be modeled with the effect-compartment approach. The effect of the growth hormone–releasing peptid (GHRP) ipamorelin on somatotropin, for example, was described by a physiologically based indirect response model [88], while the already mentioned GHRP-2 effect on somatotropin was characterized with an effect-compartment approach [79]. Similarly, effect compartment as well as indirect response models were applied for characterizing the effect of insulin on blood glucose levels. A recent comparative study, however, suggests that a mechanism-based indirect response model is a more appropriate approach for modeling the PK/PD of insulin [89].

8.6.4
Precursor Pool PK/PD Models

An extension of indirect response models are precursor pool-dependent indirect response models that include the liberation of an endogenous compound from a storage pool. These models possess the unique ability to characterize both tolerance and rebound phenomena [90]. Such a model was, for example, used to describe the effect of interferon-β1a on neopterin, an endogenous marker for cell-mediated immunity, in humans and monkeys (Fig. 8) [91, 92]. The primary elimination mechanism of interferon-β 1a was modeled as receptor-mediated endocytosis, and the pharmacodynamic model was driven by the amount of internalized drug-receptor complex DR*:

$$\frac{dP}{dt} = k_0 \cdot \left(1 + \frac{S_{max} \cdot DR^*}{SC_{50} + DR^*}\right) - k_p \cdot P$$

where P is the concentration of neopterin precursor (neopterin triphosphate), k_0 is the zero-order production rate of precursor P, and k_p is the first-order rate constant characterizing the conversion of precursor P to neopterin. The amount of internalized drug-receptor complex DR* stimulates precursor production via a stimulation function governed by the maximum effect parameter S_{max} and a sensitivity parameter SC_{50}, the concentration that results in half of S_{max}. The rate of change in neopterin concentration NP is then defined by the following expression:

$$\frac{dNP}{dt} = k_p \cdot P - k_{out} \cdot NP$$

with k_{out} as the first-order elimination rate constant of NP in the body.

8.6.5
Complex PK/PD Models

Since the effect of most biotech drugs is mediated via complex regulatory physiologic processes including feedback mechanisms and/or tolerance phenomena, some PK/PD models that have been described for biotech drugs are much more sophisticated than the four classes of models previously discussed. One example of such a complex modeling approach has been developed by Nagaraja et al. for the therapeutic effects of the LH-RH antagonist cetrorelix [93–95].

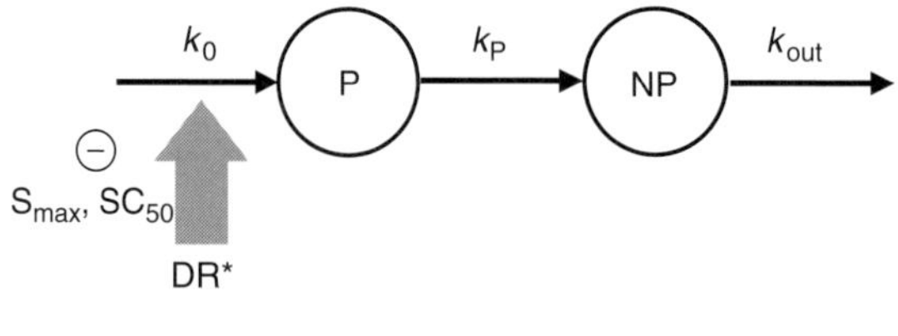

Fig. 8 Schematic representation of an indirect response model with precursor pool used to describe the effect of interferon-β 1a (represented by its internalized drug-receptor complex DR*) on endogenous neopterin concentrations (NP) via stimulation of the synthesis of its precursor neopterin triphosphate (P). See text for discussion of the details (from [91, 92]).

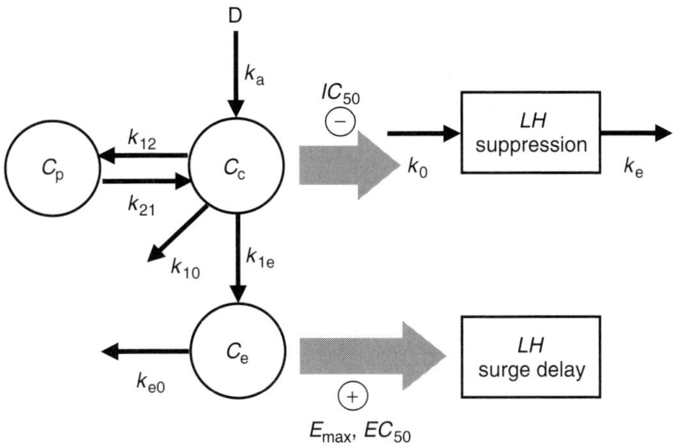

Fig. 9 PK/PD model for the LH suppression and the delay in LH surge following administration of a cetrorelix dose (D). C_C, C_P, and C_e: Concentration of drug in central, peripheral and effect compartments, respectively; k_a, k_{12}, k_{21}, k_{10}, k_{1e}, k_{e0}: pharmacokinetic first-order rate constants. See text for discussion of the details (modified from [93, 94]).

Cetrorelix is used for the prevention of premature ovulation in women undergoing controlled ovarian stimulation in in vitro fertilization protocols. LH-RH antagonists suppress the LH levels and delay the occurrence of the preovulatory LH surge, a delay in response via complex signal transduction steps of unknown mechanism of action. The combined effect of LH suppression and delaying the LH surge was described by the following relationship:

$$\frac{\mathrm{d}LH}{\mathrm{d}t} = k_0 \cdot \left(1 - \frac{C}{IC_{50} + C}\right) \cdot \left(1 + \frac{SA}{\left(\frac{t - \left(T_0 + \frac{E_{\max} \cdot C_e}{EC_{50} + C_e}\right)}{SW}\right)^N + 1}\right) - k_e \cdot LH$$

and this delay is thought to be responsible for postponing ovulation. The suppression of LH was modeled in the PK/PD approach with an indirect-response model approach directly linked to cetrorelix plasma concentrations (Fig. 9) [93]. The shift in LH surge was linked to cetrorelix concentration with a simple E_{max}-function via a hypothetical effect compartment to account for where LH is the LH concentration, k_0 and k_e are the zero-order production rate and first-order elimination rate constant for LH at baseline, C and C_e are the cetrorelix concentrations in plasma and a hypothetical effect compartment respectively, SA is the LH surge amplitude, t is time, T_0 is the time at which the peak occurs under baseline conditions, SW is the width of the

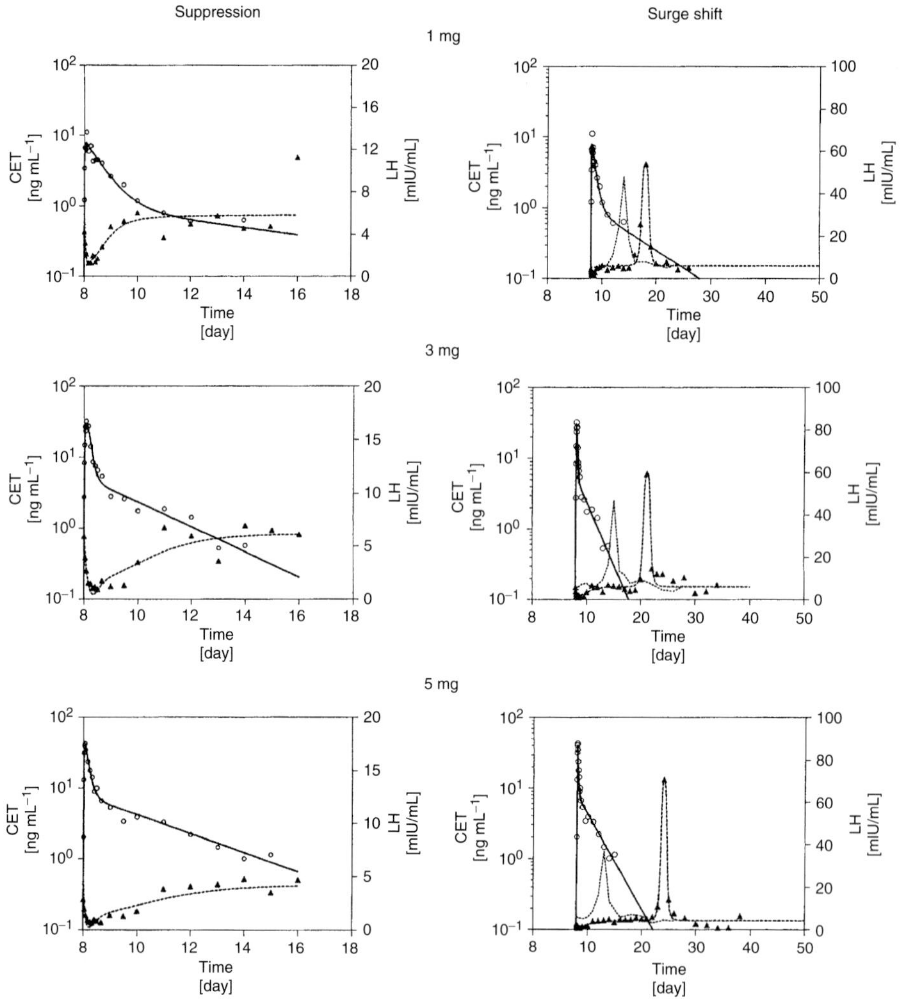

Fig. 10 Pharmacokinetic and pharmacodynamic relationship between cetrorelix (O) and LH concentrations (▲) after single doses of 1, 3, and 5 mg cetrorelix in representative subjects. Left panel: LH suppression. Right panel: LH suppression and LH surge profiles. The thick solid line represents the model-fitted cetrorelix concentration; the dashed line the model-fitted LH concentration; and the thin dotted line the pretreatment LH profile (not fitted) (from [94]).

peak in time units, IC_{50} is the cetrorelix concentration that suppresses LH levels by 50%, E_{max} is the maximum delay in LH surge and EC_{50} is the cetrorelix concentrations that produces half of E_{max}. N describes the slope of the surge peak and is an even number. Baseline data analysis indicated that N and SW were best fixed at values of 4 and 24 h, respectively [93].

Figure 10 shows the application of this PK/PD model to characterize the LH suppression and LH surge delay after subcutaneous administration of cetrorelix to groups of 12 women at different dose levels. The analysis revealed a marked dose-response relationship for the LH surge and thus predictability of drug response to cetrorelix [94].

An even more complex mechanism-based modeling approach including tolerance phenomena was used for the effect of antide, an LH-RH antagonist, on the endogenous regulatory mechanisms and plasma concentrations of LH and testosterone [96].

8.7 Summary

In general, biotech drugs underlie the same pharmacokinetic and pharmacodynamic principles as traditional, small-molecule drugs. On the basis of their similarity to endogenous compounds or nutrients, however, numerous caveats and pitfalls related to bioanalytics and pharmacokinetics have to be considered and addressed during the development process and may require additional resources. Furthermore, pharmacodynamics is frequently complicated owing to close interaction with endogenous substances and specific feedback mechanisms.

Biotech drugs, including peptides, proteins and antibodies, oligonucleotides, and DNA, are projected to cover a substantial market share in the health care systems of the future. It will be crucial for their widespread application in pharmacotherapy, however, that their respective drug development programs are successfully completed in a rapid, cost-efficient, and goal-oriented manner. A more widespread application of pharmacokinetic and pharmacodynamic concepts including exposure-response correlations has repeatedly been promoted by industry, academia, and regulatory authorities for all preclinical and clinical phases of drug development and is believed to result in a scientifically driven, evidence-based, more focused and accelerated drug product development process [72]. Thus, PK/PD concepts are likely to continue and expand their role as a decisive factor in the successful development of biotechnologically derived drug products in the future.

References

1. S. Arlington, S. Barnett, S. Hughes et al., *Pharma 2010: The Threshold to Innovation*, IBM Business Consulting Services, Somers, New York, 2002.
2. M. Hosseini, K. Nagels, *Scr. Mag.* **2001**, *December*, 22–25.
3. T. Nagle, C. Berg, R. Nassr et al., *Nat. Rev. Drug Discov.* **2003**, *2*(1), 75–79.
4. G. Levy, *J. Allergy Clin. Immunol.* **1986**, *78*(4 Pt 2), 754–761.
5. N. H. Holford, L. B. Sheiner, *Pharmacol. Ther.* **1982**, *16*(2), 143–166.
6. B. Meibohm, H. Derendorf, *Int. J. Clin. Pharmacol. Ther.* **1997**, *35*(10), 401–413.
7. H. Derendorf, B. Meibohm, *Pharm. Res.* **1999**, *16*(2), 176–185.
8. CDER/FDA, *Exposure Response Relationships: Guidance for Industry*, US Department of Health and Human Services, Food and Drug Administration, Center for Drug Evaluation and Research, Rockville, 2003.
9. International Conference on Harmonization, *ICH E4 – Dose-Response Information to Support Drug Registration*, European Agency for the Evaluation of Medicinal Products, London, 1994.
10. CDER/FDA, *Bioanalytical Method Validation: Guidance for Industry*, US Department of Health and Human Services, Food and Drug Administration, Center for Drug Evaluation and Research, Rockville, 2001.
11. K. J. Miller, R. R. Bowsher, A. Celniker et al., *Pharm. Res.* **2001**, *18*(9), 1373–1383.
12. V. Shah, K. Midha, J. Finlay et al., *Pharm. Res.* **2000**, *17*(12), 1551–1557.

13. W. Colburn, Peptide, peptoid, and protein pharmacokinetics/pharmacodynamics in *Peptides, Peptoids, and Proteins* (Eds.: P. Garzone, W. Colburn, M. Mokotoff), Harvey Whitney Books, Cincinnati, OH, 1991, pp. 94–115.
14. K. Nakao, A. Sugawara, N. Morii et al., *Eur. J. Clin. Pharmacol.* **1986**, *31*(1), 101–103.
15. F. S. Labella, J. D. Geiger, G. B. Glavin, *Peptides* **1985**, 6(4), 645–660.
16. R. J. Wills, B. L. Ferraiolo, *Clin. Pharmacokinet.* **1992**, *23*(6), 406–414.
17. A. Chen, D. Baker, B. Ferraiolo, Points to consider in correlating bioassays and immunoassays in the quantitation of peptides and proteins in *Peptides, Peptoids, and Proteins* (Eds.: P. Garzone, W. Colburn, M. Mokotoff), Harvey Whitney Books, Cincinnati, OH, 1991, pp. 54–71.
18. G. Rodes, V. Boppana, Novel derivatization approaches in the analysis of peptides in biological fluids by high performance liquid chromatography in *Peptides, Peptoids, and Proteins* (Eds.: P. Garzone, W. Colburn, M. Mokotoff), Harvey Whitney Books, Cincinnati, OH, 1991, pp. 73–79.
19. C. Mcmartin, *Adv. Drug Res.* **1992**, *22*, 39–106.
20. S. Muller, G. Hochhaus, *Pharm. Res.* **1995**, *12*(8), 1165–1170.
21. S. Muller, A. Hutson, V. Arya et al., *J. Pharm. Sci.* **1999**, *88*(9), 938–944.
22. H. C. Meissner, J. R. Groothuis, W. J. Rodriguez et al., *Antimicrob. Agents Chemother.* **1999**, *43*(5), 1183–1188.
23. A. Fasano, *J. Pharm. Sci.* **1998**, *87*(11), 1351–1356.
24. V. J. Wacher, J. A. Silverman, Y. Zhang et al., *J. Pharm. Sci.* **1998**, *87*(11), 1322–1330.
25. B. Meibohm, I. Beierle, H. Derendorf, *Clin. Pharmacokinet.* **2002**, *41*(5), 329–342.
26. L. B. Lan, J. T. Dalton, E. G. Schuetz, *Mol. Pharmacol.* **2000**, *58*(4), 863–869.
27. V. Lee, Problems and solutions in peptide and protein drug delivery in *Peptides, Peptoids, and Proteins* (Eds.: P. Garzone, W. Colburn, M. Mokotoff), Harvey Whitney Books, Cincinnati, OH, 1991, pp. 81–92.
28. R. Braeckman, Pharmacokinetics and pharmacodynamics of peptide and protein drugs in *Pharmaceutical Biotechnology* (Eds.: D. Crommelin, R. Sindelar), Harwood Academic Publishers, Amsterdam, 1997, pp. 101–122.
29. H. Berger, N. Heinrich, H. Schafer et al., *Life Sci.* **1988**, *42*(9), 985–991.
30. R. M. Reilly, J. Sandhu, T. M. Alvarez-Diez et al., *Clin. Pharmacokinet.* **1995**, *28*(2), 126–142.
31. P. Periti, T. Mazzei, E. Mini, *Clin. Pharmacokinet.* **2002**, *41*(7), 485–504.
32. D. R. Mould, C. B. Davis, E. A. Minthorn et al., *Clin. Pharmacol. Ther.* **1999**, *66*(3), 246–257.
33. S. Kageyama, H. Yamamoto, H. Nakazawa et al., *Arterioscler. Thromb. Vasc. Biol.* **2002**, *22*(1): 187–192.
34. M. Rowland, T. Tozer, *Clinical Pharmacokinetics: Concepts and Applications*, 3rd ed., Williams & Wilkins, Media, PA, 1993.
35. P. Chanson, J. Timsit, A. G. Harris, *Clin. Pharmacokinet.* **1993**, *25*(5), 375–391.
36. P. Tanswell, N. Modi, D. Combs et al., *Clin. Pharmacokinet.* **2002**, *41*(15), 1229–1245.
37. K. B. Alton, T. Kosoglou, S. Baker et al., *Clin. Ther.* **1998**, *20*(2), 307–323.
38. A. C. Tan, F. G. Russel, T. Thien et al., *Clin. Pharmacokinet.* **1993**, *24*(1), 28–45.
39. T. Kuwabara, T. Uchimura, H. Kobayashi et al., *Am. J. Physiol.* **1995**, *269*(1 Pt 1), E1–E9.
40. D. R. Clemmons, M. L. Dehoff, W. H. Busby et al., *Endocrinology* **1992**, *131*(2): 890–895.
41. Y. Taki, T. Sakane, T. Nadai et al., *J. Pharm. Pharmacol.* **1998**, *50*(9), 1013–1018.
42. R. Wills, A kinetic/dynamic perspective of a peptide and protein: GRF and rHuIFN-alpha-2a in *Peptides, Peptoids, and Proteins* (Eds.: P. Garzone, W. Colburn, M. Mokotoff), Harvey Whitney Books, Cincinnati, OH, 1991, pp. 117–127.
43. T. Maack, C. Park, M. Camargo, Renal filtration, transport and metabolism of proteins in *The Kidney* (Eds.: D. Seldin, G. Giebisch), Raven Press, New York, 1985, pp. 1773–1803.
44. S. Andersen, L. Lambrecht, S. Swan et al., *J. Clin. Pharmacol.* **1999**, *39*(10), 1015–1020.
45. A. Bartocci, D. S. Mastrogiannis, G. Migliorati et al., *Proc. Natl. Acad. Sci. U. S. A.* **1987**, *84*(17), 6179–6183.
46. R. J. Bauer, J. A. Gibbons, D. P. Bell et al., *J. Pharmacol. Exp. Ther.* **1994**, *268*(1), 152–158.
47. T. Murakami, M. Misaki, S. Masuda et al., *J. Pharm. Sci.* **1994**, *83*(10), 1400–1403.
48. S. M. Eppler, D. L. Combs, T. D. Henry et al., *Clin. Pharmacol. Ther.* **2002**, *72*(1), 20–32.

49. R. L. Dedrick, *J. Pharmacokinet. Biopharm.* **1973**, *1*(5), 435–461.
50. H. Boxenbaum, *J. Pharmacokinet. Biopharm.* **1982**, *10*(2), 201–227.
51. I. Mahmood, J. D. Balian, *Clin. Pharmacokinet.* **1999**, *36*(1), 1–11.
52. I. Mahmood, *Am. J. Ther.* **2002**, *9*(1), 35–42.
53. J. Mordenti, S. A. Chen, J. A. Moore et al., *Pharm. Res.* **1991**, *8*(11), 1351–1359.
54. S. P. Khor, K. Mccarthy, M. Dupont et al., *J. Pharmacol. Exp. Ther.* **2000**, *293*(2), 618–624.
55. S. K. Batra, M. Jain, U. A. Wittel et al., *Curr. Opin. Biotechnol.* **2002**, *13*(6), 603–608.
56. R. S. Herbst, C. J. Langer, *Semin. Oncol.* **2002**, *29*(1 Suppl. 4), 27–36.
57. P. Working, P. Cossum, Clinical and preclinical studies with recombinant human proteins: effect of antibody production in *Peptides, Peptoids, and Proteins* (Eds.: P. Garzone, W. Colburn, M. Mokotoff), Harvey Whitney Books, Cincinnati, OH, 1991, pp. 158–168.
58. B. H. Dvorchik, *Curr. Opin. Mol. Ther.* **2000**, *2*(3), 253–257.
59. ISIS Pharmaceuticals, *New data on Capsule Form of ISIS 104838 Demonstrate Antisense Drugs Can Be Administered Orally*, Press Release, Toronto, 2002.
60. S. Agrawal, J. Temsamani, W. Galbraith et al., *Clin. Pharmacokinet.* **1995**, *28*(1), 7–16.
61. B. R. Yacyshyn, W. Y. Chey, J. Goff et al., *Gut* **2002**, *51*(1), 30–36.
62. B. R. Yacyshyn, C. Barish, J. Goff et al., *Aliment Pharmacol. Ther.* **2002**, *16*(10), 1761–1770.
63. D. Sereni, R. Tubiana, C. Lascoux et al., *J. Clin. Pharmacol.* **1999**, *39*(1): 47–54.
64. A. A. Levin, *Biochim. Biophys. Acta* **1999**, *1489*(1), 69–84.
65. S. T. Crooke, M. J. Graham, J. E. Zuckerman et al., *J. Pharmacol. Exp. Ther.* **1996**, *277*(2), 923–937.
66. R. Z. Yu, R. S. Geary, J. M. Leeds et al., *J. Pharm. Sci.* **2001**, *90*(2), 182–193.
67. K. L. Sewell, R. S. Geary, B. F. Baker et al., *J. Pharmacol. Exp. Ther.* **2002**, *303*(3), 1334–1343.
68. M. Nishikawa, S. Takemura, Y. Takakura et al., *J. Pharmacol. Exp. Ther.* **1998**, *287*(1), 408–415.
69. B. E. Houk, R. Martin, G. Hochhaus et al., *Pharm. Res.* **2001**, *18*(1), 67–74.
70. Y. Takakura, M. Nishikawa, F. Yamashita et al., *Eur. J. Pharm. Sci.* **2001**, *13*(1), 71–76.
71. D. E. Mager, E. Wyska, W. J. Jusko, *Drug Metab. Dispos.* **2003**, *31*(5), 510–518.
72. B. Meibohm, H. Derendorf, *J. Pharm. Sci.* **2002**, *91*(1), 18–31.
73. G. Levy, *Clin. Pharmacol. Ther.* **1994**, *56*(4), 356–358.
74. H. Lee, H. C. Kimko, M. Rogge et al., *Clin. Pharmacol. Ther.* **2003**, *73*(4), 348–365.
75. A. Racine-Poon, L. Botta, T. W. Chang et al., *Clin. Pharmacol. Ther.* **1997**, *62*(6), 675–690.
76. E. Radwanski, A. Chakraborty, S. Van Wart et al., *Pharm. Res.* **1998**, *15*(12), 1895–1901.
77. L. B. Sheiner, D. R. Stanski, S. Vozeh et al., *Clin. Pharmacol. Ther.* **1979**, *25*(3), 358–371.
78. J. A. Gibbons, A. A. Hancock, C. R. Vitt et al., *J. Pharmacol. Exp. Ther.* **1996**, *277*(2), 885–899.
79. C. Pihoker, G. L. Kearns, D. French et al., *J. Clin. Endocrinol. Metab.* **1998**, *83*(4), 1168–1172.
80. N. L. Dayneka, V. Garg, W. J. Jusko, *J. Pharmocokinet. Biopharm.* **1993**, *21*(4), 457–478.
81. A. Sharma, W. Jusko, *Br. J. Clin. Pharmacol.* **1998**, *45*, 229–239.
82. Y. N. Sun, W. J. Jusko, *J. Pharm. Sci.* **1999**, *88*(10), 987–990.
83. F. Bressolle, M. Audran, R. Gareau et al., *J. Pharmacokinet. Biopharm.* **1997**, *25*(3), 263–275.
84. L. J. Benincosa, F. S. Chow, L. P. Tobia et al., *J. Pharmacol. Exp. Ther.* **2000**, *292*(2), 810–816.
85. F. S. Chow, L. J. Benincosa, S. B. Sheth et al., *Clin. Pharmacol. Ther.* **2002**, *71*(4), 235–245.
86. J. V. Gobburu, C. Tenhoor, M. C. Rogge et al., *J. Pharmacol. Exp. Ther.* **1998**, *286*(2): 925–930.
87. Y. N. Sun, H. J. Lee, R. R. Almon et al., *J. Pharmacol. Exp. Ther.* **1999**, *289*(3), 1523–1532.
88. J. V. Gobburu, H. Agerso, W. J. Jusko et al., *Pharm. Res.* **1999**, *16*(9), 1412–1416.
89. S. Lin, Y. W. Chien, *J. Pharm. Pharmacol.* **2002**, *54*(6), 791–800.
90. A. Sharma, W. F. Ebling, W. J. Jusko, *J. Pharm. Sci.* **1998**, *87*(12), 1577–1584.

91. D. E. Mager, B. Neuteboom, C. Efthymiopoulos et al., *J. Pharmacol. Exp. Ther.* **2003**, *26*, 26.
92. D. E. Mager, W. J. Jusko, *Pharm. Res.* **2002**, *19*(10), 1537–1543.
93. N. V. Nagaraja, B. Pechstein, K. Erb et al., *J. Clin. Pharmacol.* **2003**, *43*(3), 243–251.
94. N. V. Nagaraja, B. Pechstein, K. Erb et al., *Clin. Pharmacol. Ther.* **2000**, *68*(6), 617–625.
95. B. Pechstein, N. V. Nagaraja, R. Hermann et al., *J. Clin. Pharmacol.* **2000**, *40*(3), 266–274.
96. K. E. Fattinger, D. Verotta, H. C. Porchet et al., *Am. J. Physiol.* **1996**, *271*(4 Pt 1), E775–E787.

9
Formulation of Biotech Products

Ralph Lipp
Schering AG, Berlin, Germany

Erno Pungor
Berlex Biosciences, Richmond, CA, USA

9.1
Introduction

Today erythropoietin, insulin, and interferons belong to the 10 top drug substances on a global scale. Use of further proteins like somatotropin, for example, is rapidly growing and in future there will be an enhanced access to a growing number of therapeutically relevant proteins. This will most likely lead to a further increased importance of protein- and peptide-based drugs. Even today, drugs and drug candidates from the latter classes are being produced with high efficiency by novel biotechnological methods, and additional benefit is expected by the use of methods from the area of proteomics in the near future [1]. Novel transgenic approaches of protein and peptide production show significant advantages with respect to lower costs of goods for larger peptides in comparison to traditional recombinant methods [2]. However, in the class of small peptides that are based on less than 20 amino acids, the classical chemical synthesis still serves as a versatile method of cost-efficient drug production. In order to fully exploit the therapeutic potential of proteins, highly specific formulations are required that need to meet challenging targets from the areas of stabilization, specific application routes, and in some cases, drug-targeting aspects as well. These formulation and application route specifics will be highlighted in the subsequent paragraphs.

9.2
General Considerations on the Formulation of Proteins and Peptides

The stability, biological activity, and pharmacological activity of proteins and peptides are largely dependent on their intact primary, secondary, tertiary, and quaternary structure. Proteins and peptides can be easily modified by physical or chemical means [3]. Table 1 provides an overview of

Tab. 1 Main degradation pathways of proteins and peptides

Degradation pathways	
Physical	***Chemical***
Denaturation (ΔT, pH)	Oxidation (O_2, $h \cdot \nu$), e.g. Met, Cys, His, Trp
Noncovalent aggregation	Deamidation (ΔT, pH) e.g. Asn, Gln
Precipitation	Peptide cleavage (ΔT) e.g. Asp-X
Adsorption	Disulfide interchange (pH) Cys
	Beta-elimination (pH) e.g. Ser, Thr, Cys, Lys
	Disulfide formation (pH, O_2) Cys
	Covalent aggregation
	Cyclization (pH), e.g. Asp, Glu

the major degradation pathways that have to be considered when dealing with these molecules.

The effects of degradations can be very complex (losses in biological activity, changes in pharmacokinetics, pharmacodynamics, toxicity, biodistribution, elimination pathways, antigenecity, immunogenecity, etc.). In reality, the number of potential degradation pathways is even larger since some of the proteins are chemically modified: glycosylated, phoshorylated, etc. These nonprotein modifications may also have significant impact on the drug performance [4] (for example, glycosylation often controls circulating half-life, the mechanism of drug elimination and immunogenicity, phosphorylation in many cases contributes to biological activity and specificity of action), therefore the formulation has to protect these modifications, too. Unfortunately, there are no general rules to predict the effects of the individual degradation events; the effects of similar changes are widely variable on the pharmaceutical performances of different protein drugs. From the regulatory perspective, the degradation products are generally considered impurities, they need to be strongly controlled and listed in the specifications. These specifications have to reflect the clinically safe levels of these impurities. If a particular degradation is extensive, the presence of the degradation product on the drug safety may have to be specifically evaluated (in model systems or even in clinical studies).

As indicated in Table 1, there are several common triggers for instability of proteins and peptides, like presence of oxygen, shifts to extreme low or high pH values, elevated temperature and so on. Although there are some common approaches (e.g. using surfactants to prevent/reduce aggregation, or using antioxidants, like ascorbic acid to slow oxidation of the proteins), there is no general strategy for stabilization that will serve all drugs in an equivalent manner. In contrast, there is a tendency to develop strongly tailored manufacturing processes and formulations to serve the stabilization need of the individual compounds. Manufacturers also have to demonstrate the ability of the manufacturing processes to yield products of reproducible quality by process validation.

The variety of protein degradation pathways and the necessity to demonstrate control present a significant analytical challenge to the manufacturers of protein drugs. Besides the traditional analytical tools of protein chemistry (gel electrophoresis with protein staining or immunoblots, N-terminal sequencing, UV spectral analysis, etc.), a variety of techniques are used. These include different chromatographic (e.g. ion exchange, reverse phase, isoelectric focusing) separations of the intact proteins or enzymatically digested proteins (peptide map), separations on capillary systems, mass spectroscopic analysis of proteins, and peptide maps to assess chemical modifications. Circular dichroism is used to assess secondary and tertiary structures; light scattering techniques (sometimes in combination with chromatographic separation) and field flow fractionation are primarily applied to assess aggregations. If the protein is glycosylated, specific carbohydrate profile analysis and sequencing methods (liquid and/or gas chromatographic) are used to assess the integrity of the carbohydrates.

9.3 Application Routes for Proteins and Peptides

Directly from the beginning of the therapeutic use of proteins and peptides, parenteral formulations were of utmost importance, like sc formulations of insulin, for example. This was mainly triggered by special aspects from three different areas:

- Suboptimal physicochemical properties of the drugs for absorption through biological membranes.
- Instability of the drugs during nonparenteral administration.
- Specific therapeutic requirements with respect to onset or duration of action.

The specific physicochemical characteristics of proteins, like high molecular mass (often far beyond M_{rel} of 500), in many cases, high hydrophilicity, and the carriage of molecular charges largely cause the inability of these drugs to easily permeate biological membranes and thus diminish the bioavailability (BA) of the related compounds. Stability aspects may play a significant role during administration and absorption as well. The sensitivity of many proteins against extreme pH values has already been highlighted (see Table 1), and it is obvious that acidic pH values in the GI-tract are deleterious to the respective compounds. Furthermore, enzymatic degradation by proteases plays a significant role after oral administration and subsequent GI-passage.

In addition to the aforementioned absorption hurdles, the sometimes highly specific therapeutic requirements ask for, for example, a very rapid onset of action, like in the case of oxitocin-mediated labor induction or lactation, or the administration of insulin in the context of food intake. Some indications, on the other hand, ask for a long duration of action or even a continuous drug delivery like in the case of basal insulin administration or the application of β-interferon in the treatment of multiple sclerosis.

Today and in future, the demand for nonparenteral administration of proteins and peptides will be significantly growing. This is especially triggered by the wish of the patient to receive a convenient noninvasive treatment, rather than an invasive one. In order to fulfill this wish, two main hurdles need to be overcome. Firstly, in order to enable an effective, reliable, and safe application of proteins, specialized formulations and drug delivery systems for various application routes need to be developed. Secondly, in order to compensate for the significantly lower drug utilization, in many cases caused by lower bioavailability after nonparenteral application, the manufacturing processes for the proteins need to be optimized in order to lower the costs of goods far enough to allow for an economical overall therapy.

Owing to their overall importance, the parenteral application route and the three application routes that are currently of strong academic interest (enteral, especially oral), increasing clinical relevance (pulmonary), or therapeutic potential (nasal) are described in more detail in the subsequent paragraphs. Further, noninvasive routes of protein and peptide application like, for example, buccal [5], transdermal [6], colonic [7], and rectal [8] application will not be discussed within this chapter because of their lower significance versus the aforementioned ones, as of today.

9.4 Parenteral Application

The most widely used parenteral administration avenues are intravenous (iv), intramuscular (im), and subcutaneous (sc). In addition, there are several minor applications (e.g. intraarterial). Application of a protein drug by the different main parenteral administration routes may have profound effects on the pharmacological performances. When the drug is administered iv, it is immediately available for action in the circulation, while drugs administered im or sc need more time to reach the blood (depot effect), and consequently the pharmacokinetic (PK) profiles could be different. Besides the PK, the route of administration may have influence on the primary distribution of the drug. For example, when administered sc, smaller and hydrophillic proteins tend to enter the venous system, while larger and/or more hydrophobic proteins tend to be absorbed through the lymphatic system. The different routes of parenteral administration could also have effect on the antigenecity and immunogenecity of the drugs [9]. Several other aspects may also be taken into account when deciding on the application route. If, for example, chronic dosing is required (like with insulin and interferon therapies), sc and im administrations may offer added benefits as the patients can perform injections as opposed to the iv dosing, which is normally done in hospital settings.

Parenteral protein drugs are traditionally presented in vials with liquid, frozen, or lyophilized protein formulations. Stability and economical considerations influence the choice of formulation. Generally, the liquid formulations are less stable, many of the chemical degradation reactions are slowed down or practically eliminated in the frozen state or in the lyophilized cake with low water activity. From the manufacturing economy perspective, there are some trade-offs. Liquid and frozen formulations are less expensive to produce than the freeze-dried formulations. At the same time, many lyophilized formulations are stable at room temperature or higher, allowing for less costly shipping of the drug. Frozen formulations and some liquid formulations only stable under refrigeration conditions require a cool chain for shipping to assure stability. Liquid formulations may present additional problems in shipping and handling the vials: proteins are often highly surface active compounds capable of forming stable foam.

The manufacturers also have to assure the integrity of the container closure system used for the packaging of the drug (vials, syringes, various injector types, etc.) and the stability of the drug in the approved presentation (syringe, iv bag, etc.). As the number of protein drugs that can be self-administered by the patients is increasing, a variety of approaches have been taken to improve convenience (packaging of liquid formulation in syringes and lyophilized formulation in double chamber syringes allowing an "in-line" reconstitution to reduce manipulation required by the patients). Similarly, attempts were made to reduce injection pain (for example by using needleless injectors delivering the drug in a high-speed jet stream). The various delivery systems may create additional challenges in demonstrating drug integrity (syringes and even more so, the needleless injectors create high shear force during injection, which may cause proteins to aggregate or undergo partial denaturation). In all these cases, the manufacturers are required to demonstrate that the drug maintains the specified properties all the way to the point when it is applied to the patients in the specific systems.

9.5 Oral Application

Amongst the different routes of enteral application, the oral route is clearly the most attractive one from the patient's point of view. Upon oral application, the drug rapidly reaches the stomach and thereafter the small intestine comprising duodenum, jejunum, and ileum, and, subsequently, colon and rectum.

Several factors are promoting the drug uptake, like, for example, the high inner surface of approximately 100 m^2, the long contact time of ca. 16 h, and the presence of Peyer's patches. Factors that might decrease drug absorption on the other hand are low pH values in the stomach, the presence of endogenous proteases and bacterial enzymes, physical barriers like the mucus, and the glycocalix covering the microvilli, as well as the first liver passage.

In order to further discuss the oral route of administration, it is helpful to recall the basic epithelial transport routes in the first place (see Fig. 1).

For large molecules, it is assumed that transcytotic vesicular or – after opening of the tight junctions – paracellular transport may play a significant role, whereas transcellular transport is deemed to play a less significant role. Carrier-mediated transport so far is mainly discussed for

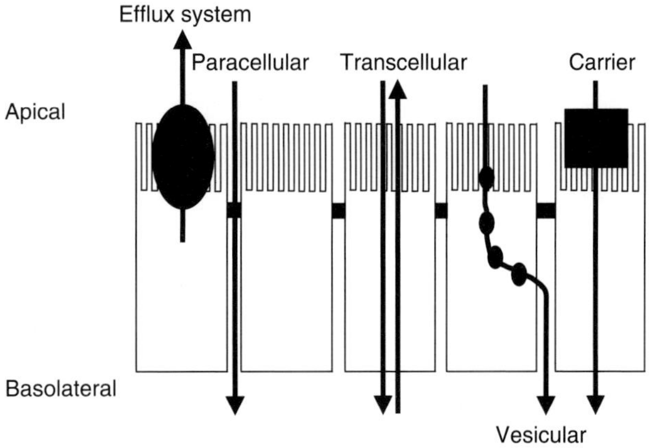

Fig. 1 Epithelial transport routes.

dipeptides and in the context of efflux systems like, for example, P-glycoprotein [10].

The uptake of proteins and peptides after enteral administration is largely prohibited by physical and enzymatic barriers [11]. The mucus and the glycocalyx that cover the microvilli of the brush border membrane need to be permeated prior to a contact of the drug with the epithelial cell and thus serve as physical barriers against protein uptake. The tight junctions that form the very close connection between adjacent to epithelial cells also build a strong physical barrier (see Fig. 2).

Enzymatic barriers against protein uptake stem from various classes of proteases. For example, gastric proteases like pepsin, and intestinal pancreatic proteases like trypsin and alpha-chymotrypsin [11]. In the microvilli, aminopeptidases as well as carboxypeptidases are located, and upon the passage of the cell membrane, the drugs are confronted with the contact of cytosolic petidases like di-tripeptidases. In addition to the aforementioned endogenous enzymes, enzymes of the intestinal bacteria have to be considered along with potential catalysts of protein degradation.

Taking into account the efficiency of the physical and biological barriers against protein uptake, it is obvious that the absorption of proteins after peroral application is typically low. One example for low bioavailability after peroral administration is the one of sal-calcitonin in dogs [12].

In Fig. 3, the bioavailability of sal-calcitonin upon application via several administration routes is displayed. Whereas infusion into the portal vein leads to a nearly full bioavailability, thus demonstrating that there is no relevant effect of the first liver passage in this specific case, and subcutaneous administration still leads to a moderate bioavailability of 50%, whereas it is zero after peroral application. Regional administration of the drug directly into either the duodenum, the ileum, or the colon led to improved bioavailability in all cases versus peroral administration. However, the uptake was still negligible and varied in the range from 0.02 to 0.06%.

In order to overcome the barriers of protein uptake, several attempts have been tried in the recent past, which will be discussed subsequently. In the area of formulation approaches, gastric resistant coatings of the protein containing dosage forms have proven to be a versatile stabilizer against the low pH value that prevails in the stomach. Owing to the fact that the drug is only released after passage of the pylorus, the deleterious effect of gastric

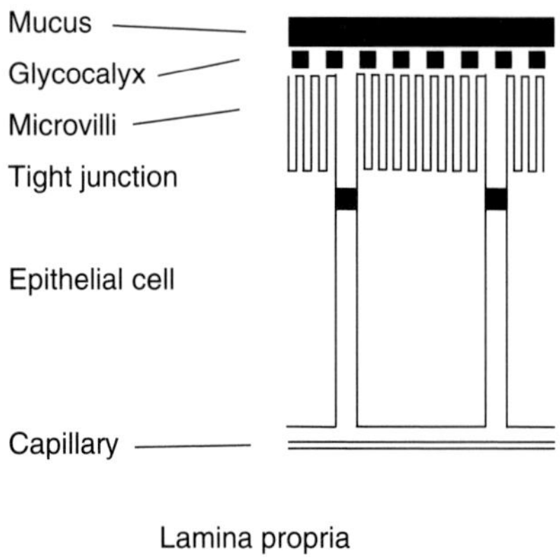

Fig. 2 Physical barriers against permeation.

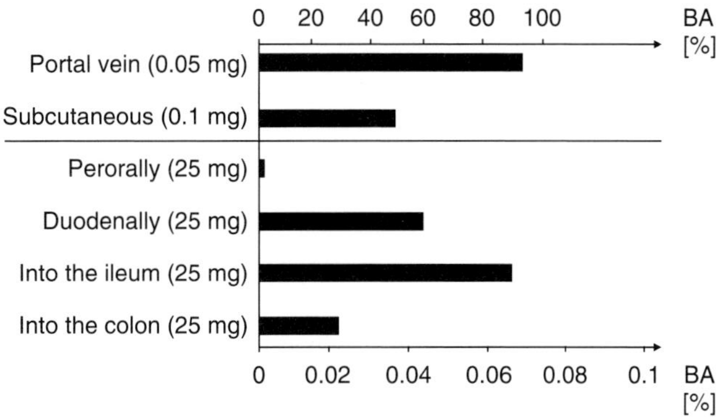

Fig. 3 Bioavailability of sal-calcitonin in dogs.

proteases like pepsin is also minimized via a gastric resistant coating [13]. Inhibition of catabolic proteases can be achieved in different ways, and is mainly performed in the large intestine. Firstly, protease inhibitors like, for example, puromycin [14] may be added to the formulation. Secondly, a less specific approach may be followed via shifting the pH to lower values where some proteases exhibit less activity. In that respect, compounds like, for example, citric acid are useful formulation additives. A practical example of a formulation approach using the combined effect of a gastric resistant coating and the addition of citric acid to a capsule-based sal-calcitonin formulation is discussed below.

Microencapsulation is a further formulation tool that helps overcome the absorption barriers against proteins. Firstly, encapsulation minimizes the susceptibility of the drugs against proteolysis, and secondly, particle uptake is discussed by several authors as well [15]. Mathiowitz et al. reported the in vivo effect of PLGA/FA-encapsulated insulin on fed rats after peroral administration. The latter formulation of 20 I.U. insulin led to unchanged glucose levels upon feeding, whereas the same dose administered as a simple solution could not prevent an increase in blood glucose by ca. 40 mg dL^{-1}, 1.5 h upon feeding.

Bioadhesion, mediated either by means of polymer particles or lectines, is discussed to prolong the gastrointestinal transit time and thereby enhance the absorption potential of the drugs [16]. Penetration enhancers are derived from various compound classes such as bile salts and fatty acids [13], for example. Some of these compounds are deemed to interact with the lipid bilayer of cell membranes, thus increasing their fluidity and decreasing their resistance against drug permeation. A very specific permeation enhancer on the other hand is the zona occludens toxin, which opens the tight junctions, and thereby allows higher penetration rates [13].

Besides the means to increase bioavailability through optimization of formulations, some efforts concentrate on the optimization of the proteins themselves with respect to use of analogues, like, for example, sal-calcitonin instead of human calcitonin due to the long in vivo half-life of the nonhuman analogue. Pegylation is a further means of creating analogues with optimized pharmacokinetical characteristics. In the area of prodrug formation, vitamin B12 derivatives play an important role because of their susceptibility to the vitamin B12-carrier system, which allows for higher (pro)drug uptake [17].

In order to substantially increase the bioavailability of proteins, a combination of elements from the aforementioned enhancement technologies is often applied. Lee et al. [12] report a study where sal-calcitonin in formulations containing up to 570 mg of citric acid in a hard gelatine capsule with a gastric-resistant coat were administered, thus combining three approaches: stable analogue, gastric acid protection, and protease inhibition. Trypsin, for example, is known to exhibit its maximum activity at a pH of 5 to 6, whereas it displays only 15% activity at pH 3.5. Capsules loaded with 1.2-mg sal-calcitonin were administered perorally to beagle dogs. The pH value in the vicinity of the drug-carrying formulation was monitored via a Heidelberger capsule. The outcome of this study was that in individual dogs the area under the curve (AUC) of the drug was increased up to 70-fold.

An additional route of uptake upon peroral administration is the passage of the M-cells of the Peyer's patches [18]. Although this route in essence has only a low transport capacity, it is of importance for mucosal (peroral) vaccination, for example. The Peyer's patches are located in the small intestine where M-cells, which neither possess a mucus nor a glycocalix layer in comparison to the adjacent enterocytes, allow the drug uptake via transcytosis. The drug is then transported into the lymphatic system and causes an immune response in the case of the vaccination approach. Absorption via M-cells in essence is possible due to the decreased enzymatic activity and their rather high permeability.

Although there is a rather high number of research initiatives ongoing in the area of peroral application of pharmacologically active compounds, no breakthrough with respect to reasonably high bioavailability rates has been made so far.

9.6 Nasal Application

The nose is characterized by a four-chambered structure, with the ostries dividing the front chamber from the lower, middle, and upper chamber [19]. One of the characteristics of the upper chamber is its coverage with cilliars. The cilliars' function is to clean the chambers from particles that stem from the air by a directed movement toward the throat. They are covered by a protective mucus that is fully renewed every 20 min owing to the aforementioned transport mechanism. A factor that is accountable for increased absorption potential after nasal application is the availability of certain "pores" that specifically increase the absorption of small molecules of M_{rel} up to 300, especially hydrophilic ones [20]. Furthermore, the nasal cavity is easily accessible by medications, for example, in form of droplets of 20 to 30 μm in diameter. Furthermore, there will be obviously no deleterious effects from liver first pass metabolism nor degradation from gastric or pancreatic enzymes upon nasal application.

Factors limiting the drug uptake after nasal application are the limited absorption area of 160 cm^2 and the short contact time of 20 min due to cilliaric transport (sa). Furthermore, there are proteases and peptidases located in the mucus of the nasal tissue, however, at concentrations that are easily saturable in many cases.

Several peptide products are in the marketplace today, which achieve high acceptance by the patients, predominantly due to their ease of use. These drugs stem from the class of oligopeptides such as Luteinizing Hormone Releasing Hormone (LHRH) and octreoid, for example [21]. Formulations for drugs with higher molecular weight, such as calcitonin or insulin, for example, are being developed as well.

In order to increase bioavailability after nasal application, several approaches were reported. Enzyme inhibition using puromycine or bile salts is one means [22]. The use of middle-chain phospholipids as penetration enhancers is another one. Powder formulations based on dextran or chitosan, for example, are described as well [23]. Besides the formulation approaches, the prodrug approach is followed, for example, by forming acyloxymethyl-derivatives of amines [24]. Furthermore, studies on the absorption of larger molecules like, for example, insulin have been reported as well. Hussain shows how the addition of the enzyme inhibitor puromycine, for example, inhibits the cleavage of Leu-enkephalin to des-tyrosin Leu-enkephalin during its passage through nasal tissue in a concentration-dependent manner [22].

Figure 4 provides absolute bioavailability values of several proteins and peptides after nasal administration in man.

Although rather high bioavailability values have been reported for drugs such as Leuprorelin and Insulin, for example, it has to be pointed out that the variability of the drug uptake is sometimes pronounced. For calcitonin, for example, it has been reported to vary from 0.3 to 30% with a mean of 3% in one study. High variance, however, will not be acceptable for drugs with a narrow therapeutic window.

Formulations for nasal application need to be sterile and free of cilliotoxic substances. Solutions may be applied via pump dispensers. Nasal powders on the other side require special application systems like, for example, the Jetilizer® [25]. This system follows a twin construction pattern with two nozzles. The capsule containing the particle-based formulation is opened by needles. A pump-activated air stream aerosolizes the powder in an equilibrium chamber. After passing a conical tube, the powder is applied with a high deposition rate to the absorption area within the nose.

Owing to the benefits mentioned above, the nasal route of peptide application already plays a significant role in the clinical practice of today. Depending on the outcome of the development activities

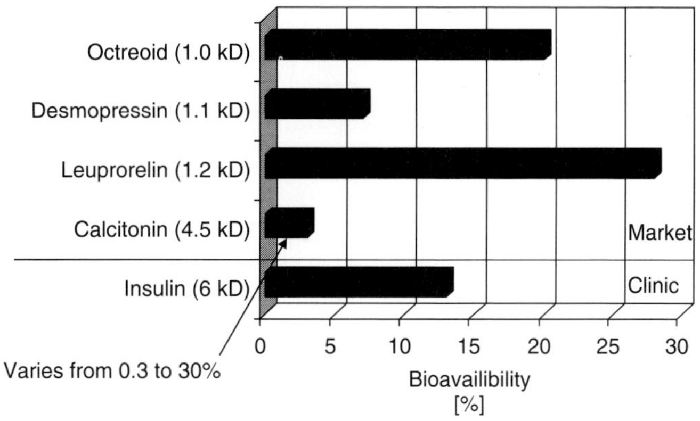

Fig. 4 Bioavailability of proteins and peptides after nasal administration in man.

in the areas of formulations and devices, this role will be growing in the near future, eventually even reaching importance for the application of larger compounds.

9.7 Pulmonary Application

With regard to the aim to achieve a high absolute bioavailability, the pulmonary route of administration bears several beneficial aspects. First of all, the large absorption surface of approximately 100 m^2 has to be mentioned. This surface is provided by the alveoli of the deep lung, which are characterized by a layer consisting predominantly of very thin so-called type I-cells, which are characterized by a height of only ca. 0.2 μm. This layer forms, together with the adjacent layer of capillary endothelium cells, the main part of the barrier between the air-filled space of the lung and the bloodstream in the capillaries [26]. Furthermore, the existence of pores within this barrier is discussed. Obviously, there will be no metabolic effect of a liver first pass effect after pulmonary application. However, factors potentially decreasing the drug uptake after pulmonary administration are potentially low alveolar deposition rates, which might be due to the fact that large particles or droplets will not reach the deep lung but might be deposited in the segments of the upper lung, or, in case of very small particles or droplets, exhalation might take place even after arriving at the deep lung in the first place. Furthermore, the so-called surfactant, which is a layer consisting of amphiphilic substances [26] covering the apical side of the alveolar type I-cell layer might hamper the drug uptake as well as macrophages, which are physiologically cleaning the alveoli from exogenic particles and germs. Furthermore, it has to be pointed out that the cell layer of the alveoli is rather tight, and does not easily allow for paracellular transport.

Patton proposed [27] the existence of different transport routes for macromolecules depending on their size. Compounds smaller than 40 kD are supposed to utilize paracellular and transcytotic routes in parallel, whereas larger compounds should utilize the latter only, leading to an reduced uptake into the bloodstream of the capillaries, but to an relatively elevated uptake into the lymph system, and the venoles as well.

The proteolytic activity of the alveolar epithelium has been studied by, for example, Yang et al. [28]. They showed that the half-life of LHRH in type-I cell cultures in vitro was 5 h, and that the stability could be significantly increased by substituting the Gly^6 by a $D\text{-}Ala^6$, leading to an approximately threefold more stable analogue of the drug. Even larger proteins are reported to exhibit good absorption rates upon pulmonary administration like, for example, parathyroid hormone (PTH) [29], and an overview of the bioavailability of proteins and peptides after pulmonary administration is given in Fig. 5.

The high bioavailability values ranging in the area of up to 40% as well as the fact that rather large compounds such as Somatotropin (22 kD) demonstrate moderate to good bioavailability are very impressive.

In order to apply proteins to the lung with the target to achieve high bioavailability, it is important to formulate them in ways that allow for a high amount of deep lung deposition in the first place. Two main formulation approaches have to be differentiated, the particle-based and the droplet-based one.

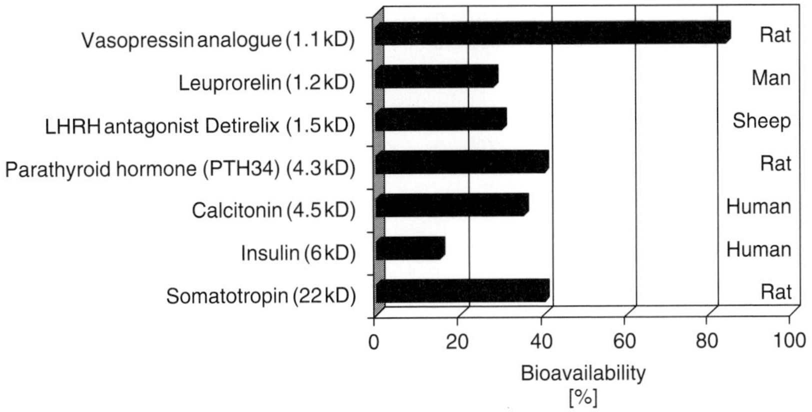

Fig. 5 Bioavailability of proteins and peptides upon pulmonary administration.

Novel particle-based formulations are designed to achieve high deep lung deposition as well as low clearance by the phagocytotic activity of the macrophages of the lung [30]. These special particles were engineered to be too large for a significant uptake into the macrophages, which clean the lung epithelium from airborne particles up to a diameter of approximately 5 µm. However, particles above this size typically exhibit a low percentage of deep lung deposition. The aforementioned particles, however, are manufactured in a tailor-made spray-drying process based on lactose as an excipient. They are designed to be porous, and because of their low density, they display aerodynamic properties of particles of significantly lower particle diameters. Thus they combine the advantages of elevated deep lung deposition with low macrophage clearance.

Particle-based formulations for pulmonary protein delivery request novel types of powder inhalers, which combine several attributes like the ability to apply rather high dosages of up to the milligram range. They need to exhibit high dose accuracy as well as to be even less dependent on the breath rate of the patient.

Drug solutions for pulmonary applications combine several benefits over particulate formulations, but there are severe drawbacks as well. Beneficial is the ease of access to liquid formulations, which are typically available for parenteral administration in the first place, thereby allowing for an immediate start of development activities. One further benefit is the rather low stress, which is exposed on the proteins during manufacture. One drawback versus dry, particle-based formulation, however, is the rather low drug load of the water-based formulations ranging predominantly from 1 to 2%, thereby requiring the aerosolization of large quantities of formulation.

So far, Pulmozyme® is the first pulmonary drug based on a protein. It contains rhDNase, an enzyme active in the treatment of cystic fibrosis. This, however, is a locally active drug, which does not require absorption but only sufficient deep lung deposition. The water-based solution is aerosolized by a large nebulizer with application phases of 15 to 30 min, which strongly restricts the use of the drug. In general, development of handy carry-on nebulizers is of utmost importance for

Tab. 2 Absolute bioavailibility of proteins and peptides after noninvasive administration in comparison

Route of administration	***Class of drug***	
	Proteins	***Peptides***
Oral	Up to ~ 1%	
Nasal	Up to ~ 10%	Up to ~ 30%
Pulmonary	Up to ~ 40%	

the success and the future acceptance of inhalative protein drugs formulated as solutions. An example of an effective, small device, which provides effective nebulization by the passage of the drug solution through micro nozzles is given in [30].

In summary, pulmonary application of proteins and peptides seems to be a promising approach to enable rather high bioavailabilites of the drugs applied. Main tasks of the near future remain to clarify the long-term safety of pulmonary protein application on the lung physiology and function by means of, for example, forced expiratory volume measurement and chest X rays. This is especially important, since all drugs under current development are targeting at chronic administration.

9.8 Conclusion

Table 2 provides a condensed information on the bioavailability ranges that have been reported so far for the oral, nasal, and pulmonary application of proteins and peptides, clearly demonstrating the advantage of pulmonary application over nasal and especially oral application in this respect.

In the area of oral delivery, a large number of preclinical and even a few early clinical trials are still being carried out. In the area of nasal delivery, there are various activities in research and development ongoing. Most important, some products based on oligo-peptides are successful in the market for years. Pulmonary drug delivery is characterized by various activities in preclinic and clinic. Especially, several insulin-based drug developments are in a late stage of phase III trials [31], and with the locally acting Pulmozyme® one protein-based pulmonary drug, although acting locally, is being marketed already.

References

1. A. Dove, *Nat. Biotechnol.* **1999**, *17*, 233–236.
2. P. W. Latham, *Nat. Biotechnol.* **1999**, *17*, 755–757.
3. Basic review of formulation development of proteins for parenteral use. J. A. Bontempo, (Ed.), *Development of Biopharmaceutical Parenteral Dosage Forms*, Marcel Dekker, New York, Basel, Hong Kong, 1994.
4. Basic review of protein modifications and their effects. D. J. Graves, B. L. Martin, J. H. Wang, *Co- and Post-Translational Modification of Proteins. Chemical Principles and Biological Effects*, Oxford University Press, New York, Oxford, 1994.
5. H. E. Junginger, J. A. Hoogstrate, J. C. Verhoef, *J. Control. Release* **1999**, *62*, 149–159.
6. J. E. Riviere, M. C. Heit, *Pharm. Res.* **1997**, *14*(6), 686–697.
7. H. Tozaki, J. Nishioka, J. Komoike et al., *J. Pharm. Sci.* **2001**, *90*(1), 89–97.

8. T. Uchida, S. Sakakibara, Y. Toida et al., *Pharm. Pharmacol. Commun.* **1999**, *5*, 523–527.
9. S. E. Grossberg, Human antibody development to therapeutic response modifiers in *Progress in Oncology, Update on Cytokines* (Ed.: E. C. Borden), Mediscript, London, UK, 1990, 17–23.
10. V. H. L. Lee, C. Chu, E. D. Mahlin, *J. Control. Release* **1999**, *62*, 129–140.
11. A thorough review of oral protein delivery. W. Wang, *J. Drug Target* **1996**, *4*(4), 195–232.
12. Y. H. Lee, P. Sinko, *Adv. Drug Deliv. Rev.* **2000**, *36*, 225–238.
13. Review on insulin delivery highlighting several enhancement strategies for oral bioavailability. G. P. Carino, E. Mathiowitz, *Adv. Drug Deliv. Rev.* **1999**, *35*, 249–257.
14. A. Bernkop-Schnürch, *J. Control. Release* **1998**, 1–16.
15. S. P. Baldwin, W. M. Saltzman, *Adv. Drug Deliv. Rev.* **1998**, *33*, 71–86.
16. G. Ponchel, J. M. Irache, *Adv. Drug Deliv. Rev.* **1998**, 191–219.
17. J. Alsenz, G. J. Russel-Jones, S. Westwood et al., *Pharm. Res.* **2000**, *17*(7), 825–832.
18. F. Niedergang, J. P. Kraehenbuhl, *Trends Cell Biol.* **2000**, *10*, 137–141.
19. Basic reference on anatomy and physiology of the nose. N. Mygind, R. Dahl, *Adv. Drug Deliv. Rev.* **1998**, *29*, 3–12.
20. V. Agarwal, B. Mishra, *Ind. J. Exp. Biol.* **1999**, *37*, 6–16.
21. Thorough review on the nasal administration of peptide hormones. A. E. Pontiroli, *Adv. Drug Deliv. Rev.* **1998**, *29*, 81–87.
22. Basic review on nasal drug delivery. A. A. Hussain, *Adv. Drug Deliv. Rev.* **1998**, *29*, 39–49.
23. L. Illum, N. F. Farraj, S. S. Davis, *Pharm. Res.* **1994**, *11*(8), 1186–1189.
24. R. Krishnamoorthy, A. K. Mitra, *Adv. Drug Deliv. Rev.* **1998**, *29*, 135–146.
25. Detailed comparison of devices for nasal application. M. Nomura, A. Yanagawa, O. Tokomo et al., *Pharm. Technol. Eur.* **1998**, *10*(10), 48–58.
26. Thorough review of absorption routes of the lung. J. S. Patton, *Adv. Drug Deliv. Rev.* **1996**, *19*, 3–36.
27. J. S. Patton, *Nat. Biotechnol.* **1998**, *16*, 141–143.
28. X. Yang, J. K. A. Ma, C. J. Malanga et al., *Int. J. Pharm.* **2000**, *195*, 93–101.
29. J. S. Patton, *Adv. Drug Del. Rev.* **2000**, *42*, 239–248.
30. D. A. Edwards, A. Ben-Jeriba, R. Langer, *J. Appl. Physiol.* **1998**, *85*(2), 379–385.
31. Comparison of different pulmonary protein delivery platforms. K. Haan, *Biocentury* **2002**, *10*(51), A1–A6.

10
Patents in the Pharmaceutical Biotechnology Industry: Legal and Ethical Issues

David B. Resnik
East Carolina University, Greenville, NC, USA

10.1
Introduction

This chapter will provide the reader with an overview of patenting in the pharmaceutical biotechnology industry and summarize some of the key legal and ethical issues related to the patenting of biomedical products and processes. It will examine the legal aspects of patenting before considering the ethical and policy issues. This essay will focus primarily on the US patent laws, which are very similar to the European patent laws. The essay will note some differences between the US and European laws, and it will mention some relevant international intellectual property treaties.

10.2
Patent Law

10.2.1
What is a Patent?

A patent is a type of intellectual property. All properties can be understood as a collection of rights to control a particular thing. Tangible properties give the property holder rights to control tangible things, such as cars or land. Intellectual properties, on the other hand, give the property holder rights to control intangible things, such as inventions, poems, or computer programs. Tangible things have a particular location in space and time, whereas intangible things do not. The main types of intellectual property are patents, copyrights, trademarks, and trade secrets [1].

A patent is a private right granted by the government to someone who invents a new and useful product or process. The initial patent holder, the inventor, has the right to exclude others from making, using, or commercializing his invention. The patent holder may transfer all or part of his rights to another party, including another individual or a corporation. Researchers who work for biotechnology companies usually assign their patent rights to the company in exchange for a salary, a fee, or a share of royalties. Assignment of patent rights transfers all the rights to the assignee, who becomes the new patent holder. Patent holders may also grant licenses to other parties in exchange for royalties or a fee.

For example, a biotechnology company with a patent on a gene therapy technique could grant individuals or companies licenses to use the technique [1].

In the United States, a patent holder has the right to refrain from making, using, or licensing his invention, if he or she so desires. In the United States, a patent confers rights to make, use, or commercialize a thing but implies no corresponding obligations. As a result, some companies in the United States use patents to block technological development to gain an advantage over their competitors. Some European countries, however, have compulsory licensing, which requires the patent holder to make, use, or commercialize his or her invention or license others to do so [1].

The term of patent in the United States and most countries that belong to the European Union (EU) lasts for 20 years from the time the inventor submits his application. A patent is not renewable. Once the patent expires, the invention becomes part of the public domain, and anyone can make, use, or commercialize the invention without the permission from the inventor [1]. In the pharmaceutical industry, the average interval between the discovery of a new drug and its final approval by the Food and Drug Administration (F.D.A.) for human consumption is 10 years, which includes the time required to conduct clinical research, product development, as well as an F.D.A. review. Thus, most pharmaceutical companies can expect that they will have about 10 years to recoup the money they have invested in a new drug before the patent expires. Once the patent expires, the name of the drug may still have trademark protection, but other companies can manufacture and market a generic version of the drug without obtaining permission from the company [2].

The main policy rationale for patent laws is that they promote the progress of science, technology, and industry by providing financial incentives for inventors, entrepreneurs, and investors [1]. By granting property rights over inventions, the patent system gives inventors, private companies, and other organizations the opportunity to profit from their investments of time and money in research and development. Since most new scientific discoveries and technological innovations benefit the society, the public benefits from granting private rights over intellectual property. However, excessive private control over intellectual property can impede access to science and technology. Thus, patent laws attempt to strike an appropriate balance between the public and private control of inventions. A good example of this balancing is the length of a patent: if the term of a patent is too short, companies and researchers will not have enough time to obtain a fair return on their investment; if the term is too long, the public will not have adequate access to the technology.

10.2.2 How Does One Obtain a Patent?

To obtain a patent, one must submit a patent application to the patent office. In the United States, the Patent and Trademark Office (PTO) examines the patent applications. The application must provide a description of the invention that would allow someone trained in the relevant practical art to make and use the invention. One or more individuals may be listed as inventors on the patent application. The application need not include a sample or model of the invention; a written description will suffice. The application will contain information about the invention, background references, data, as well as one or more claims pertaining to the invention. The claims stated on the patent application will determine the scope of the inventor's patent rights.

If the PTO rejects a patent application, the inventor may submit a revised application. The process of submission/revision/resubmission, otherwise known as "prosecuting" a patent, may continue for months or even years. If the PTO rejects the patent, the applicant may appeal the decision to a federal court [1].

The PTO will award a patent to an inventor only if he or she provides evidence that his or her invention satisfies all of the following conditions (EU countries have similar requirements [1, 3]):

1. *Originality*: The invention is new and original; it has not been previously disclosed in the prior art. The rationale for this condition is that the public does not benefit when the patent office grants a patent on something that has already been invented. Thus, if someone else has already submitted an application for the same invention, this would qualify as a prior disclosure. Also, disclosure could occur if a significant part of the invention has been published or used [1].
2. *Nonobviousness*: The invention is not obvious to someone who is trained in the relevant practical art.
3. *Usefulness*: The invention has some definite, practical utility. The utility of the invention should not be merely hypothetical, abstract, or contrived. The rationale for this condition is self-explanatory: the public does not benefit from useless patents. Recently, the U.S. PTO raised the bar for proving the utility of patents on DNA in response to concerns that it was granting patents on DNA sequences when the inventors did not even know the biological functions of those sequences [4].

In addition to satisfying these three conditions, to obtain a patent in the United States, the inventor must exhibit due diligence in submitting an application and developing the invention. In the United States, the person who is the first to conceive an invention will be awarded the patent provided he exhibits due diligence. If the first inventor does not exhibit due diligence, the PTO may award the patent to a second inventor, if that inventor reduces the invention to practice and submits an application before the first inventor [1].

Once the PTO awards the patent, the application becomes part of the public domain, and other inventors and researchers may use the knowledge contained in the application. Indeed, the "patent bargain" is an agreement between the government and a private party in which the party agrees to disclose the knowledge related to his invention to the public in exchange for a limited monopoly on the invention [1]. The public benefits from this bargain because it encourages the inventor to avoid protecting his knowledge through trade secrecy. A great deal of the world's scientific and technical information is disclosed in patent applications [5]. For example, the PTO has a large online, searchable database of patent applications [6].

10.2.3 What is the Proper Subject Matter for a Patent?

Under the US law, the PTO can award patents on articles of manufacture, compositions of matter, machines, or techniques or improvements thereof [7]. The EU countries allow patents on similar types of things [3]. Although different patent laws use different terms to describe the subject matter of patents, there are three basic types of subjects for patents: (1) products (or materials), (2) processes (or methods), and (3) improvements. For example, one could patent a mousetrap (a product), a method for making a mousetrap (a process), or a more efficient and humane mousetrap (an improvement) [1].

One of the most important doctrines in patent law is that patents only apply to products or processes that result from human ingenuity (or inventiveness). Thus, the US courts have held that one may not patent laws of nature or natural phenomena, since these would be patents on a product of nature. Over two decades ago, a landmark US Supreme Court case, *Diamond v. Chakrabarty*, set the legal precedent in the United States for patents on life forms [8]. Chakrabarty had used recombinant DNA techniques to create a type of bacteria that metabolizes crude oil. The PTO had rejected his patent application on the grounds that the bacteria did not result from human ingenuity, but the Supreme Court vacated this ruling and held that Chakrabarty could patent his genetically engineered life forms [8, 9]. This decision helped to establish the legal precedent for other patents on life forms, such as patents on laboratory mice, cell lines, and bioengineered tissues and organs [10]. The EU countries have followed the United States in allowing patents on life forms that result from human ingenuity [11].

Before Chakrabarty received his patent, the PTO had also granted inventors patents on DNA, proteins, and recombinant DNA techniques [12–14]. In granting patents on organic compounds that occur in living organisms, such as animals or plants, patent agencies have distinguished between naturally occurring compounds and isolated and purified compounds [15]. For example, DNA in its natural state occurs in virtually all organisms and is unpatentable in its natural state. However, scientists can use various chemical and biological techniques to create isolated and purified samples of DNA, which are patentable. The reason why patent agencies allow patents on isolated and purified compounds is that they result from human ingenuity [16].

Another important doctrine in patent law is that patents apply to applications, not to ideas. Ideas are part of the public domain. For example, courts in the United States have ruled that mathematical algorithms are unpatentable ideas but that the computer programs that use algorithms to perform practical functions are patentable [17].

10.2.4
Types of Patents in Pharmaceutical Biotechnology

There are many different types of patents that may be available to researchers and companies in the field of pharmaceutical biotechnology. Following the distinction in Sect. 10.2.3 between products and processes, potential patents might include the following:

1. Patents on pharmaceutical and biomedical products, such as bioengineered drugs, proteins, receptors, neurotransmitters, oligonucleotides, hormones, genes, DNA, DNA microchips, RNA, cell lines, bioengineered tissues and organs, and genetically modified bacteria, viruses, animals, and plants.

2. Patents on pharmaceutical and biotechnological processes, such as methods for genetic testing, gene therapy procedures, DNA cloning techniques, methods for culturing cells and tissues, DNA and RNA sequencing methods, and xenotransplantation procedures.
3. Patents on improvements of pharmaceutical, biomedical, and biotechnological products and processes.

For any of these products or processes to be patentable, they would need to result from human ingenuity.

10.2.5
Patent Infringement

Patent infringement occurs when someone uses, makes, or commercializes an invention without the permission of the patent holder. In the United States, the patent holder has the responsibility of bringing an infringement claim against a potential infringer and proving that infringement occurred [1]. A court may issue an injunction to stop the infringement or award the patent holder damages for loss of income due to infringement. There are three types of infringement: direct infringement, indirect infringement, and contributory infringement. Patent holders may also settle infringement claims out of court. Researchers, corporations, and universities usually try to avoid any involvement in an infringement lawsuit, since patent infringement litigation is expensive and time consuming [16].

Many EU countries have a defense to patent infringement known as the research exemption [3]. The United States also has a research exemption (also known as the experimental use exemption), which has been used very infrequently [18]. Under this exemption, someone who uses or makes a patented invention for pure research with no commercial intent can assert this defense in an infringement lawsuit to avoid an adverse legal decision. The research exemption is similar to the "fair use" exemption in copyright law insofar as it permits some unconsented uses of intellectual property [18]. There are some problems with the exemption, however. First, the research exemption is not well publicized. Second, the research exemption is not well defined [18]. Indeed, in the United States the research exemption has no statutory basis but is a creation of case law. Some commentators have argued that countries should clarify and strengthen the research exemption in order to promote research and innovation in biotechnology and avoid excessive private control of inventions [3].

10.2.6
International Patent Law

Every country has the authority to make and enforce its own patent laws and to award its own patents. Thus, a patent holder must apply for a patent in every country where he wants patent protection. For example, a corporation that patents a new drug in the United States must also apply for a patent in Germany, if it desires patent protection in Germany. Furthermore, complex matters relating to jurisdiction can arise when someone infringes a patent that is protected in one country but not in another. For example, if someone infringes a US patent in Germany, but the invention is not protected by the German patent laws, then the patent holder will need to bring a lawsuit in a court in the United States, which may or may not have jurisdiction.

To deal with international disputes about intellectual property and to harmonize intellectual property laws, many countries have signed intellectual property treaties. Most of these treaties define minimum standards for intellectual property protection and obligate signatories to cooperate in the international enforcement of property rights. The most important treaty related to patents is the Trade Related Aspects of Intellectual Properties agreement (TRIPS), which has been developed and negotiated by the World Trade Organization (WTO). The TRIPS agreement defines minimum standards for patent rights. For example, it requires that patents last for 20 years. Countries that have signed the agreement agree to adopt patent laws that provide at least the minimum level of protection under the agreement. Countries must also agree to cooperate in the enforcement of patent rights. TRIPS allows countries to override patents rights to deal with national emergencies, such as public health crisis [1].

10.3 Ethical and Policy Issues in Biotechnology Patents

Having provided the reader with some background information on patenting in biotechnology, this chapter will briefly review some important ethical and policy issues.

10.3.1 No Patents on Nature

In the 1990s, a variety of writers, political activists, theologians, ethicists, and professional organizations opposed patents on biotechnological products and processes for a variety of reasons. Many of these critics argued that patents on living bodies, as well as patents on body parts, are unethical because they are patents on natural things [19]. They argued that it is immoral and ought to be illegal to patent organisms, tissues, DNA, proteins, and other biological materials. Some of these critics based their opposition to biotechnology patents on religious convictions [20], while others based their opposition on a general distrust of biotechnology and the biotechnology industry [21, 22]. Some of the more thoughtful critics of biotechnology patents accepted some types of patents on biological materials, but objected to patents on other types of biological materials, such as patents on genes or cell lines, on the grounds that these types of patents attempt to patent nature [23, 24].

As noted in Sect. 10.2.3, patents on products of nature are illegal; a product or process must have resulted from human ingenuity to be patentable. But how much human ingenuity should be required to transform something from an unpatentable product of nature to a patentable, human invention? Defining the boundaries between the products of nature and human inventions is a fundamental issue in patent law and policy that parallels the tenuous distinction between the natural and artificial [25]. While most people can agree on paradigmatic cases of things that are natural, such as gold, and things that are artificial, such as gold jewelry, it difficult to reach an agreement on borderline cases, such as DNA sequences. On the one hand, DNA sequences exist in nature and can therefore be regarded as natural. On the other hand, isolated and purified DNA sequences do not exist in nature and are produced only under laboratory conditions. They are, in some sense, human artifacts. However, the nucleotide sequences in isolated and purified DNA are virtually identical to the sequences in naturally occurring DNA.

There is probably no objective (i.e. scientific) basis for distinguishing between naturally occurring DNA and isolated and purified DNA. Likewise, there is probably no objective basis for distinctions between natural cell lines versus artificial cell lines, natural proteins versus artificial proteins, and natural organisms versus artificial organisms.

If the distinction between a product of nature and a human invention is not objective, then it depends, in large part, on human values and interests. It is like other controversial distinctions in biomedical law and ethics, such as human versus nonhuman and alive versus dead. The best way to deal with these controversial distinctions is to carefully consider, negotiate, and balance competing values and interests in light of the particular facts and circumstances. Laws and policies that define patentable subject matter should also attempt to promote an optimal balance between competing interests and values and should carefully consider the facts and circumstances relating to each item of technology [25]. Policies adopted by the United States and the European Union with respect to the patenting of DNA appear to strike an optimal balance between competing interests and values because these policies disallow the patenting of DNA in its natural state but allow the patenting of isolated and purified DNA [11, 15].

10.3.2
Threats to Human Dignity

Critics of biotechnology patents have also claimed that patents on human body parts, such as genes, cell lines, and DNA, are unethical because they treat people as marketable commodities [19, 21, 22, 26]. Some have even compared patents on human genes to slavery [27]. The issues concerning the commercialization of human body parts are complex and emotionally charged. They also have implications for many different social policies, including organ transplantation, surrogate parenting, and prenatal genetic testing. This chapter will give only a brief overview of this debate.

According to several different ethical theories, including Kantianism and the Judeo–Christian tradition, human beings have intrinsic moral value (or dignity) and should not be treated as if they have only extrinsic value. An entity (or thing) has intrinsic value if it is valuable for its own sake and not merely for the sake of some other thing. A commodity is a thing that has a value – a market value or price – which serves as a basis for exchanging it for some other thing. For example, one can exchange a barrel of oil for $30 or exchange a visit to the dentist for $50. Treating an entity as a commodity is treating it as if it has only extrinsic value and not intrinsic value. Thus, it would be unethical to treat a human being as a commodity because this would be treating that person as if they have only extrinsic value and no intrinsic value. Slavery is therefore unethical because it involves the buying and selling of whole human beings. People are not property [28, 29].

Even though treating a whole human being as a commodity violates human dignity, one might argue that treating a human body part as a commodity does not violate human dignity. Human beings have billions of different body parts, ranging from DNA, RNA, proteins, and lipids to membranes, organelles, cells, tissues, and organs. Properties that we ascribe to the parts of a thing do not necessarily transfer to the whole thing; inferences from parts to wholes are logically invalid. For example, the fact that a part of an automobile, such as the front tire, is made of rubber does not imply that the whole car is made of rubber [28]. Likewise, treatment of a part of human being, such as blood or hair, as a commodity does not imply treatment of the whole human being as a commodity. It is possible to commodify (or commercialize) a human body part without commodifying the whole human being.

This argument proves that buying and selling hair, blood, or even a kidney is not equivalent to slavery. Even so, one might argue that treating human body parts as commodities constitutes incomplete commodification of human beings; partial commodification of human beings can threaten human dignity even if it does not violate human dignity [26]. Incomplete commodification can threaten human dignity because it can lead to exploitation, harm, and injustice, as well as complete commodification of human beings. For example, in the now famous case of *Moore v. Regents of University of California*, the desire to patent a valuable cell line played an important role in the exploitation of a cancer patient [24]. The researchers took cells from Moore's body that overexpress cytokines. The researchers did not tell Moore what they planned to do with the tissue samples they took from him or that the samples could be worth millions of dollars [30]. One might argue that treating human body parts as commodities inevitably leads to the abuse of human rights and dignity as in the Moore case. Although incomplete commodification of human beings is not intrinsically immoral, it can lead the society down a slippery slope toward various types of immorality and injustices. In order to stop the slide down this slippery slope, society should forbid activities that constitute incomplete commodification of human beings, such as the patenting of cell lines and DNA, a market in human organs, surrogate pregnancy contracts, cloning for reproduction, and selling human gametes [31].

One could reply to this argument by acknowledging that the slippery slope poses a genuine threat to human dignity but maintain that it may be possible to prevent exploitation, injustice, and other abuses by developing clear and comprehensive regulations on practices that commodify human body parts. Regulations should require informed consent to tissue donation, gamete donation, and organ donation, as well as fair compensation for subjects that contribute biological materials to research and product development activities. Regulations should also protect the welfare and privacy of human research subjects and patients [28, 32]. These regulations should also state that some human biological materials, such as embryos, should not be treated as commodities because they pose an especially worrisome threat to human dignity. Although an embryo is not a human being, it should be illegal to buy, sell, or patent a human embryo. However, it should be legal to buy or patent embryonic stem cells, provided that the society has appropriate regulations [33]. Although selling organs is illegal in many countries, including the United States and many European nations, some have argued that organs could be bought and sold, provided that appropriate regulations are in place [34, 35].

10.3.3 Access to Technology

One of the most important ethical and policy concerns raised by the critics of biotechnology patenting is that patenting will have an adverse impact on access to materials and methods that are vital to research and innovation in biotechnology as well as medical tests and treatments. The negative effects of patenting on science, industry, and medicine will constitute a great social cost rather than a social benefit. In Sect. 10.2.1, we noted that the primary rationale for the patent system is that it benefits the society by encouraging progress in science, technology, and industry. However, this argument loses its force when patenting has the opposite effect. If patenting does more harm than good, then we should forbid or greatly restrict patenting [24, 36]. The issue of access to materials and methods in biotechnology, like the issues discussed in Sects. 10.3.1 and 10.3.2, is very complex and controversial. This chapter will not attempt to explore these issues in great depth, but it will attempt to provide the reader with an outline of the arguments on both sides.

Concerns about access to materials and methods stem from potential problems with the licensing of patents on products and processes that are useful in research and innovation of biomedicine and biotechnology [37]. First, if a researcher or a company wants to develop a new product or process in biotechnology and biomedicine, then he or she may need to negotiate and obtain dozens of different licenses from various patent holders in order to avoid patent infringement. The researcher or company might need to fight through a "patent thicket" in order to develop a new and useful invention. For example, DNA chip devices test for thousands of different genes in one assay. If dozens of companies hold patents on these different genes, then one may need to obtain dozens of different licenses to develop this new product. Although larger biotechnology and pharmaceutical companies are prepared to absorb the legal and administrative transaction costs associated with licensing, smaller companies and universities may find it difficult to navigate the "patent thicket" [38].

Second, "blocking patents" in biotechnology could prevent the development of downstream products and processes [39]. In industries with many different interdependent products and processes, someone who holds a particular invention may be able to affect or control the development of subsequent inventions that depend on that prior invention. These prior inventions are also known as "upstream" inventions, and the subsequent inventions are also known as "downstream" inventions. Some companies may obtain patents for the sole purpose of preventing competitors from developing useful inventions in biotechnology. In the United States, these companies would have no obligation to use, make, market, or license such inventions. They could use their inventions to block the development of downstream products and processes. In countries that have compulsory licensing, companies would have a legal duty to make, use, commercialize, or license their inventions, but they could still use other means to prevent the development of downstream technologies, such as setting very high licensing fees.

Third, high licensing fees could impose a heavy toll on research and innovation in biotechnology and biomedicine [37]. Companies with patents on upstream inventions might issue licenses on the condition that they receive a percentage of profits from downstream inventions. While downstream patent holders have no legal obligation to share their profits with upstream patent holders, upstream patent holders may try to acquire a portion of downstream profits by issuing these "reach through" licenses. Even companies that do not issue "reach through" licenses may still set high licensing fees. For example, many commentators have claimed that Myriad Genetics' high licensing fees for its tests for BRCA1 and BRCA2 mutations, which increase the risk of breast and ovarian cancer, have had a negative impact on research and innovation, and diagnostic and predictive testing [40].

These aforementioned problems related to licensing – the patent thicket, blocking patents, and high licensing fees – could undermine not only research and innovation but could also have an adverse impact on health care by undermining the access to new medical products and services, such as genetic tests. For example, if a company is unable to develop a genetic test, due to licensing problems, then the patients will not benefit from that test. If a company develops a genetic test but charges a high fee to conduct the test or charges a high fee to license the test, then many patients may not be able to afford the test. In either case, problems related to the licensing of biotechnology products and processes could prevent the public from benefiting from new developments in biomedicine.

On the other hand, many commentators and industry leaders have rebutted these criticisms of biotechnology patenting by arguing that the free market, patent offices, and the legal system will keep potential licensing problems in check [41–43]. Companies will not have any major difficulties in negotiating and obtaining licenses because they will all understand the importance of cooperation in the biotechnology industry. Few companies will develop blocking patents because these patents will usually prove to be unprofitable: one can make much more money from marketing or licensing a new invention than from keeping it on the shelf. Finally, high licensing costs will decline in response to lower consumer demands, especially if competitors are able to enter the market by developing new inventions that work around existing ones. (A "work around" invention is an improvement on a patented invention or an alternative to a patented invention.) Industry leaders also point out that the potential licensing problems faced by the biotechnology industry are not new because many other industries have faced – and solved – similar problems [41]. For example, many different companies in the semiconductor industry have worked together to develop licensing agreements [44]. There are many interdependent products and processes in the semiconductor industry and many different patent holders, but companies have managed to avoid licensing problems and the industry has thrived. Indeed, the semiconductor industry is one of the most successful and innovative industries the world has ever known.

Some commentators have argued that societies should reform the patent system to prevent licensing problems from occurring and to ensure that new biomedical technologies are affordable and accessible. These proposed reforms, some of which have been mentioned above, include the following:

1. Banning patents on particular kinds of products or processes, such as patents on genes that are associated with diseases or patents on genetic tests [23].
2. Expanding and clarifying the research exemption in biotechnology [3, 16].
3. Raising the bar for the various conditions for awarding patents, such as novelty and utility [3, 16].
4. Restricting the scope of biotechnology patents in order to allow for "work around" inventions and to promote competition [3, 16].

5. Applying antitrust laws to the biotechnology industry to promote fair competition [16].
6. Conducting an ethical review of patent applications to address ethical and policy issues before awarding patents [3, 45].
7. Developing a patent pool in the biotechnology industry to promote efficient licensing [46].

Most of these proposed reforms, with the exception of banning some types of biotechnology patents, would probably promote research and innovation in biotechnology and biomedicine without undermining the financial incentives for researchers and companies. Many of these reforms could be enacted without any additional legislation, since patent offices and the courts already have a great deal of authority to shape patent law and policy through their interpretation and application of existing statutes [47].

10.3.4
Benefit Sharing

The final issue this chapter will consider involves the sharing of the benefits of research and innovation in biotechnology. Some critics of biotechnology patents have claimed that the distribution of the benefits of research and innovation is often unfair [22, 24, 48, 49]. According to these critics, pharmaceutical and biotechnology companies benefit greatly from research and innovation by earning large profits, but individual patients or research subjects, populations, or communities benefit very little. For example, to study a genetic disease, researchers need to take tissue samples from patients/subjects. Very often, researchers do not offer to pay subjects any money for their tissue samples or promise them any royalties from the commercialization of their research or its applications. If a company develops a profitable genetic test from free genetic samples, patients/subjects could argue that the company is not sharing the benefits fairly. Unequal distributions of benefits could also occur between companies and entire communities or countries. For example, some pharmaceutical and biotechnology companies are now developing drugs on the basis of the knowledge obtained from indigenous populations concerning their medicinal plants. If a company develops a profitable medication from this indigenous knowledge and does not offer the population any compensation, the population could argue that the company has not shared the benefits of research fairly. Unequal distributions of benefits could also take place between developed nations and developing nations. For example, if researchers, patients, and companies from the developed world benefit a great deal from biotechnology, but people in the developing world do not, one might argue that the benefits of biotechnology have been distributed unfairly.

Several writers and organizations have called for the fair distribution of the benefits of research in biotechnology [50–53]. Some writers appeal directly to theories of justices, such as utilitarianism, egalitarianism, or social contract theory, to argue for a fair distribution of research benefits [47, 54]. Others appeal to the concept of a common heritage relating to human biological materials, such as DNA [52, 53]. Regardless of how one justifies a general principle of benefit sharing in biotechnology, the most important practical problems involve determining how the benefits should be shared. What would be a fair sharing of benefits between researchers and companies and subjects/populations/communities? Should researchers and companies offer to give subjects/populations/communities financial compensation for providing research materials and methods, such as tissue samples of indigenous knowledge? Should researchers and companies offer to pay royalties for the commercialization of research to subjects/populations/communities? Although financial compensation might be useful and appropriate in some situations, such as giving communities royalties for indigenous knowledge or providing some subjects with compensation for their valuable tissues (as in the Moore case, discussed in Sect. 10.3.2), in other situations, direct financial compensation may not be very useful or appropriate. For example, if a company collects thousands of tissue samples from subjects and uses the knowledge gained from those samples to develop a commercial product, the financial benefit offered to any particular subject might be miniscule, since the benefits would need to be divided among thousands of subjects. Moreover, it may be impossible to estimate the potential benefits to subjects prior to the development of the product, since most new products are not profitable. Furthermore, subjects in some cultures might not be interested in financial rewards for participation. Perhaps the best way to share the benefits in situations like these would be to offer to provide the population or community with nonfinancial benefits, such as improvements in health care, education, or infrastructure. In any case, these are complex questions that cannot be addressed in depth in this chapter. To answer questions about the fair distribution of research benefits in any particular case, one needs to apply the theories and concepts of distributive justice.

Even though there is little consensus about how to distribute the benefits of research and innovation in biotechnology, almost everyone with an interest in the issue agrees that subjects should be informed about the plans for benefit sharing (if there are any) [55]. For example, the researchers in the Moore case should have told Moore that they planned to develop a cell line from his tissue and that they were not planning to offer him any financial compensation. If researchers conduct a study that involves an entire population or community, they should discuss the benefit-sharing plans with the representatives of the community or population [56]. Indeed, respect for human dignity requires nothing less than fully informing the subjects of the material facts related to their research participation, including facts pertaining to the commercialization of research [57, 58].

10.4 Conclusion

This essay has provided the reader with an overview of the legal, ethical, and policy issues relating to the patenting of products and processes used in pharmaceutical biotechnology. Although this essay has attempted to provide the reader with up-to-date information, it is possible that some of this information may soon be out-of-date, due to the changes in technology, case law, legislation, and international treaties. Since most of these issues are very complex and constantly changing, those who are interested in learning more about this topic should review the relevant documents, guidelines, and policies relating to their particular areas of research and development.

References

1. A useful overview of intellectual property law. A. Miller, M. Davis, *Intellectual Property*, West Publishing, St. Paul, MN, 2000.
2. A useful overview of legal and ethical issues involving DNA patents. A. Goldhammer, *Accountability Res.* **2001**, *8*, 283–292.
3. The Nuffield Council on Bioethics, *The Ethics of Patenting DNA*, Nuffield Council, London, 2002.
4. U.S. Patent and Trademark Office, *Fed. Register* **1999**, *64*(244), 71440–71442.
5. Derwent Information, Frequently asked questions, http://www.derwent.com/, Accessed: December 31, 2002.
6. U.S. Patent and Trademark Office, www.uspto.gov, Accessed: December 31, 2002.
7. U.S. Patent Act. 35 U.S.C. 101, 1995.
8. A landmark court case relating to the patenting of biological materials. Diamond v. Chakrabarty, 447 U.S. 303, 1980.
9. A. Chakrabarty, US 4259444, 1980.
10. R. Eisenberg, *Nat. Genet.* **1997**, *15*(2), 125–130.
11. A clear statement of the E.U.'s opinions on some ethical issues in biotechnology. European Commission, *Opinions of the Group of Advisors on the Ethical Implications of Biotechnology of the European Commission*, European Commission, Brussels, 1998.
12. A useful overview of legal issues relating to the patenting of DNA. R. Eisenberg, *Emory Law J.* **1990**, *39*, 721–745.
13. D. Kevles, A. Berkowitz, *Brooklyn Law Rev.* **2001**, *67*, 233–248.
14. S. Cohen, H. Boyer, US 4237224, 1980.
15. Outlines the U.S. PTO's DNA patent policy. J. Doll, *Science* **1998**, *280*, 689–690.
16. A useful overview of legal and ethical issues involving DNA patents. D. Resnik, *Sci. Eng. Ethics* **2001**, *7*(1), 29–62.
17. Parker v. Flook, 437 U.S. 584, 1978.
18. J. Karp, *Yale Law J.* **1991**, *100*, 2169–2188.
19. A useful summary of religious objections to patents on biological materials. M. Hanson, *Hastings Cent. Rep.* **1997**, *27*(6) (Special Supplement), 1–21.
20. R. Land, B. Mitchell, *First Things* **1996**, *63*(2), 20–22.
21. A. Kimbrell, *The Human Body Shop*, Gateway, Washington, 1997.
22. J. . Rifkin, *The Biotech Century*, Penguin Putnam, New York, 1998.
23. J. Merz, M. Cho, *Camb. Q. Healthc. Ethics* **1998**, *7*, 425–428.
24. L. Andrews, D. Nelkin, *Body Bazaar*, Crown Publishers, New York, 2000.
25. D. Resnik, Discoveries, inventions, and gene patents in *Who Owns Life?* (Eds.: D. Magnus, A. Caplan, G. McGee), Prometheus Press, New York, 2002, pp. 171–181.
26. M. Radin, *Contested Commodities*, Harvard University Press, Cambridge, MA, 1996.
27. Joint Appeal Against Human and Animal Patenting, Press Conference Text, May 17, 1995. Board of Church and Society of the United Methodist Church, Washington, DC, 1995.
28. A useful discussion of some ethical issues in patenting human DNA. D. Resnik, *J. Law, Med. Ethics* **2001**, *29*(2), 152–162.
29. R. Green, *Kennedy Inst. Ethics J.* **2001**, *11*(3), 247–261.
30. A key court case involving informed consent and the commercialization of research. Moore vs. Regents of the University of California, 793 P.2d 479, Cal. 1990.
31. M. Hanson, *J. Med. Philos.* **1999**, *24*(3), 267–287.
32. D. Resnik, *Bioethics* **2001**, *15*(1), 1–25.
33. D. Resnik, *Health Care Anal.* **2002**, *10*, 127–54.
34. S. Wilkinson, *Health Care Anal.* **2000**, *8*, 352–61.
35. American Medical Association, Council on Ethical and Judicial Affairs, *Arch. Intern. Med.* **1995**, *155*, 581–89.
36. A. Caplan, J. Merz, *Br. Med. J.* **1996**, *312*, 926.
37. An influential essay outlining some potential problems with licensing in biotechnology. M. Heller, R. Eisenberg, *Science* **1998**, *280*, 698–701.
38. C. Shapiro, Navigating the patent thicket: cross-licenses, patent pools, and standard setting in *Innovation Policy and the Economy* (Eds.: A. Jaffe, J. Lerner, S. Stern), MIT Press, Cambridge, MA, 2000, pp. 119–150.
39. L. Guenin, *Theor. Med. Bioethics* **1996**, *17*, 279–314.
40. J. Merz, M. Cho, M. Robertson, D. Leonard, *Mol. Diagn.* **1997**, *2*(4), 299–304.
41. R. Scott, Testimony before the House Judiciary Subcommittee on Courts and Intellectual Property (July 13, 2000).

42. J. Tribble, *Camb. Q. Healthc. Ethics* **1998**, *7*, 429–432.
43. G. Woolett, O. Hammond, An industry perspective on the gene patenting debate in *Perspectives on Gene Patenting* (Ed.: A. Chapman), American Association for the Advancement of Science, Washington, 1999, pp. 43–50.
44. H. Hall, R. Ziedonis, *Rand J. Econ.* **2001**, *32*, 101–128.
45. T. Caulfield, R. Gold, *Clin. Genet.* **2000**, *57*, 370–375.
46. Discuss patents pools in biotechnology. U.S. PTO, *Patent Pools: A Solution to the Problem of Access in Biotechnology Patents?* U.S. PTO, Washington, 2000.
47. Thorough discussion of legal, ethical, and policy issues related to DNA patenting. D. Resnik, *Owning the Genome: A Moral Analysis of DNA Patenting*, S.U.N.Y. Press, Albany, New York, 2004.
48. M. Knoppers, M. Hirtle, K. Glass, *Science* **1999**, *286*, 2277–2278.
49. V. Shiva, *Biopiracy: The Plunder of Nature and Knowledge*, South End Press, Boston, 1996.
50. R. Crespi, *Sci. Eng. Ethics* **2000**, *6*(2), 157–180.
51. Human Genome Organization (HUGO), *Statement on Benefit Sharing*, HUGO, Bethesda, MD, 2000.
52. M. Sturges, *Am. Univ. Int. Rev.* **1997**, *13*, 219–261.
53. A useful discussion of the common heritage argument. P. Ossario, Common heritage arguments and the patenting of DNA in *Perspectives on Gene Patenting* (Ed.: A. Chapman), American Association for the Advancement of Science, Washington, DC, 1999, pp. 89–110.
54. D. Resnik, *Health Policy* **2003**, *65*, 181–197.
55. E. Clayton, K. Steinberg, M. Khoury, E. Thomson, L. Andrews, M. Kahn, L. Kopelman, J. Weiss, *J. Am. Med. Assoc.* **1995**, *274*, 1786–1792.
56. R. Sharp, M. Foster, *J. Law, Med. Ethics* **2000**, *28*(1), 41–49.
57. World Medical Association, *J. Am. Med. Assoc.* **2000**, *284*, 3043–3046.
58. A useful discussion of ethical and policy issues relating research involving human biological materials. National Bioethics Advisory Commission (NBAC), *Research Involving Human Biological Materials: Ethical Issues and Policy Guidance*, NBAC, Washington, DC, 1998.

11
Drug Approval in the European Union and the United States

Gary Walsh
University of Limerick, Limerick City, Ireland

11.1
Introduction

The pharmaceutical sector is arguably the most highly regulated industry in existence. Legislators in virtually all the regions of the world continue to enact/update legislation, controlling every aspect of pharmaceutical activity. Interpretation, implementation, and enforcement of these laws is generally delegated by the lawmakers to dedicated agencies. The relevant agencies within the European Union (EU) and the United States (USA) are the European Medicines Evaluation Agency (EMEA) and the US Food and Drug Administration (FDA), respectively. This monograph focuses upon the structure, remit, and operation of both these organizations, specifically in the context of biopharmaceutical products.

11.2
Regulation within the European Union

11.2.1
The EU Regulatory Framework

The founding principles of what we now call the European Union are enshrined in the treaty of Rome, initially adopted by six countries in 1957. While this treaty committed its signatories to a range of cooperation and harmonization measures, it largely deferred health care–related issues to individual member states. As a consequence, each member state drafted and adopted its own set of pharmaceutical laws, enforced by its own national regulatory authority. Although the main principles underpinning elements of national legislation were substantially similar throughout all European countries, details did differ from country to country. As a result, pharmaceutical companies seeking product-marketing authorizations were forced to apply separately to each member state. Uniformity of regulatory response was not guaranteed and each country enforced its own language requirements, scale of fees, processing times, and so on. This approach created enormous duplication of effort, for companies and regulators alike.

In response, the European Commission (EC, Brussels) began a determined effort to introduce European-wide pharmaceutical legislation in the mid-1980s. The commission represents the EU body with responsibility for drafting (and subsequently ensuring the implementation) of EU law, including pharmaceutical law. In pursuing this objective, it has at its disposal two legal instruments, "regulations" and "directives". Upon approval, a regulation must be enforced immediately and without alteration by all EU member states. A directive, in contrast, is a "softer" legal instrument, requiring member states only to introduce its "essence" or "spirit" into national law.

By the early 1990s, some 8 regulations and 18 directives had been introduced, which effectively harmonized pharmaceutical law throughout the European Union. In addition to making available the legislative text, the European Commission has also facilitated the preparation and publication of several adjunct documents designed to assist industry and other interested parties to interpret and conform to the legislative requirements. Collectively these documents are known as "the rules governing medicinal products in the European union" and they make compulsory reading for those involved in any aspect of pharmaceutical regulation. The nine-volume (Table 1) publication is regularly updated and hard copies may be purchased from the commission's publication office [1] or may be consulted/downloaded (for free) from the relevant EU website [2].

11.2.2
The EMEA

Harmonization of pharmaceutical law made possible the implementation of an EU wide system for the authorization and subsequent supervision of medicinal products. The EMEA was set up to coordinate and manage the new system [3]. Based in Canary Wharf, London, the agency became operational in 1995. The EMEA mission statement is "to contribute to the protection and promotion of public and animal health". It seeks to achieve this by

- providing high-quality evaluation of medicinal products;
- advising on relevant R & D programs;
- providing a source of drug and other relevant information to health care professionals/users;
- controlling the safety of medicines for humans and animals.

An outline structure of the EMEA is provided in Fig. 1. From a technical standpoint, the most significant organizational

Tab. 1 The volumes comprising "the rules governing medicinal products within the European union"

Volume	*Title*
1	Pharmaceutical legislation: Medicinal products for human use
2	Notice to applicants: Medicinal products for human use
3	Guidelines. Medicinal products for human use
4	Good manufacturing practices. Medicinal products for human and veterinary use
5	Pharmaceutical legislation: Veterinary medicinal products
6	Notice to applicants. Veterinary medicinal products
7	Guidelines. Veterinary medicinal products
8	Maximum residue limits. Veterinary medicinal products
9	Pharmacovigilance. Medicinal products for human and veterinary use

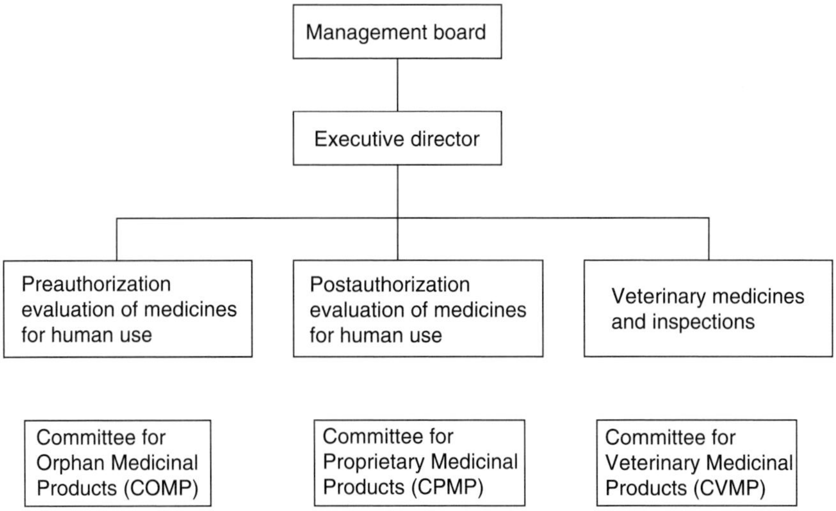

Fig. 1 Simplified structural overview of the EMEA. Refer to [3] for further details.

structures are the following:

- The unit for preauthorization evaluation of medicines for human use.
- The unit for post authorization evaluation of medicines for human use.
- The unit for veterinary medicines and inspections.

A more detailed description of these units and their responsibilities is available on the EMEA home page [3]. Two additional structural units also exist: administration, and communications and networking.

From a drug approval perspective, at the heart of the functioning of the EMEA are three key scientific committees:

- The Committee for Proprietary Medicinal Products (CPMP)
- The Committee for Veterinary Medicinal Products (CVMP)
- The Committee for Orphan Medicinal Products (COMP).

Each committee is composed of a number of (mainly technical) experts, the majority of whom are drawn from the national drug regulatory authorities of each EU member state (Table 2). The function of these committees in the context of new biotechnology drug approvals will be discussed in the next section. In addition to these three committees, the EMEA has at its disposal a bank of some 3000 European technical experts (the majority of whom, again, are drawn from the national regulatory authorities). The EMEA draws upon this expert advice as required.

11.2.3 New Drug Approval Routes

The rules governing medicinal products in the European Union provide for two independent routes by which new potential medicines may be evaluated. These are termed the "centralized" and "decentralized" procedures, respectively, and the EMEA plays a role in both [4]. The centralized procedure is compulsory for biotech medicines and as such is described in greatest detail below. This route may also be used to evaluate new chemical entities.

Tab. 2 National drug regulatory authorities of the 15 current EU member states. Internet addresses given if listed in the EMEA annual report, 2002

Austria Federal ministry for labour, health and social affairs, Wien.	**Belgium** Ministere des affaires socials de la sante publique et de l'environment, Brussles http://www.afigp.fogv.be
Denmark Danish medicines agency, Bronshoj. http://www.dkma.dk	**Finland** National agency for medicines, Helsinki
France Agence Francaise de securite sanitaire des produits de sante, Saint Denis, Cedex http://www.afssaps.sante.fr	**Germany** BfArM, Bonn http://www.bfarm.de
Greece National organization for medicines, Athens	**Ireland** Irish medicines board, Dublin http://www.imb.ie
Italy Dipartimento della valutazione dei medicinali e della farmacovigilanza, Rome http://www.sanita.it/farmaci	**Luxembourg** Division de la pharmacie et des medicaments, Luxembourg
The Netherlands Medicines evaluation board, Den Haag http://www.cbg-meb.nl	**Portugal** Infarmed, Lisbona http://www.infarmed.pt
Spain Agencia Espanola del medicamento, Madrid http://www.agemed.es	**Sweden** Medical products agency, Uppsala http://www.mpa.se
UK Medicines control agency, London. http://www.gov.uk/mca	

11.2.3.1 **The Centralized Procedure**

Under the centralized route, marketing authorization applications (dossiers) are submitted directly to the EMEA. Before evaluation begins, the EMEA staff first validate the application, by scanning through it to ensure that all necessary information is present and presented in the correct format. This procedure usually takes one to two working weeks to complete. Biotech-based dossiers are termed "part A applications", whereas new chemical entities are termed "part B applications".

The validated application is then presented at the next meeting of the CPMP (human medicine applications) or CVMP (veterinary medicines). This committee then appoints one of its members to act as "rapporteur" for the application. The rapporteur organizes technical evaluation of the application (product safety, quality, and efficacy), and this evaluation is often carried out in the rapporteur's home national regulatory agency. Another member of the committee (a corapporteur) is often also appointed to assist in this process. Upon completion of the evaluation phase, the rapporteurs draw up a report, which they present, along with a recommendation, at the next CPMP (or CVMP) meeting. After discussion, the committee issues a scientific opinion on the product, either recommending acceptance or rejection of the marketing application. The EMEA then transmits this scientific opinion to the European Commission in Brussels (who represent the only body with the legal authority to actually grant marketing authorizations). The Commission, in turn, issues a final decision on the product (Fig. 2).

Regulatory evaluation of marketing authorization applications must be completed within strict time limits. The EMEA is given a 210-day window to evaluate an application and provide a scientific opinion. However, during the application process, if the EMEA officials seek further information/clarification on any aspect of the application, this 210-day "clock" stops until the sponsoring company provides satisfactory answers. The average duration of active EMEA evaluation of biotech-based product applications is in the region of 175 days, well within this 210-day time frame. The duration of clock stops can vary widely – from 0 days to well over 300 days. Most applications, however, incur clock stops of the order of 30 to 80 days. Upon receipt of the EMEA opinion, the commission is given a maximum of 90 days in which to translate this opinion into a final decision. Overall therefore, the centralized process should take a maximum of 300 "active" evaluation days. EMEA opinions provided in 2002 for both human- and veterinary-based biotech drugs are listed in Table 3.

11.2.3.2 **Mutual Recognition**

The second route facilitating product authorization is termed "mutual recognition" or the "decentralized procedure". This is open to non-biotechnology products, and the procedure entails the initial submission of an authorization application to a single national regulatory agency of an EU member state (Table 2). This agency then assesses the application (within 210 days), formulates an opinion, and either grants or rejects the application. If authorization is granted, the sponsoring company may then apply via "mutual recognition" to extend the market authorization to the remaining EU states.

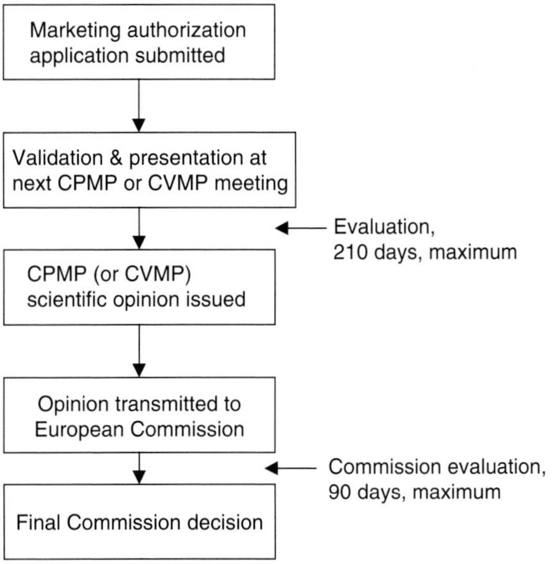

Fig. 2 Overview of the EU centralized procedure. Refer to text for details.

Tab. 3 Products of pharmaceutical biotechnology that were evaluated by the EMEA in 2002

Product brand name	***Indication***	***Sponsoring company***
Human medicines		
Pegasys (PEGylated alpha interferon)	Hepatitis C	Roche
Velosulin/monotard/ultratard/ protphane/actrapane/mixtard/ insulatard/actrapid (various formulations of recombinant human insulin)	Diabetes	Novo Nordisk
Neupopeg & neulasta (recombinant, pegylated colony-stimulating factor)	Neutropenia	Amgen
Xigiris (recombinant activated protein C)	Severe sepsis	Eli Lilly
InductOs (recombinant bone morphogenic protein)	Bone (tibia) fractures	Genetics institute
Ambirix (contains recombinant hepatitis B surface antigen)	Vaccination against hepatitis A and B	Glaxo SmithKline
Somavert (recombinant human growth hormone antagonist)	Acromegaly	Pharmacia
Veterinary medicines		
Eurifel RCP Fe LV (multicomponent biotech vaccine)	Vaccination against various feline viruses	Merial
Porcilis porcoli diluvac forte (biotech vaccine)	Vaccination of pigs against neonatal diarrhea	Merial
Proteqflu (biotech vaccine)	Vaccination against equine influenza	Merial
Proteqflu Te (biotech vaccine)	Vaccination against equine influenza and tetanus	Merial

Theoretically, awarding of authorization in these remaining countries should follow almost automatically as the authorization requirements (dictated by pharmaceutical law) are harmonized throughout the EU. Should disputes arise, the EMEA acts as an arbitrator, itself forming a scientific opinion, which it transmits to the European Commission that issues a final binding decision.

Tab. 4 Product categories regulated by the FDA

Foods, nutritional supplements
Drugs: chemical & biotech based
The blood supply & blood products
Cosmetics & toiletries
Medical devices
All radioactivity-emitting substances
Microwave ovens

11.3 Regulation in the United States

The FDA is the US regulatory authority [5]. Its mission is simply to protect public health. In addition to pharmaceuticals and cosmetics, food as well as medical and a range of other devices come under its auspices (Table 4). Founded in 1930, it now forms part of the US Department of Health and Human Services, and its commissioner is appointed directly by the US president.

The FDA derives its legal authority from the federal food, drug, and cosmetic (FD&C) act. Originally passed into law in 1930, the act has been updated/amended several times since. The FDA interprets and enforces these laws. Although there are many parallels between the FDA and the EMEA, its scope is far broader than that of the EMEA and its organizational structure is significantly different. Overall, the FDA now directly employs some 9000 people, has an annual budget in the region of US$1 billion and regulates over US$1 trillion worth of products annually (Table 4). A partial organizational structure of the FDA is presented in Fig. 3. In the context of pharmaceutical biotechnology, the centre for Drug Evaluation and Research (CDER) and, in particular, the Centre for Biologics Evaluation and Research (CBER) are the most relevant FDA bodies.

11.3.1 CDER and CBER

A major activity of CDER is to evaluate new drugs and decide if market authorization should be granted or not. Additionally, CDER also monitors the safety and efficacy of drugs already approved (i.e. post marketing surveillance and related activities). CDER predominantly regulates "chemical"-based drugs (i.e. drugs which are usually of lower molecular weight and often manufactured by direct chemical synthesis). Included are prescription, generic, and over-the-counter drugs. CDER also regulates some products of pharmaceutical biotechnology, including recombinant hormones (e.g. recombinant insulins and gonadotrophins) and certain cytokines (e.g. recombinant interferons).

The CBER undertakes many activities similar to that of CDER, but it focuses upon biologics and related products. The term "biologic" historically has a specific meaning, relating to "a virus, therapeutic serum, toxin, antitoxin, vaccine, blood, blood components or derivatives, or allergenic products that are used in the prevention, treatment, or cure of diseases of human beings" [6]. CDER therefore regulates products such as vaccines and blood factors, whether they are produced by traditional or modern biotechnological means (i.e. by nonrecombinant or recombinant means). Additional "biological products", including cell, gene therapy,

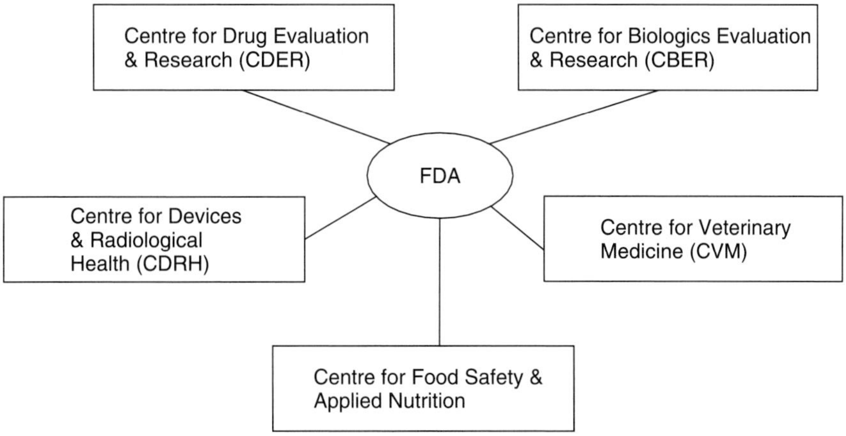

Fig. 3 Partial organizational structure of the FDA.

and tissue-based products also fall under the auspices of CBER.

11.3.2
The Approvals Procedure

The overall procedure by which biotechnology and other drugs are evaluated and approved by CDER or CBER are predictably very similar, although some of the regulatory terminology used by these two centers differ. A summary overview of the main points along the drug development/approval road where CDER/CBER play key regulatory roles is provided in Fig. 4.

Once a sponsor (company, research institute, etc.) has completed the preclinical evaluation of a proposed new drug, it must gain FDA's approval before instituting clinical trials. The sponsor seeks this approval by submitting an investigational new drug (IND) application to either CDER or CBER, as appropriate. The application, which is a multivolume work of several thousand pages, contains information detailing preclinical findings, methods of product manufacture, and proposed protocols for initial clinical trials. The regulatory officials then assess the data provided and may seek more information/clarification from the sponsor if necessary. Evaluation is followed by a decision to either permit or block clinical trials. Should clinical trials commence, the sponsor and regulatory officials hold regular meetings in order to keep the FDA appraised of trial findings. Upon successful completion of clinical trials, the sponsor then usually applies for marketing authorization. In CDER speak, this application is termed a new drug application (an NDA). NDAs usually consist of several hundred volumes containing over 100 000 pages in total. The NDA contains all the preclinical as well as clinical findings and other pertinent data/information. Upon receipt of an NDA, the CDER officials check through the document ensuring completeness (a process similar to the EMEA's validation phase). Once satisfied, they "file" the application and evaluation begins.

The NDA is reviewed by various regulatory experts, generally under topic headings such as "medical," "pharmacology," "chemistry," "biopharmaceutical," "statistical," and "microbiology" reviews. Reviewers may seek additional information/clarification from the sponsor as they feel necessary. Upon review completion, the application is either approved or rejected. If approved, the product may go on sale but regulatory officials continue to monitor its performance (postmarketing surveillance). Should unexpected/adverse events be noted, the regulatory authority has the legal power (and responsibility) to

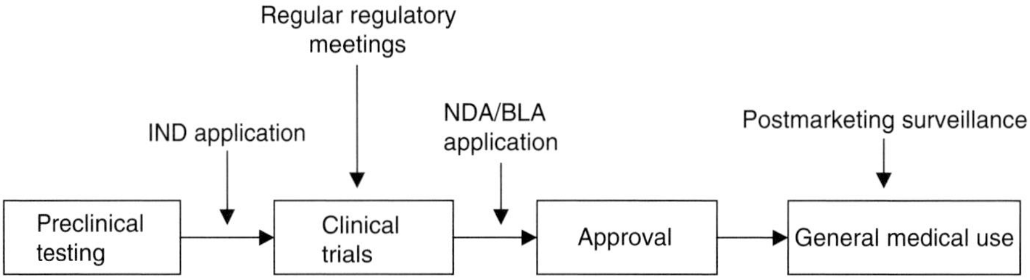

Fig. 4 Summary overview of the main points during a drug's lifetime at which the FDA plays a key regulatory role. Refer to text and Ref. [5] for further details.

suspend/revoke/modify the approval, as appropriate.

The review process undertaken by CBER officials upon biologic and related products is quite similar to that described above for CDER-regulated product. CBER-regulated investigational drugs may enter clinical trials subject to gaining IND status. The application process for marketing authorization undertaken by the sponsor subsequent to completion of successful clinical trials is termed the licensure phase in CBER terminology. The actual product application is known as a biologics licence application (BLA). Overall, the content and review process for a BLA is not dissimilar to that of the analogous CDER NDA process, as discussed above. The bottom line is that the application must support the thesis that the product is both safe and effective and that it is manufactured and tested to the highest quality standards. Overall, the median time between submission and approval of product marketing application to CBER/CDER stands at approximately 12 months.

While the majority of biotech-based drugs are regulated in the United States by either CBER or CDER, it is worth noting that some such products fall outside their auspices. Bone morphogenic proteins (BMPs) function to stimulate bone formation. As such, several have been approved for the treatment of slow-healing bone fractures. Product "administration" requires surgical implantation of the BMP in the immediate vicinity of the fracture, usually as part of a supporting device. As such, in the United States, these products are regulated by the FDA's Centre for Devices and Radiological Health (CDRH) [7]. Drugs (both biotech and nonbiotech) destined for veterinary use also fall outside the regulation of CBER or CDER. Most such veterinary products are regulated by the FDA's Centre for Veterinary Medicine (CVM), although veterinary vaccines (and related products) are regulated not by the FDA but by the Centre for Veterinary Biologics (CVB), which is part of the US Department of Agriculture [8].

11.4 International Regulatory Harmonization

Europe, the United States, and Japan represent the three main global pharmaceutical markets. As such, pharmaceutical companies usually aim to register most new drugs in these three key regions. Although the underlining principles are similar, detailed regulatory product authorization requirements differ in these different regions, making necessary some duplication of registration effort. The international conference on harmonization of technical requirements for registration of pharmaceuticals for human use (the ICH process) is an initiative aimed at harmonizing regulatory requirements for new drug approvals in these regions. The project was established in 1990 and brings together both regulatory and industry representatives from Europe, the United States, and Japan. ICH is administered by a steering committee consisting of representatives of the above-mentioned groupings. The steering committee in turn is supported by an ICH secretariat, based in Geneva, Switzerland [9]. The main technical workings of ICH are undertaken by expert working groups charged with developing harmonizing guidelines. The guidelines are grouped under one of the following headings:

- Efficacy (clinical testing and safety monitoring–related issues)

Tab. 5 Finalized ICH guidelines that specifically focus upon products of pharmaceutical biotechnology

Guideline number	***Guideline title***
Q5A	Viral safety evaluation of biotechnology products
Q5B	Quality of biotechnology products: analysis of the expression construct in cells used for the production of rDNA-derived products
Q5C	Quality of biotechnological products: stability testing of biotechnological/biological products
Q5D	Quality of biotechnological products: derivation and characterization of cell substrates used for production of biotechnological/biological products
Q6B	Specifications: test procedures and acceptance criteria for biotechnological/biological substances
S6	Preclinical safety evaluation of biotechnology-derived pharmaceuticals

- Quality (pharmaceutical development and specifications)
- Safety (preclinical toxicity and related issues)
- Multidisciplinary (topics not fitting the above descriptions).

Thus far, 37 guidelines aimed at both traditional and biotechnology-based products have been produced and are being implemented (Table 5). One of the ICHs most ambitious initiatives to date has been the development of the common technical document. This provides a harmonized format and content for new product authorization applications within the European Union, the United States, and Japan. When this and the other guidelines are fully implemented, considerable streamlining of the drug development and, in particular, registration process will be evident. This will make more economical use of both the company's and regulatory authorities' time, will reduce the cost of drug development and speed up the drug development procedure, ensuring faster public access to new drugs.

References

1. Office for Official Publications of the European Communities, http://www.publications.eu.int.
2. http://www.pharmacos.eudra.org/.
3. http://www.emea.eu.int/.
4. G. Walsh, *Nat. Biotechnol.* **1999**, *17*, 237–240.
5. http://www.fda.gov/.
6. http://www.fda.gov/cber/index.html.
7. http://www.fda.gov/cdrh/index.html.
8. http://aphis.usda.gov/vs/cvb/.
9. http://www.ich.org/.

Part IV
Biotech 21 – Into the Next Decade

12
Rituximab: Clinical Development of the First Therapeutic Antibody for Cancer

Antonio J. Grillo-López
Neoplastic and Autoimmune Diseases Research Institute, Rancho Santa Fe, CA, USA

12.1
Introduction

After many years of research, monoclonal antibodies (Mabs) were a source of disappointment to many but a small core group of investigators [1]. In about one year, between 1991 and 1992, rituximab was engineered as a chimeric anti-CD20 Mab. Then, in late 1992, the investigational new drug (IND) application for rituximab (Rituxan, MabThera) was submitted and clinical trials were initiated in February 1993 (Table 1) [2]. In the year 2002, rituximab became the number one, brand name, cancer therapeutic product in the world (approximately US$1.3 billion in sales). Over 300 000 patients have been treated with rituximab. The events of the intervening 10 years changed the course of history. The clinical development phase was completed in record time for a lymphoma agent (three years from the first to the last patient enrolled) and included combination trials with chemotherapy, biologicals, and radioimmunotherapy. The International Workshop Response Criteria for non-Hodgkin's lymphoma (NHL) had their origin in the criteria used for the rituximab clinical trials and were validated using the rituximab database. The dossier filed simultaneously with the FDA in the United States and with the EMEA in Europe was available in electronic format. This was the first electronic Biologics License Application (BLA) filed under the FDA's developing electronic submission standards. For the first time, a Mab used as a single agent showed sufficient clinical activity to warrant worldwide approvals for a cancer indication. For the first time, a Mab was approved specifically for the treatment of patients with NHLand an agent (combined with Cyclophosphamide Hydroxydaunorubiein Oncovin Preonisone (CHOP) chemotherapy) was shown to be superior to CHOP alone in patients with aggressive NHL – a new "gold standard" was established. Rituximab + CHOP produced a significant increase in overall survival (OS) as compared to CHOP alone. The Zevalin treatment regimen includes rituximab as the cold antibody. In February 2001, Zevalin became the first radioimmunotherapy approved for the treatment of NHL. Rituximab has surpassed its original indications, and today, it is being investigated in clinical trials for a variety of autoimmune disorders. Enthusiasm for

Tab. 1 Milestones in the history of rituximab

Date	*Milestone*
January 1991	Mice immunized with human CD20 antigen. Anti-CD20 antibodies isolated. Parent antibody for rituximab (murine anti-CD20 antibody, IDEC-2B8) identified and characterized.
June 1991	Chimeric antibody IDEC-C2B8 engineered with murine variable and human constant regions (IgG1 kappa isotype).
August 1991	Vector engineered, introduced by electroporation into CHO cells, antibody produced by fermentation process.
December 1992	IND filed with the US FDA – 1st Phase I trial protocol submitted.
February 1993	First patient treated with rituximab (single dose Phase I trial)
April 1994	First patient treated on the 1st combination study – CHOP + rituximab in LG/NHL.
March 1995	First patient treated on the Phase III (pivotal) trial that led to the approval of rituximab by regulatory authorities in the United States and Europe.
March 1996	Last patient entered on the Phase III (pivotal trial).
February 1997	Biologics License Application (BLA) filed with the US FDA and simultaneous filing of the European dossier with the EMEA.
July 1997	Biologics Response Modifiers Advisory Committee of the US FDA meets in Washington DC and recommends approval.
November 1997	Rituximab approved in the United States by the FDA for patients with LG/NHL – 1st Mab approved for the treatment of cancer, 1st Mab approved for the treatment of NHL.
June 1998	Rituximab approved in Europe by the EMEA for LG/NHL.
December 2000	Results of the Grupe Dej Etudes De Lymphome D'adultes (GELA) study presented at the American Society of Hematology meeting showing statistically significant superiority of CHOP + rituximab over CHOP alone in patients with Intermediate Grade NHL.
October 2001	Committee for Proprietary Medicinal Products (CPMP) of the EMEA recommends approval of rituximab for patients with Intermediate Grade NHL.
December 2002	Rituximab named the number one selling, brand name, cancer therapeutic product in the world (approximately US$1.3 billion).

clinical and laboratory research in the area of antibody therapeutics was renewed and grew in an accelerated fashion yielding numerous new Mabs for cancer and autoimmune diseases, some of which are already approved [3]. This chapter provides information and insight as to how these historical events came about.

12.2 Clinical Development and Regulatory Approvals

The IND for rituximab was submitted to the US FDA in December 1992 [4]. The anti-CD20 Mab had been purposely engineered with human constant regions (IgG1 kappa isotype) to ensure that it would effectively bind complement and, through Fc receptors, effector cells so that it could effect complement-dependent cytotoxicity (CDC) and antibody-dependent cellular cytotoxicity (ADCC) [5, 6] (Fig. 1). This was confirmed in vitro. Additionally, in vivo experiments in monkeys revealed immediate, profound, and specific B-cell depletion with recovery within 100 days. The effects of the antibody on the human immune system were unknown. It was expected to produce B-cell depletion and

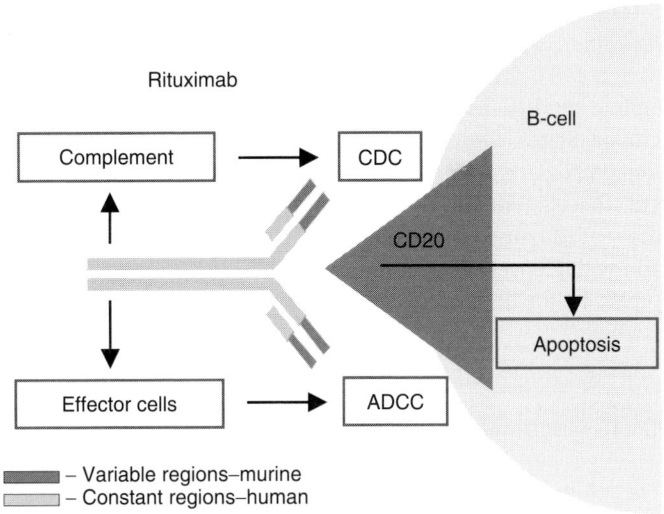

Fig. 1 Mechanism of action of rituximab. The chimeric (mouse/human) antibody, rituximab, binds to the CD20 antigen on B-cells and (a) activates complement to effect CDC, (b) attracts effector cells via Fc receptors to effect ADCC, and (c) transmits a signal into the cell to induce apoptosis. (See Color Plate p. xxii).

to decrease lymphomatous nodes, masses, and infiltrates. Its effects on immunoglobulin levels were unknown. The timing and duration of B-cell depletion could not be accurately predicted from the animal studies. Thus, the initiation of the first Phase I, single dose, clinical trial was delayed for 2 months (beyond the usual 30-day wait) because of the FDA's safety concerns and specifically their concerns regarding effects on immunoglobulins and on B-cells. Eventually, an agreement was reached on a starting dose of 10 mg m^{-2}. Today, we know that this dose represents less than 1% of the total dose that patients receive over four infusions (about 3 gm for the average patient). Clinical trials were initiated in February 1993.

12.2.1 Clinical Development

The clinical development of rituximab was conducted entirely by IDEC Pharmaceuticals Corporation, San Diego, California. The Division of Medical Research and Regulatory Affairs (M&RA) at IDEC was responsible for all aspects of clinical development and regulatory interactions worldwide including preparation and defence of the BLA and European submissions through the review process and to approvals. The clinical trials were all designed, implemented, conducted, analyzed, interpreted, and reported by a small group of professionals (staff of seven in 1992) at IDEC's M&RA division. Investigators were chosen from important academic institutions in the United States and Canada. A Clinical Research and Development Agreement (CRADA) was initiated with the US NCI around 1996. However, the US NCI did not participate in the development studies. The first study conducted under US NCI sponsorship was the CHOP versus R + CHOP intergroup study that started in 1997.

IDEC began collaborations with Genentech Inc., San Francisco, California, in March 1995, for the manufacturing, marketing, and sales of rituximab in the United States. Shortly thereafter, collaborations began with F. Hoffmann-La Roche Ltd., Basel, Switzerland, for the development of rituximab in the European Union, and with Zenyaku Kogyo Co., Ltd, Tokyo, Japan, for the development of rituximab in Japan.

12.2.2
The Clinical Development Plan

The clinical development plan was designed to achieve an early approval based on single-agent efficacy in patients with relapsed or refractory LG/F NHL. Resources were limited and it was not possible to conduct large randomized trials or to pursue an additional indication (such as aggressive NHL). Thus, the plan relied on single-arm studies that utilized surrogate endpoints (e.g. response rates) and qualified for approval under "accelerated approval" guidelines. However, three pilot studies of different combinations were carried out (combination trials with chemotherapy, biologicals, and radioimmunotherapy). These were considered important as it was clear that the Mab would eventually be used as part of a combination or multimodality therapy and not just as a single agent. If successful, these pilot studies could lead to larger randomized trials. The single-agent studies in patients with LG/F NHL included Phase I single dose, Phase I/II multiple dose, Phase II in patients with bulky disease, Phase II re-treatment, Phase II 8 infusion, and Phase III (pivotal) studies. The first patient was treated in February 1993 and the last patient (included in the regulatory dossiers) was enrolled in February 1996. Completion of enrolment in all of these studies in a three-year period established a record in NHL where most cooperative group studies take years to complete. An aggressive but realistic clinical development plan, including both single-agent and combination studies, set the pace that made these achievements possible.

12.2.3
Clinical Trials Methodology

The methodology utilized in the implementation, conduct, analysis, and interpretation of the clinical development plan

Tab. 2 Key investigators in rituximab clinical trials

Investigator	*Site*	*Study participation*
David Maloney, MD	Stanford U. Med. C.	102–01, 02, 05
Thomas Davis, MD	Stanford U. Med. C.	102–07, 08
Peter McLaughlin, MD	M.D. Anderson C.C.	102–05, PK
Myron Czuczman, MD	Roswell Park C.C.	102–03, 05, 06
Neil Berinstein, MD	Toronto-Sunnybrook C.C.	102–05, PK
Larry Piro, MD	Scripps Clinic	102–06, PK

Notes: U: University; C: Center; CC: Cancer Center;
Many other investigators and staff at investigational sites made important contributions and are not listed because of space constraints.

was critical to its success. A limited number of academic institutions (about 30 in the United States and Canada) were chosen to participate in the clinical trials and most enrolled patients in two or more studies (Table 2). This served several purposes: (a) the staff at these sites were instructed on clinical trials methodology only once as it was consistent across all studies, (b) the investigator's meetings could address several studies and the overall number of meetings was decreased, (c) the investigators and the company staff interacted more frequently and more efficiently, (d) the site staff became experts at studying drug administration as well as safety and efficacy monitoring and reporting. The required bureaucracy (including clinical trials agreements, confidentiality agreements, adverse event reporting, queries and audit trails, accounting for experimental drug, etc.) was consistent and thus could be simplified and minimized. Protocols, Case Report forms (CRFs), and data collection were standardized. Importantly, a peer-level relationship based on mutual professional respect was established between the investigators and their staff and the company clinician and staff.

Clinical trials conducted during development (as described below in Sects. 12.2.6 and 12.2.7) were similar in a number of ways. Inclusion and exclusion criteria were almost identical across studies. These studies focused on the treatment of patients with relapsed or refractory LG/F NHL. Patients were dosed at 375 mg m^{-2} of rituximab by intravenous infusion weekly for a total of four doses. This consistency resulted in a degree of homogeneity that allowed for analyses across studies as well as comparisons between studies. Any exceptions to these general rules are noted in the individual study descriptions below.

12.2.4 Response Criteria

Defining a set of clinical response criteria was a difficult task. At the time the clinical trials started in 1993, the WHO and Eastern Cooperative Oncology Group (ECOG) criteria were being utilized [7, 8]. Historically, these criteria had been developed for the efficacy evaluation of patients with solid tumors. There were no standard criteria for NHL. The WHO and ECOG criteria were inadequate for the evaluation of NHL patients as they were based on the disappearance of a tumor mass, whereas in lymphoma the "tumor mass" is in part a normal anatomical structure, a lymph node, that may decrease in size but will not disappear. We convened a panel of NHL experts from the United States to draft lymphoma-specific criteria for the rituximab clinical trials [9]. These criteria were subsequently endorsed by a group of European NHL experts [10]. In October 1996, these rituximab NHL response criteria were reviewed and approved by the Biologic Response Modifiers Advisory Committee of the US FDA [11, 12]. A third-party blinded panel of NHL experts (Lymphoma Experts Confirmation of Response, LEXCOR) evaluated patients in the rituximab Phase III (pivotal) trial by applying these criteria [13, 14, 15]. The criteria were accepted by the US FDA in 1997 [16]. In February 1998, we collaborated with the US NCI to convene an international working group in order to reach a consensus on new response criteria for NHL that could be accepted and applied worldwide. We invited, in addition to the US NHL experts, a number of international experts from Europe and other areas including: Coiffier B (France), Connors JM (Canada), Lister TA (United Kingdom), Hagenbeek A (Netherlands), Hiddemann

W (Germany), and others. The committee, at a meeting in Washington DC, drafted a set of criteria. These criteria were tested by application to the rituximab clinical trials database. This database included raw data from all patients treated with rituximab in the clinical trials conducted during development. The tumor measurements for these patients had been collected initially at the investigational sites by the principal investigators, radiologists, and their staff. All CT scans were collected on an ongoing basis and were subsequently subject to a centralized and blinded review by an independent (third party) panel of NHL experts (oncologists and radiologists) termed the *LEXCOR panel.* The LEXCOR panel included the following: Hematologists/Oncologists – Cheson B (US NCI), Horning S (Stanford U), Just R (San Diego), Kossman C (San Diego), Morrison V (U Minn.), Peterson B (CALGB), and Rosen P (UCLA); and Radiologists – Carter W (San Diego), Klippenstein D (Roswell Park), and Kortman K (San Diego). These experts measured all the lesions on each CT scan for every patient. Some patients had more than 50 measurable lesions. To our knowledge, the resulting database (with bidimensional measurements of all lesions) is the only one of its kind as investigators usually measure only 6 to 10 "sentinel" lesions.

These International Working Group Response Criteria for NHL (IWRC), published in 1999, have become the standard criteria for response evaluation in NHL [17] and have been applied to the rituximab studies [18, 19].

12.2.5
The Medical Research and Regulatory Affairs Staff

The company staff included a core group of experienced professionals: clinical scientists, clinical research associates (site monitors), statisticians, medical writers, regulatory specialists, and others (Table 3). This core staff had years of experience in cancer drug development in pharmaceutical industry as well as in academic centers. In the biotech world of 1992, having an experienced core clinical staff was the exception rather than the rule. Many biotech companies were relatively small with inadequate funds and resources and their staff had limited clinical trials experience. Frequently that experience was at the level of clinical trials of institutional or cooperative group type and not the highly regimented studies required by regulatory agencies. Those who have not had the experience of conducting a clinical trial that must meet worldwide regulatory requirements do not comprehend the degree of rigor, detail, accuracy, specificity, and clarity demanded of such studies. Such studies represent the best clinical science and are not just designed to meet, the sometimes arbitrary, regulatory requirements. In the academic world, "peer review" is considered to be the highest-level test that a manuscript must undergo in order to be published. This usually entails review by two or three anonymous reviewers with varying degrees of expertise who will not have access to the raw data. "Peer review" of a regulatory dossier, the clinical trials, results, and interpretation, is a much more detailed and rigorous process. The reviewers have access to the raw data and will review it in detail. A representative sample of the sites participating in the clinical trials will be audited. When a manuscript fails peer review, it can be rewritten and resubmitted. When a regulatory dossier fails peer review, the consequences have a more significant impact, as the work of many years may have to be repeated. Thus,

Tab. 3 Key IDEC Pharmaceuticals staff in rituximab clinical trials

Name and title	*Responsibilities*
Antonio J. Grillo-López, MD, Chief Medical Officer and Senior VP, Medical and Regulatory Affairs Division	Chief Medical Officer and Project Clinician 1992–2001.
Brian K. Dallaire, Pharm D, Senior Director, Clinical Operations	Clinical Scientist and Divisional operations, plans, and resources 1993–2001.
Christine White, MD, Senior Director, Hematology and Oncology	Safety Officer and Clinical Scientist 1995–2001.
Chester Varns, Director, Clinical Trials Monitoring	Clinical trials monitoring, study implementation, data acquisition.
Anne McClure, MS, Director, Medical Writing	Medical writing.
David Shen, PhD, Senior Director, Biometrics	Biostatistics, data entry and analysis.
Jay Rosenberg, PhD,	Clinical Immunology Laboratory
John Leonard, PhD, Senior Director, Project Planning and Regulatory Affairs	Project planning and regulatory affairs.
Alice Wei, Senior Director, Regulatory Affairs	Regulatory filings and interactions.

Notes: Clinical trials with rituximab began in 1993. The BLA was filed with the US FDA in February 1997 and the MAA was filed simultaneously with the EMEA in Europe. Approval was granted in the United States in November 1997 and in Europe in June 1998. Many others at IDEC Pharmaceuticals made important contributions and are not listed because of space constraints.

having an experienced, professional, and dedicated clinical staff is invaluable.

Some companies chose to conduct their work through the use of consultants and contractors. The so-called virtual company has been justified by the expected fiscal efficiency and the lower overhead. However, consultants and contractors can be more expensive than in-house personnel and will never have the degree of loyalty and dedication. Continuity is a major problem as the outside staff is usually subject to greater turnover and changes of assignment. The critical issue is loss of control. Someone other than you is having daily contact with the investigators and sites. Someone else is interacting with the FDA. Importantly, the database is not held by the company and the consistency and quality of the data is at risk. All of these factors constitute the real price of having a virtual company. Virtual companies many times generate costly "virtual data". It is important for the small biotech companies to have their own clinical development staff and thus hold in their own hands the reigns to their ultimate success.

12.2.6
Phase I and I/II Clinical Trials

FDA and EMEA approvals of rituximab were based on five single-agent studies conducted primarily in patients with relapsed or refractory, low-grade or follicular, CD20+, B-cell NHL. Clinical trial results are listed in Table 4. Two of these were Phase I (single dose) or I/II (multiple dose) studies.

The first Phase I study, single rituximab infusions ranging from 10 to 500 mg m^{-2} in 15 patients, reached the highest dose without dose-limiting toxicity [20]. The maximum tolerated dose (MTD) was not

Tab. 4 Rituximab clinical development – single-agent trials

Study description	***N***	***ORR [%]***	***CR [%]***	***PR [%]***	***TTP mo.***	***References***
Phase I – Single dose	15	20	0	20	9	20
Phase I/II – Multiple dose, PI part	18	33	0	33	6.4	24
Phase I/II – Multiple dose, PII part	37	50	9	41	13.2	26
Phase III – Pivotal trial	166	50	6	44	13.2	29
Phase II – Bulky disease	28	43	4	39	8.1	30
Phase II – Eight infusions	35	60	14	46	19.4+	32
Phase II – Re-treatment	57	40	11	30	17.8+	36

Notes: N: patients treated and evaluable; ORR: overall response rate; CR: complete response rate; PR: partial response rate; TTP: median time to progression for responders (+ indicates a Kaplan Meier projection where true median has not been reached).
(a) Dosing in all studies, except the first two listed above, was at 375 mg m^{-2} weekly × 4 doses.
(b) All response rates are based on the "evaluable/treated" patient population (*N*) and on response criteria as reported by authors (not the new IWRC).

reached. However, the length of infusion time at higher doses was not considered feasible for outpatient therapy. This study served to provide the first safety experiences with the Mab. Infusion-related adverse events (including fever, chills, nausea, headache, myalgia, bronchospasm, hypotension, and others) were observed. The very first patient treated (single dose, 10 mg m^{-2}) experienced fever, chills, and bronchospasm. The benefits of premedication with antipyretics (acetaminophen) and antihistaminics (diphenhydramine) became evident. The first observation of the relationship between higher B-cell (CD20+) counts and more significant adverse events was made in the course of this study. The overall response rate (ORR) was 20%. Two patients had partial responses (PR) that lasted eight and nine months. A third patient, the first one treated and mentioned above, had a delayed response reaching PR after seven months. This response lasted about a year. This is an important observation. Delayed responses have been observed in many patients. The importance of observation in patients who are stable Stable Disease (SD) has been stressed. These patients may show progressive tumor shrinkage over time and eventually reach a PR. Likewise, some patients with PR may, with time, become CRs [21–23].

The Phase I/II study consisted of two parts: a multiple dose, dose escalation part (Phase I) and a Phase II part. In the dose escalation part of the study, patients were treated with four infusions of rituximab at 125, 250, or 375 mg m^{-2}. An MTD was not reached and the highest dose, 375 mg m^{-2}, was chosen for further studies. This eventually became the approved standard dose. The dose could have been higher, but at the time of this study, it was limited not by adverse events but by the limited supplies of the Mab. In fact, this study was designed considering the minimum number of patients necessary to perform dose escalation and the Phase II part of the study versus the total amount of rituximab that was available [25]. Higher doses have been studied as discussed below (eight-infusion study). In the Phase II part of the study, 34 patients were treated at the chosen dose in an effort to establish a response rate within the reasonable 95% confidence intervals (CI) [26]. Pharmacokinetic studies revealed a mean half-life of 225 h for the free antibody in serum. Mean serum Cmax was 500 μg mL^{-1}. One patient with a PR developed a transient detectable (not quantifiable) HACA seven months post treatment. There were no patients with quantifiable (>100 ng mL^{-1}) HAMA or HACA. Seventeen of 34 patients responded (50% ORR, 36 to 67% CI). The median TTP was reported as 10.2 months. This was a Kaplan Meier projected median and was later revised when the true median was reached at 13.2 months. These results have stood the test of time and have been duplicated time and again in different studies [27–29].

12.2.7 Phase II and III Clinical Trials

Phase II clinical trials performed during development included three single-agent trials: a study in patients with bulky disease, an eight-infusion study, and a re-treatment study (Table 4). Additionally, three combination studies were conducted: a study in combination with chemotherapy, a study in combination with biologicals, and a study in combination with radioimmunotherapy (Table 5).

12.2.7.1 Rituximab in Bulky Disease

It was important to conduct a study in patients with bulky disease because a decade ago the bias was that Mabs would not be active in such patients. Additionally, there was no experience using rituximab in patients with bulky disease as a lesion greater than 10 cm in diameter constituted an exclusion criterion in the development studies. The Phase II trial in patients with bulky disease was designed to include only those patients who had at least one lesion that was 10 cm or greater in its largest diameter [30]. The 28 patients treated had multiple characteristics indicative of poor prognosis, as one would expect given their bulky disease. About a third had International Working Formulation (IWF) A Histology (small lymphocytic) that is known to respond poorly to rituximab. They were heavily pretreated (median 3 prior regimens, range 1–13) and had a progressive/clinically aggressive disease at study entry. In spite of this, the ORR was 43% (4% CR and 39% PR) with a median TTP of 8.1 months (range 4.5 to 18.6+ months). In a historical comparison with the 166-patient pivotal trial, there

Tab. 5 Rituximab clinical development – combination studies

Study description	*N*	*ORR [%]*	*CR [%]*	*PR [%]*	*TTP mo.*	*References*
Phase II – R + CHOP	38	100	58	42	72.0+	45
Phase II – R + Interferon	37	45	11	34	25.2	48
Phase III – R + ^{90}Y Zevalin	73	80	34	45	15.4+	54

Notes: N: patients treated and evaluable; ORR: overall response rate; CR: complete response rate; PR: partial response rate; TTP: median time to progression for responders (+ indicates a Kaplan Meier projection where true median has not been reached).
All response rates are based on the "evaluable/treated" patient population (*N*) and on response criteria as reported by authors except Zevalin (based on the new IWRC).

were no significant differences in ORR between patients with bulky disease and patients with lesions of 5 to 7 cm, or with the general pivotal trial population [30]. Rituximab was shown to be active in patients with bulky disease. This clinical trial remains the only reported study of rituximab in patients with bulky disease.

12.2.7.2 Optimizing the Dose and Schedule

The eight-infusion study was necessary as patients in the four-infusion dosing experience (at 375 mg m^{-2}) had not reached either MTD or steady state/plateau. The dose and schedule of administration had not been optimized. Given the positive correlation between higher serum levels of antibody and response, it was important to explore a higher total dose administered over eight doses (375 mg m^{-2} weekly × 8 doses) [31]. The eight-infusion study was significant in showing a numerically higher ORR and CR than the previous four-infusion studies [32]. The ORR in 35 treated patients was 60% (14% CR and 46% PR). The median TTP exceeded 19.4 months and was also longer than historical controls using four infusions. Pharmacokinetic studies revealed a progressive increase in serum concentration levels of rituximab beyond the fourth infusion with a possible plateau following the seventh and eighth infusions [32, 33]. Aviles et al. have reported on a six-infusion study in which they also achieved better ORR and longer TTP than with four infusions [34]. O'Brien et al., in a Phase I dose escalation trial in patients with Chronic Lymphocytic Leukemia (CLL), have also shown the benefit of higher doses of rituximab. A controlled, randomized study will probably never be carried out to formally settle the issue of whether or not more rituximab is better as these pilot studies suggest. Nevertheless, there is a pharmacokinetic and biologic rationale and the results reported to date have sufficed for the US FDA to include the option of eight-infusion dosing in the package insert [35].

12.2.7.3 Repeated Treatment – as Maintenance or Following Disease Progression

A number of issues regarding repeated treatment with rituximab had to be addressed. Can rituximab treatment be repeated safely? What are the long-term effects of sustained B-cell depletion? Do patients continue to respond and for how long? Should treatment be repeated upon relapse or is maintenance therapy feasible and preferable? The timing of relapse for the individual patient cannot be predicted. We know that the median TTP for responders is about one year, but there is a wide range with some patients relapsing early on and others having prolonged sustained remissions with no other therapy (Fig. 2) [37, 38]. It is also clear that B-cell recovery in peripheral blood is not a marker for disease progression (PD) as some patients relapse before (during depletion) and many patients remain in remission beyond the point of B-cell recovery. Also, it has been shown that the median tumor volume for responding patients, as measured by the sum of the products of the perpendicular diameters (SPD), continues to decrease even after B-cells have recovered in peripheral blood (Fig. 3). It would have been nice to have a simple marker, such as the B-cell count, to indicate PD. However, it is clear that normal B-cell recovery cannot be equated with lymphomatous B-cell recovery or with PD [39].

Patients enrolled on the re-treatment study had been previously treated with rituximab, responded and later relapsed. They were required to have PD and

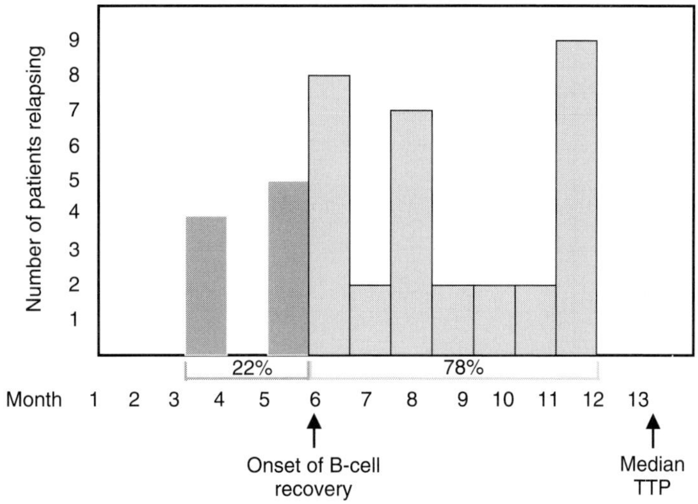

Fig. 2 Analysis of rituximab responders relapsing prior to median TTP (pivotal trial). Patients who respond to rituximab show a median TTP of 13.2 months. Of those who relapse prior to that median, 22% will relapse in the first 6 months and 78% between 6 and 13 months. Most responders will relapse beyond the point that marks the onset of median B-cell recovery in peripheral blood (6 months) [38, 39].

remain CD20-positive upon entering the study. The ORR in 57 patients treated was 40% (11% CR and 30% PR) and the TTP exceeded 17.8 months. Historical comparison to patients' prior TTP showed a significant increase upon re-treatment. There was a numerically higher TTP in this study as compared to the TTP for patients treated in the pivotal trial. Re-treatment was feasible, well tolerated, and had significant clinical activity. Development of HACA was not detected in any patient participating in this study. Successful maintenance treatment has been reported by Hainsworth et al. [40] and by Ghielmini et al. [41]. In the later study, patients initially treated with rituximab were randomized to maintenance or observation. Progression rates were significantly different with only 20% of the patients progressing on the maintenance arm, while 44% progressed on the observation arm in the first 12 months. Although the optimal schedule and dose for maintenance has not been defined, it is clear that maintenance therapy with rituximab is beneficial in increasing the response rate over time and in prolonging the remission duration.

12.2.7.4 **Combinations with Chemotherapy: The R + CHOP Combination**

No clinical trial has had the impact on the treatment of lymphoma that the study by Coiffier et al. has had [42]. This randomized trial showed that the combination of rituximab and CHOP resulted in a significant increase in overall survival as well as in ORR, CR, and Event Tree Survival (EFS) as compared to CHOP chemotherapy alone. It was the first time since the initial experiences with CHOP over 25 years ago that any combination was shown to be statistically superior to CHOP for patients with aggressive NHL. Importantly, through this study, rituximab was elevated to the rank

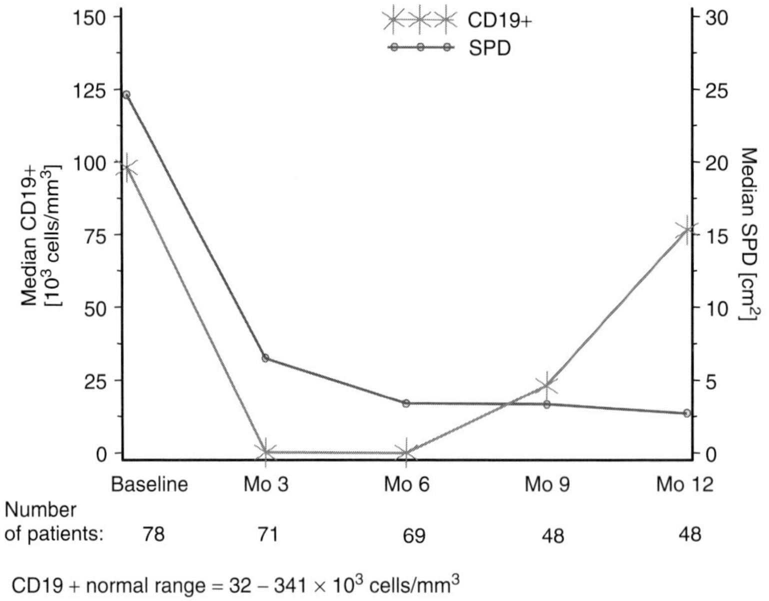

Fig. 3 B-cell recovery and SPD in all responders (pivotal trial). Median B-cell counts in peripheral blood (as measured by CD19 positivity on Fluorescence Activated Cell Sorter (FACS) analysis) drop to zero and start recovering by the sixth month. Tumor volume (as measured by SPD) for responders continues to decrease beyond nine months despite normalization of B-cell counts [38, 39]. This figure is used by permission from the copyright holders – Grillo-Lopez AJ and Idec Pharmaceuticals.

of a curative therapy for NHL. Many other combinations have been evaluated [43] and have shown promise, as for example R + EPOCH [44]. None to date can surpass R + CHOP in either aggressive or indolent NHL [3]. R + CHOP is today the "gold standard" curative therapy for aggressive NHL.

The first trial of the R + CHOP combination was a Phase II study in patients (mostly frontline, some relapsed) with Indolent NHL initiated in April 1994 (Table 5) [45]. Patients received six infusions of the Mab and six cycles of CHOP. This study was intended as a pilot study to define the safety and tolerability of the combination in these patients prior to initiating clinical trials in aggressive NHL. It paved the way for a subsequent pilot study in aggressive NHL [46] and eventually led to the Coiffier et al. [42], and the currently ongoing US Intergroup, randomized studies. The results of this first study were also remarkable for the prolonged TTP that was observed. When last reported, a median had not yet been reached (could not be projected by Kaplan Meier methodology), and the median observation time exceeded six years [47].

12.2.7.5 Combinations with Biologicals: Rituximab + Interferon

The first biological to be combined with rituximab was interferon alpha 2a. In 1993, Interferon was the only biological with some clinical activity in NHL. It was also an immunostimulant with a variety of different effects on the immune system.

Patients on the combination study received 12 doses of interferon on a weekly basis and 4 infusions of rituximab on weeks 5 through 8 [48]. The combination was safe and well tolerated. The ORR was 45% (CR 11% and PR 34%) and was lower than expected. However, the TTP for responders was 25.2 months. Interferon is currently approved for the prolongation of chemotherapy-induced remissions. Even though it did not increase the response rate of rituximab, it appears to have prolonged the TTP. Despite these encouraging results (confirmed by others [49]), investigators have not shown much interest in utilizing interferon during maintenance therapy following rituximab. Other promising rituximab combinations include interleukin 2, GM-CSF, G-CSF, alemtuzumab, epratuzumab, and so on.

12.2.7.6 Combinations with Radioimmunotherapy: Rituximab + Zevalin

Rituximab has been shown to synergise with chemotherapeutic agents [50]. There is also some early data that anti-CD20 Mabs may synergise with radiation. Although the clinical development of Zevalin (^{90}Yttrium-labeled ibritumomab tiuxetan) began utilizing murine IDEC-2B8 (ibritumomab) as the cold antibody in the treatment regimen, we had always planned to switch to rituximab as soon as feasible [51 to 54]. The activity of the Zevalin treatment regimen, including rituximab, was established during clinical development. In the Phase III randomized study, Zevalin had an ORR of 80% (34% CR and 45% PR) with a TTP of 15.4+ months in responders. In February 2001, Zevalin became the first radioimmunotherapy approved for the treatment of NHL. Other radioimmunotherapies are under development (radiolabeled tositumomab, epratuzumab, etc.).

12.2.8 Regulatory Dossiers, Review, and Approvals

Communication and coordination with the US FDA was a very important factor in carrying out the clinical development plan and proceeding on to the dossier submission, review, and approval processes. A professional and peer-level relationship based on openness and mutual trust was established between the clinical and regulatory staff at IDEC Pharmaceuticals and the FDA reviewers. This served to preempt or expedite resolution of the many issues that always arise during development. The IND for rituximab was submitted to the US FDA in December 1992 [4]. Clinical trials were initiated in February 1993 and enrolment completed three years later in February 1996. The dossier, filed simultaneously with the FDA in the United States and with the EMEA in Europe in February 1997, was available in electronic format. This was the first electronic BLA filed under the FDA's developing electronic submission standards and IDEC Pharmaceuticals' first BLA. At the time, the US FDA had published a draft guidance manual for Computer-Assisted Product License Applications (CAPLAs); however, final standards and requirements were still under development. The BLA was a new entity, and no guidelines had been published specifically for electronic submission of this type of application. Therefore, IDEC worked with the Center for Biologics Evaluation and Research (CBER) to design a user-friendly e-BLA while simultaneously laying the groundwork for future standards. On the basis of the success of the first e-BLA submission, IDEC geared up for a second electronic filing three years later. This second product developed by IDEC Pharmaceuticals, Zevalin (^{90}Y ibritumomab tiuxetan), for the treatment of non-Hodgkin's lymphoma, ultimately became the first radioimmunotherapy approved by the US FDA (dossier under review by EMEA in Europe).

The rituximab presentation to the Biologics Response Modifiers Advisory Committee (BRMAC) of the FDA took place on July 25, 1997 and the final approval was granted on November 26, 1997 (Fig. 4) [55–58]. The EMEA granted approval to rituximab in Europe in June 1998. Rituximab became the first monoclonal antibody approved for the treatment of cancer and specifically for patients with NHL.

12.3 Rituximab: Other Indications/Applications

Rituximab has been approved in the United States for relapsed or refractory LG/F NHL and in Europe for relapsed or refractory LG/F NHL and for aggressive NHL. Multiple other indications have been explored including hematologic malignancies – CLL, multiple myeloma, and so on; autoimmune disorders – Immune Thrombocytopenic Purpura (ITP), heumatoid arthritis, Systemic Lupus Erythematosus (SLE), hemolytic anemias, and so on (Figs. 5 and 6) [4, 59, 60].

12.4 Conclusions: Achievements, Current Role, and Future Applications of Rituximab

Rituximab represents the most important scientific achievement of the past decade. It was the first therapeutic antibody approved for the treatment of cancer and specifically for NHL. The current IWRC for NHL had their origin with the rituximab response criteria. Clinical development was completed in record time. Dossiers were filed simultaneously in the United States and in Europe. The US filing utilized an electronic format (computer-aided product license application, CAPLA). These and many other achievements during clinical development served to provide tremendous impetus to the monoclonal antibody research area. The renewed enthusiasm in this area has yielded many new Mabs with activity in both hematologic malignancies and in autoimmune diseases. Several of these have been approved (e.g. herceptin, alemtuzumab, mylotarg, others) and many others are under active investigation.

Rituximab is approved for the treatment of patients with relapsed or refractory LG/F NHL. In Europe, it is also approved for aggressive NHL (R + CHOP). It is used as a single agent, in combinations

Rituximab: BRMAC presentation
25 July 1997

IDEC pharmaceuticals staff

Augusta Cerny	C. David Shen, Ph.D
Chet Varns	Christine A. White, M.D.
Alice Wei	Brian K. Dallaire, Pharm.D.
William Hauser	Susan K. Langley

Antonio J. Grillo-López, M.D.

Fig. 4 Rituximab: BRMAC presentation – 25 July 1997. The IDEC Pharmaceuticals staff's presentation and the rituximab data were commended by the BRMAC advisory committee members for the excellent organization, clarity, scientific quality, and methodological rigor. The videotaped presentation is now used by pharmaceutical companies as a teaching tool for advisory committee presentations [55–58]. Presenters: Alice Wei – regulatory, Antonio Grillo-Lopez – medical and scientific. (This figure is used by permission from the copyright holders – Grillo-Lopez AJ and IDEC Pharmaceuticals).

Fig. 5 Rituximab: literature reports (neoplastic diseases). The indications and clinical applications for rituximab are expanding as research in the therapy of multiple neoplastic diseases is reported. This list is a sample of the literature reports on the many indications currently under investigation.

Rituximab: Literature reports
Neoplastic diseases

- Low-grade NHL
- Intermediate–and high-grade NHL
- Waldenstrom's macroglobulinemia
- Chronic lymphocytic leukemia
- Acute prolymphocytic leukemia
- Acute lymphoblastic leukemia
- Hodgkin's disease
- Cutaneous B-cell lymphoma
- Colon cancer and other solid tumors

Fig. 6 Rituximab: literature reports (autoimmune diseases). The indications and clinical applications for rituximab are expanding as research in the therapy of multiple autoimmune diseases is reported. This list is a sample of the literature reports on the many indications currently under investigation.

Rituximab: Literature reports
Autoimmune diseases

- Idiopathic thrombocytopenic purpura
- Acquired Factor VIII inhibitors
- Pure red cell aplasia
- Systemic lupus erythematosus
- Rheumatoid arthritis
- Hemolytic anemias
- Posttransplant lymphoproliferative disease
- Paraneoplastic pemphigus
- IgM polyneuropathies
- Myasthenia gravis
- Graft versus host disease

(with chemotherapy, biologicals, radioimmunotherapies), and as part of myeloablative regimens. Its part (as cold antibody) in the Zevalin treatment regimen led to the approval of this new therapeutic as the first radioimmunotherapy for the treatment of cancer. A wide variety of other indications are being explored.

In 2002, rituximab became the number one, brand name, cancer therapeutic in the world. It is an important component of the current curative combination for aggressive NHL (R + CHOP). In the future, rituximab may become a part of other curative treatment regimens for hematologic malignancies and will also find utility in the treatment of major diseases such as rheumatoid arthritis and other autoimmune disorders.

Acknowledgments

None of the research described in this chapter could have been performed without the selfless, courageous, and generous participation of the numerous lymphoma patients who volunteered for the initial clinical trials. This is particularly true of those who participated in our Phase I trials, at a time when the efficacy and safety of the antibody were unknown. They had hope, faith, and courage. They trusted us with their lives. Without these heroes, Rituxan would not exist today. We (the researchers), and the many patients who will benefit from their participation in these studies, owe them a debt of gratitude.

References

1. A. J. Grillo-López, *Oncol. Spec.* **2001**, *2*, 700–705.
2. Review of the clinical data from the initial development studies of rituximab. A. J. Grillo-López, C. A. White, C. Varns et al., *Semin. Oncol.* **1999**, *26* (Suppl. 5, 14), 66–73.
3. Presents new therapeutic paradigms in the treatment of NHL. A. J. Grillo-López, *Expert Rev. Anticancer Ther.* **2002**, *2*(3), 323–329.
4. A candid review of historical events during the development of rituximab. A. J. Grillo-López, *Semin. Oncol.* **2000**, *27* (Suppl. 6, 12), 9–16.
5. Provides an insight into the key laboratory and in vivo studies that helped elucidate the mechanisms of action of rituximab. D. R. Anderson, A. Grillo-López, C. Varns et al., *Biochem. Soc. Trans.* **1997**, *2*, 705–708.
6. Review of the preclinical data on rituximab. M. E. Reff, K. Carner, K. S. Chambers et al., *Blood* **1994**, *83*(2), 435–445.
7. World Health Organization, *WHO Handbook for Reporting Results of Cancer Treatment*, Offset Publication, Geneva, 1979, p. 48.
8. M. Oken, R. Creech, D. Tormey et al., *Am. J. Clin. Oncol.* **1982**, *5*, 649–655.
9. S. Horning, B. Cheson, B. Peterson et al., *Proc. Am. Soc. Clin. Oncol.* **1997**, *16*, 18a (#62).
10. A. J. Grillo-López, S. Horning, B. D. Cheson et al., *Exp. Hematol.* **1997**, *25*(8), 732 (#17).
11. A. J. Grillo-López, *Response Evaluation in Patients with Low-Grade or Follicular NHL*, Presentation to the Biological Response Modifiers Advisory Committee of the FDA, Washington, DC, 1996.
12. A. J. Grillo-López, Response criteria and QA auditing of responses in low-grade or follicular lymphomas: experience in the IDEC-C2B8 single arm pivotal trial in *CBER, FDA (Chair), Clinical Trials in Biotechnology* (Ed.: K. Weiss), Workshop Sponsored by the Drug Information Association, Dana Point, CA, 1997.
13. A. J. Grillo-López, B. Cheson, S. Horning et al., *Blood* **1998**, *92*(10), 412a (#1701).
14. Important to all who evaluate NHL patients for response and also to those who would like to compare clinical results from one publication in the medical literature with others. A. J. Grillo-López, B. D. Cheson, S. J. Horning et al., *Ann. Oncol.* **2000**, *11*, 399–408.
15. A. J. Grillo-López, B. D. Cheson, S. J. Horning et al., *Ann. Oncol.* **2000**, *11*, 399–408.
16. A. J. Grillo-López *Scientific and Medical Summary on IDEC-C2B8: Rituxan (rituximab)*, Biologics License Application Presentation to the Biological Response Modifiers Advisory Committee of the FDA, Washington, DC, 1997.
17. Current response criteria for lymphoma are discussed and presented in this important paper. B. D. Cheson, S. J. Horning, B. Coiffier et al., *J. Clin. Oncol.* **1999**, *17*(4), 1244–1253.
18. A. J. Grillo-López, P. Mclaughlin, B. Cheson et al., *Proc. Am. Soc. Clin. Oncol.* **1999**, *18*, 13a (#44).
19. A. J. Grillo-López, C. Varns, D. Shen et al., *Ann. Oncol.* **1999**, *10*(3), 178. (#660).
20. Report on the first clinical trial with rituximab. D. G. Maloney, T. M. Liles, D. K. Czerwinski et al., *Blood* **1994**, *84*(8), 2457–2466.
21. A. J. Grillo-López, D. Shen, D. Lee et al., *Blood* **2000**, *96*(11), 238b (#4760).
22. M. S. Czuczman, A. J. Grillo-López, C. A. White et al., *Blood* **2001**, *98*(11), 601a (#2519).
23. E. E. Hedrick, R. I. Fisher, B. K. Link et al., *Blood* *98*(11, Part 2), 229b (#4636).
24. D. G. Maloney, A. J. Grillo-López, D. J. Bodkin et al., *J. Clin. Oncol.* **1997**, *15*(10), 3266–3274.
25. A. J. Grillo-López, *Semin. Oncol.* **2000**, *27* (Suppl. 6, 12), 9–16.
26. D. G. Maloney, A. J. Grillo-López, C. A. White et al., *Blood* **1997**, *90*(6), 2188–2195.
27. J. M. Foran, R. K. Gupta, D. Cunningham et al., *Br. J. Haematol.* **2000**, *109*(1), 81–88.
28. M. Ghielmini, S. F. Schmitz, K. Burki et al., *Ann. Oncol.* **2000**, *11* (Suppl. 1), S123–S126.
29. Report on the pivotal trial that lead to the approval of rituximab. P. Mclaughlin, A. J. Grillo-López, B. K. Link et al. *J. Clin. Oncol.* **1998**, *16*, 2825–2833.
30. T. A. Davis, C. A. White, A. J. Grillo-López et al., *J. Clin. Oncol.* **1999**, *17*(6), 1851–1857.
31. Key article on the pharmacokinetics of rituximab. N. L. Berinstein, A. J. Grillo-López, C. A. White et al., *Ann. Oncol.* **1998**, *9*, 995–1001.
32. L. D. Piro, C. A. White, A. J. Grillo-López et al., *Ann. Oncol.* **1999**, *10*, 655–661.

33. A. Saven, A. J. Grillo-López, N. Janakiraman et al., *Blood* **2000**, *96*(11), 730a (#3155).
34. A. Aviles, M. I. Leon, J. C. Diaz-Maqueo et al., *J. Hematother. Stem Cell Res.* **2001**, *10*(2), 313–316.
35. Rituximab (Rituxan, MabThera) Package Insert, FDA Website at http://www.fda.gov/.
36. T. A. Davis, A. J. Grillo-López, C. A. White et al., *J. Clin. Oncol.* **2000**, *18*(17), 3135–3143.
37. A. J. Grillo-López, D. Shen, D. Lee et al., *Blood* **2000**, *96*(11), 238b (#4760).
38. T. A. Davis, A. J. Grillo-López, P. Mclaughlin et al., *Blood* **2000**, *96*(11), 733a (#3171).
39. A. J. Grillo-López, P. Mclaughlin, K. Wey et al., *Blood* **2000**, *96*(11), 332a (#14342).
40. J. D. Hainsworth, *Semin. Oncol.* **2002**, *29*(1) (Suppl. 2), 25–29.
41. M. Ghielmini, S. -F. Hsu Schmitz, S. Cogliatti et al., *Blood* **2002**, *100*, 161a. (abstract #604).
42. Key article on the new "gold standard" in the therapy of aggressive NHL. B. Coiffier, E. Lepage, J. Briére et al., *N. Engl. J. Med.* **2002**, *346*(4), 235–242.
43. J. Boye, T. Elter, A. Engert, *Ann. Oncol.* **2003**, *14*, 520–535.
44. W. H. Wilson, S. R. Frankel, N. Drbohlav et al., *Proc. Am. Soc. Clin. Oncol.* **2000**, *20*(1), 290a. (#1158).
45. First study of the rituximab + CHOP combination. M. S. Czuczman, A. J. Grillo-López, C. A. White et al., *J. Clin. Oncol.* **1999**, *17*(1), 268–276.
46. J. M. Vose, B. K. Link, M. L. Grossboard et al., *J. Clin. Oncol.* **2001**, *19*, 389–397.
47. M. S. Czuczman, A. J. Grillo-López, C. A. White et al., *Blood* **2001**, *98*(11), 601a. (#2519).
48. T. A. Davis, D. G. Maloney, A. J. Grillo-López et al., *Clin. Cancer Res.* **2000**, *6*, 2644–2652.
49. S. Sacchi, M. Frederico, U. Vitolo et al., *Haematologica* **2001**, *86*, 951–958.
50. A. Demidem, T. Lam, S. Alas et al., *Cancer Biother. Radiopharm.* **1997**, *12*(3), 177–185.
51. A. J. Grillo-López, P. Chinn, R. Morena, et al., *Blood* **1995**, *86*(10), 55a (#207).
52. S. J. Knox, M. L. Goris, K. Trisler et al., *Clin. Cancer Res.* **1996**, *2*, 457–470.
53. T. E. Witzig, C. A. White, G. A. Wiseman et al., *J. Clin. Oncol.* **1999**, *17*(12), 3793–3803.
54. A. J. Grillo-López, *Expert Rev. Anticancer Ther.* **2002**, *2*(5), 485–493.
55. A. J. Grillo-López, *IDEC-C2B8: Rituxan™ (rituximab)*, Oral/slide Presentation – Biological Response Modifiers Advisory Committee, Washington, DC, July 1997, (Data on file – IDEC Pharmaceuticals and US FDA).
56. A. J. Grillo-López, C. A. White, B. K. Dallaire et al., *Curr. Pharm. Biotechnol.* **2000**, *1*(1), 1–9.
57. A. Grillo-López, Presentation to BR-MAC, Scientific and Medical Summary of IDEC-C2B8, July 25, 1997, www.fda.gov/OHRMS/dockets/ac/97/transcpt/3311t2.rtf.
58. A. Wei, A. J. Grillo-López, Rituximab Presentation to the US FDA Biologic Response Modifiers Advisory Committee, FDA Advisory Committee Meetings on Video Tape, F-D-C Reports, Special Projects Department (Phone: 301-657-9830), July 25, 1997, http://www.fdaadvisorycommittee.com/FDC/advisorycommittee/toc.htm.
59. A. J. Grillo-López, B. K. Dallaire, A. Mcclure et al., *Curr. Pharm. Biotechnol.* **2001**, *2*, 301–311.
60. A. J. Grillo-López, *Rituxan: The First Decade and the Future*, Grand Rounds Presentation at Teacher's Hospital, San Juan, Puerto Rico, February 2003.

13
Somatic Gene Therapy – Advanced Biotechnology Products in Clinical Development

Matthias Schweizer, Egbert Flory, Carsten Muenk and Klaus Cichutek
Paul-Ehrlich-Institut, Langen, Germany

Uwe Gottschalk
Pharma-Biotechnology, Wuppertal, Germany

13.1
Introduction

Innovative biopharmaceuticals of the future include gene transfer medicinal products [1, 2]. It can be assumed that by mid-2003, approximately 4000 patients or healthy individuals have been treated within a clinical gene therapy trial, approximately 600 of those in Europe and approximately 260 in Germany. Most of the clinical trials are currently in phase I or II because, due to a great diversity of ongoing developments, clinical experience must first be gained before target-orientated product development and phase III clinical trials can be initiated. In this regard, investigator-driven gene therapy strategies developed by biomedical laboratories together with special clinical teams are very distinct from those developed by the pharmaceutical industry. Investigator-driven gene therapy strategies are being invented by teams of biomedical researchers and physicians while developing a new approach for the treatment of a special disease in a defined stage. This is used, for the first time, on a selected group of patients in first clinical trials of phase I/II and is aimed at proving the safety of the medicinal product. In clinical trials sponsored by the pharmaceutical industry, this phase of orientation has often already been completed and further development in phase II or III is aimed at dose finding or proving efficacy. Concerning product development, there are no standard approaches because, at this stage of development, little experience has been gained and the types of gene transfer medicinal products are very diverse [3]. Therefore, in the following sections the main current clinical developments will be described while a brief outline of a single example of a manufacturing process, also due to manifold diversity, is given.

Gene transfer medicinal products for human use are medicinal products used for in vivo diagnosis, prophylaxis, or therapy (Fig. 1). They contain or consist of

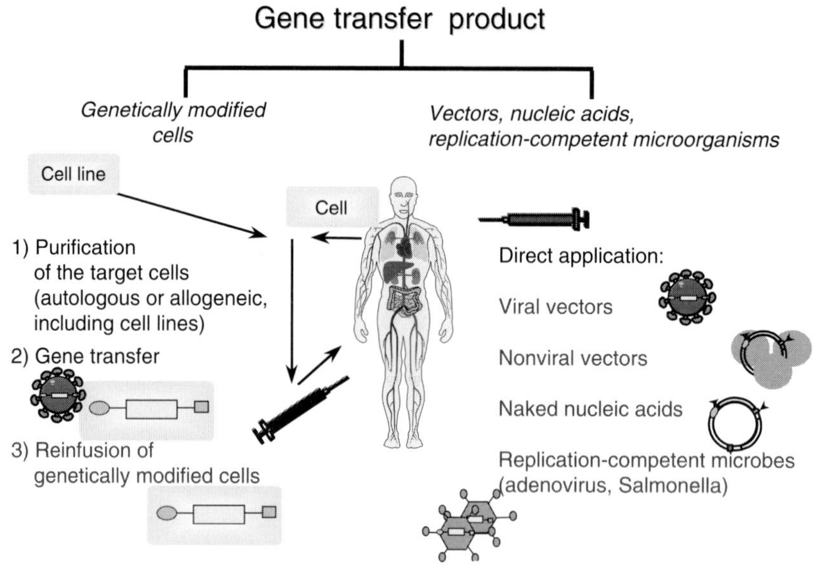

Fig. 1 Gene transfer medicinal products. The gene transfer medicinal products mentioned here are identical with those described in Table 1 of the European "Note for guidance on the quality, preclinical, and clinical aspects of gene transfer medicinal products (CPMP/BWP/3088/99)". The definition given is in compliance with the legally binding definition of gene therapeutics in Part IV, Annex I of Directive 2003/63/EC amending Directive 2001/83/EC.

1. genetically modified cells,
2. viral vectors, nonviral vectors or so-called naked nucleic acids, or
3. recombinant replication-competent microorganisms used for purposes other than the prevention or therapy of the infectious diseases that they cause.

The aim of the nucleic acid or gene transfer is the genetic modification of human somatic cells, either in the human body, that is, in vivo, or outside the human body, that is, ex vivo, in the latter case followed by transfer of the modified cells to the human body [4, 5]. The simplest case of genetic modification of a cell results from the addition of a therapeutic gene encompassed by an expression vector [6]. At least in theory, nucleic acid transfer may also be aimed at exchange of individual point mutations or other minimal genetic aberrations. Scientifically, this process is termed homologous recombination with the aim of repairing a defective endogenous gene at its locus. In principle, this can be achieved by a so-called homologous recombination achieved by transferring oligonucleotides, where – owing to 5′ and 3′ flanking homology regions – the new correct DNA sequence is replacing the existing defective one. In practice, homologous recombination is technically not yet achievable with the efficiency that will be required for clinical use.

Normally, genetic modification of cells is nowadays achieved by the transfer of an expression vector on which the therapeutic gene is located. The vector is transferred to cells via a delivery system (Fig. 2) such as a viral vector particle, a

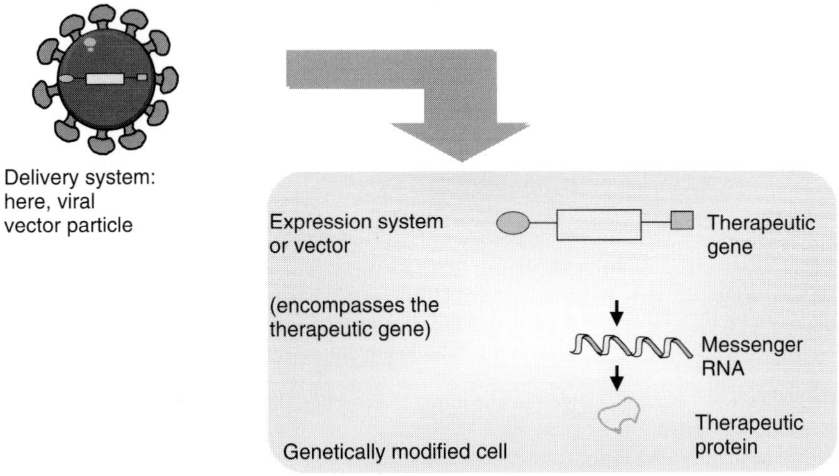

Fig. 2 Delivery system and expression vector used as gene transfer medicinal products. The terminology complies with the definition of gene therapeutics in Part IV, Annex I of Directive 2003/63/EC amending Directive 2001/83/EC.

nonviral vector complex, or a plasmid. In the latter case, the expression vector is inserted into and is therefore part of a bacterial plasmid, which allows its manufacture and amplification in bacteria. Viral expression vectors contain the sequence signals (nucleic acid sequences) required for transfer by a particular viral vector particle. For retroviral vectors, for example, such signals are encompassed by the flanking "Long Terminal Repeat" (LTR) sequences, the packaging signal Psi (Ψ) required for incorporation of the expression vector by the retroviral vector particle, and other sequence signals. For nonviral vector complexes and naked nucleic acid, the expression vector is part of a bacterial carrier, the so-called plasmid DNA. Nonviral vectors are, for example, plasmid DNA mixed with a transfection reagent, whereas naked DNA does not contain a transfection reagent.

Another example of a gene transfer medicinal product is recombinant microorganisms such as conditionally replication-competent adenoviruses for tumor therapy [7]. Here, neither an endogenous cellular gene is repaired by homologous recombination nor is a non-adenoviral therapeutic gene transferred. The transfer of conditional replicating adenoviruses to the malignant tumor cells induces cell lysis and local tumor ablation. The entire genome of the adenovirus is transferred without an additional therapeutic gene. The adenoviral genome may therefore be considered as the therapeutic gene.

Gene transfer efficiency plays a central role in gene transfer. It depends on a number of factors, for example, target cell, type of application (ex vivo or in vivo strategy), the tissue or organ containing the target cells, the physiological situation, and the

Tab. 1 Gene transfer methods (vectors/delivery systems)

Delivery system	***Description***	***Chromosomal integration***
Naked nucleic acid	Plasmid DNA, in the absence of transfection reagents	No (after im inoculation)
Nonviral vector	Plasmid DNA/transfection reagent mixture	No (application dependent)
Viral vector		
Retroviral vector	Derived from murine leukemia virus (MLV)	Yes
Lentiviral vector	Derived from HIV-1	Yes
Adenoviral vector	Deletions in the virus genes E1, E3 or E4, E2ts, combinations thereof, or "gutted" (gene-depleted)	No
Conditionally replication-competent adenovirus	No therapeutic gene except for the virus genome	No
Adeno-associated virus (AAV) vector	Wild-type AAV-derived	Yes/no (application dependent)
Smallpox virus vector	MVA (Modified Vaccinia Ancara)	No
	ALVAC (Avian Vaccinia)	No
	Vaccinia	No
Alphavirus vector	Semliki Forest virus (SFV)	No
Herpes-viral vector	Herpes simplex virus	No

disease and disease stage. Table 1 shows the most common viral vectors currently in clinical use. The vectors shown are replication incompetent and only transfer the expression vector void of any viral genes as much as possible. So-called integrating vectors mediate chromosomal integration of the expression vector (e.g. retroviral vectors) [8], whereas nonintegrating vectors lead to an episomal status of the expression vector in the cell (e.g. adenoviral vectors), or to its cytoplasmatic replication (e.g. alphavirus-derived vectors, vaccinia). Vectors derived from vaccinia, for example, used for tumor vaccination, may be replication incompetent such as Modified Vaccinia Ankara (MVA) or Avian Vaccinia (ALVAC), or replication competent, but attenuated like vaccinia.

After uptake by human somatic cells, the expression vector is transcribed like a normal cell gene. The resulting messenger RNA (mRNA) is translated and the therapeutic protein is synthesized by the cellular machinery. When so-called ribozyme genes are used, the mRNA acts like a catalytic enzyme and is itself the therapeutic gene product. As already mentioned, when a recombinant microorganism such as a conditionally replication-competent adenovirus (RCA) is used, the genome of the microorganism may be seen as the therapeutic gene.

13.2 Gene Transfer Methods

The objective of clinical gene transfer is the transfer of nucleic acids for the purpose of genetically modifying human cells (Fig. 3).

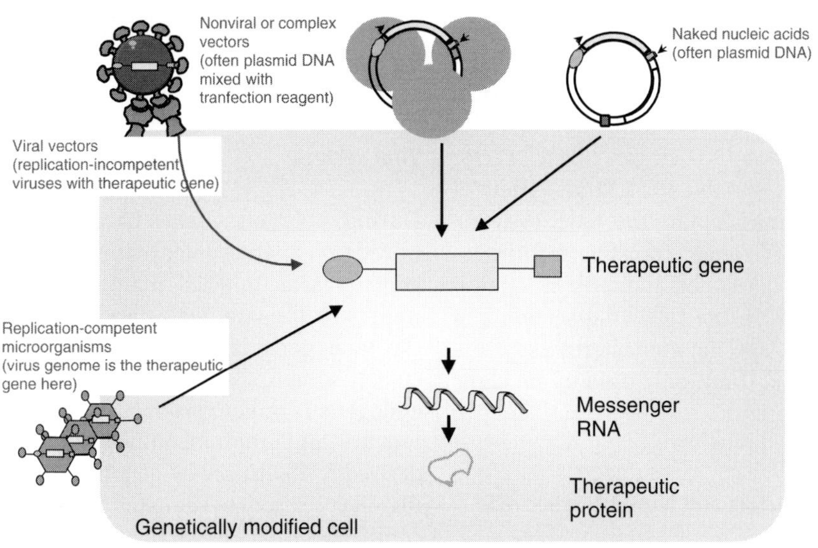

Fig. 3 Delivery systems used in clinical gene transfer. During gene therapy, an expression vector (therapeutic gene) is transferred to somatic cells via a delivery system, for example, a viral or nonviral vector (replication-incompetent), a naked nucleic acid, or a recombinant, mostly conditionally replication-competent microorganism. The gene transfer, termed transfection when a viral vector is used or when naked DNA or a nonviral vector is used, leads to genetic modification of the cell. The gene transfer can be carried out in vivo, that is, directly in or on the human body, or ex vivo, that is, in cell culture followed by the transfer of the modified cells to the human body.

Whether a viral, a nonviral vector, or naked plasmid DNA is used depends on the target cell of the genetic modification and whether an in vivo modification of the cell is at all possible. For a monogeneic disease affecting immune cells, it is, for example, possible to purify CD34-positive cells or lymphocytes from the peripheral blood (e.g. by leukapharesis) to genetically modify the cells in culture, and to return the treated cells. Before reapplication, the treated cells may or may not be enriched. Currently, long-term correction of cells is only possible when integrating vectors such as retroviral or lentiviral vectors are used. Owing to the chromosomal integration of the expression vector, the genetic modification is passed on to the daughter cells during cell division and it persists. Only long-term expression may still be a problem. For therapy of a monogeneic disease such as cystic fibrosis, the target cells are primarily the endothelial cells of the broncho-pulmonary tract, which can only be subjected to in vivo modification attempts. Although long-term correction would be desirable, in vivo modification using adenoviral vectors appeared to be more promising because the target cells were largely in a resting state of the cell cycle amenable to adenoviral gene transfer due to expression of the cell surface receptors used by adenoviruses for cell entry. In addition, the amount and titers of adenoviral vectors seemed suitable. These examples illustrate that a number of factors contribute to the choice of the treatment strategy, the vector, and the route of administration. No single "ideal" vector is therefore suitable for a large variety of gene therapies [9]. In the past 15 years, many novel gene transfer techniques have been developed and used in clinical studies [10]. In the following section and in Table 1, specific characteristics of the vectors most commonly used in the clinic are summarized.

13.2.1 Nonviral Vectors and Naked Nucleic Acid

The advantage of nonviral gene transfer systems compared with viral gene transfer systems is the smaller size limitations for the genes to be transferred. The expression vectors are nowadays usually part of a bacterial plasmid that can easily be amplified and grown in bacterial cultures. Plasmid DNA of up to 20 kb pairs encompassing an expression vector of up to 17 kb pairs can easily be manufactured. Promising methods for the in vivo administration of plasmid DNA include intradermal or intramuscular injection for the so-called naked nucleic acid transfer. Needle injection or application by medical devices such as gene guns can be used for this purpose. For so-called nonviral vectors, such as synthetic liposomes or other transfection reagents mixed with plasmid DNA, the DNA-binding liposomes mediate contact with the cellular plasma membrane, thus releasing the DNA into the cytoplasma of the cell where uptake by the nucleus has to occur subsequently [11]. During receptor-mediated uptake of nonviral vectors, cell surface proteins (receptors), such as asialoglycoprotein or the transferrin receptor, mediate cellular uptake of the DNA complex containing a specific receptor ligand [12, 13].

13.2.2 Viral Vectors

During evolution, viruses have been optimized to efficiently enter mammalian cells and replicate. Infected mammalian cells transcribe the viral genes and synthesize the viral gene products with high efficiency, sometimes to the disadvantage of endogenous protein production. Viral vectors are replication-incompetent particles derived from viruses by genetic engineering that no longer transfer to cells the complete set or any viral genes. Instead, an expression vector with one or more therapeutic genes is transferred to cells. Since no complete viral genome is transferred, virus replication is impossible or, in some cases, impaired as with first- or second-generation adenoviral vectors. The following section briefly describes the properties of the currently frequently used viral vectors.

13.2.2.1 Retroviral Vectors

The retroviral vectors in clinical use have mainly been derived from murine leukemia virus (MLV) [14]. MLV causes leukemia in mice and replication-competent retrovirus (RCR) in a contaminated vector preparation was shown to cause leukemia in severely immunosuppressed monkeys. RCR absence therefore has to be verified before human use of retrovirally modified cells; MLV vector use in vivo has been very rare. The genome of the retroviral vectors consists of two copies of single-stranded RNA, which contains one or more coding regions flanked by the viral control elements, the so-called "long terminal repeat" (LTR) regions. In the infected cells, the RNA is translated into double-stranded viral DNA and integrated into the cell. The integrated vector DNA is the expression vector. MLV vectors allow efficient genetic modification of proliferating cells by chromosomally integrating the expression vector.

Advantages of retroviral vectors include high gene transfer (transduction) efficiency, and long-term modification of cells due to stable integration of the expression vector into the chromosome of the cells. In addition, the MLV envelope proteins can be exchanged against those from other viruses (which is termed "vector pseudotyping"). This allows preparation of MLV vectors with improved transduction efficiency for certain cell types. Disadvantages of retroviral vectors include the small size of the coding region (approximately 9 kb pairs or less), the restriction of transduction to proliferating cells only, insertional mutagenesis due to integration, and the low titer of usually not more than 10^8 transducing units per milliliter of vector preparation. Although chromosomal integration occurs generally at random, it may lead to activation of cellular cancer genes, so-called proto-oncogenes, or, theoretically, to inactivation of tumor suppressor cells. In conjunction with additional genetic mutations, this may result in very low frequency in malignant cell transformation. Hundreds of patients who have been treated with retrovirally modified hematopoietic cells years ago have not shown any signs of cancer related to the gene transfer except for two patients treated during a *SCID-X1* gene therapy trial in France. In the latter two leukemia cases, the vector-mediated overexpression of the proto-oncogene *LMO2*, possibly in conjunction with the therapeutic γc chain gene (which may influence cell proliferation and signal transduction) and the SCID-X1 disease, is the probable cause of leukemia.

13.2.2.2 **Lentiviral Vectors**

Lentiviral vectors have been derived from human immunodeficiency virus type 1 (HIV-1), simian immunodeficiency virus (SIV) isolated from various Old World monkeys, feline lentivirus (FIV), and equine infectious aneamia virus (EIAIV) isolated from horses [15, 16]. Lentiviruses cause an acquired immunodeficiency syndrome and replication-competent virus has therefore to be excluded before human use by batch-to-batch analysis and verification of replication-competent lentivirus (RCL) absence. Lentiviral vectors may transfer coding regions of up to 9 kb pairs and allow pseudotyping just like MLV vectors. Their advantage is the dual capacity to transfer therapeutic genes into nonproliferating cells in conjunction with persistent genetic modification due to chromosomal integration. This could be useful for ex vivo modification of stem cells and in vivo modification of neuronal cells. Most lentiviral vectors have been pseudotyped with the G protein of vesicular stomatitis virus (VSV-G) or the envelope proteins of Gibbon ape leukemia virus. The first clinical study using lentiviral vectors has started in 2003 and involves the ex vivo modification of autologous lymphocytes of HIV infected patients with a therapeutic ribozyme gene shown in vitro to inhibit HIV-1 replication.

13.2.2.3 **Adenoviral Vectors**

The adenoviral genome consists of double-stranded DNA that persists episomally, that is, inside the nucleus, but not integrated into the chromosome of the cell [17]. Therefore, the genetic modification may be lost during cell proliferation. Adenoviral vectors are the currently preferred vectors for the in vivo transduction of a variety of human somatic cells including nonproliferating cells. In contrast to lentiviral vectors, they allow insertion of larger coding regions of therapeutic genes above 10 kb pairs and are not associated with a detectable risk of insertional oncogenesis. In addition, vector titers above 10^{11} transducing units per milliliter can usually be achieved. The lack of long-term expression is in part due to the fact that certain adenovirus genes have been kept on first- or second-generation adenoviral expression vectors, and because of the frequent generation of RCA during production. So-called gutless vectors are void of any adenoviral genes, but have to be purified from RCA after production.

Some wildtype (replicating) adenovirus strains cause inflammations of the airways and the conjunctivae. Adenoviral vectors may therefore also be transferred by inhalation of aerosols, and inflammations observed following vector applications are mainly local, transient, and associated with very high titer applications. High-titer adenoviral vectors are no longer systemically administrated because one patient had died during systemic administration of a maximum dose of approximately 10^{13} vector particles during gene therapy of the monogeneic disease OTC (Ornithine Transcarbamylase) deficiency, a life-threatening metabolic disorder.

13.2.2.4 AAV (Adeno-associated Viral) Vectors

Adeno-associated viruses (AAVs) belong to the family of parvoviruses [18, 19]. Their genome consists of single-stranded DNA. Wild-type AVV can only replicate in the presence of helper viruses like adenovirus or herpesvirus and has not been associated with any disease. AAV can infect hematopoietic cells including nonproliferating cells. Integration in infected human somatic cells is often confined to a distinct locus on human chromosome 19. AAV-derived vectors are usually classified as integrating vectors, although vector integration is unfortunately no longer confined to chromosome 19, but absence of integration may be observed, for example, following intramuscular administration The size of the coding region is very limited (approximately 4 kb pairs).

13.2.2.5 Poxvirus Vectors

Poxvirus vectors encompass vaccinia derived from the smallpox vaccine and more attenuated variants like ALVAC or MVA (Modified Vaccinia Ankara). Their genome consists of single-stranded DNA of 130 to 300 kb pairs. Replication is restricted to the cytoplasm of cells and high amounts of protein are synthesized by the cell following transduction. Most applications therefore involve intramuscular vaccination.

13.3 Clinical Use

13.3.1 Overview on Clinical Gene Therapy Trials

A number of clinical trials show promising results (see Table 2). In the past few years, it has become increasingly clear that for each disease, the development of a particular and specific gene transfer method in connection with a particular treatment approach will probably be necessary. The first standard use of an approved gene transfer medicinal product is to be expected within the next seven years since approximately 1% of the clinical gene therapy studies are in an advanced stage of phase II or phase III clinical trial.

Tab. 2 Promising clinical gene therapy trials

Disease	*Therapeutic gene*	*Pharmaceutical form/vector*	*Target cell*	*Remarks*
Severe combined immunodeficiency (SCID-X1)	γc chain gene (e.g. interleukin-2-receptor part)	MLV vector	Bone marrow stem cells ex vivo	4 of 1 patient cured, 2 leukemias
PAOD (Peripheral Artery Occlusive Disease)	*VEGF* gene (Vascular Endothelial Growth Factor)	Plasmid DNA	Muscle/ endothelial cells in vivo (i.m.)	Improved blood flow
Head and neck tumor	Adenovirus genome (cell lysis/apoptosis)	Tumor cell specific replicating adenovirus	p53-negative tumor cells in vivo	Local tumor remission in combination with chemotherapy
Graft versus Host Disease (GvHD) in donor lymphocyte transfer for leukemia treatment	Thymidin kinase gene of the herpes simplex virus, followed by treatment with Ganciclovir	T-cells, MLV vectors	T-lymphocytes ex vivo	Successful treatment of host-versus-graft disease
Hemophilia B	Coagulation factor IX-gene	AAV (Adeno-associated virus) vector	Muscle cells in vivo (i.m.)	Improved coagulation factor concentration

Clinical gene therapy studies had been performed initially in North America and Europe. About 50 clinical gene transfer studies have been registered in Germany, with slightly more than 250 patients treated (http://www.pei.de, http://www.zks.uni-freiburg.de/dereg.html). A general overview on registered studies is listed on the following websites: http://www.wiley.co.uk/genetherapy or www.pei.de. In Germany, a public registry will be available in 2004.

Target diseases in most clinical gene therapy trials have been cancer, cardiovascular diseases, infectious diseases such as AIDS or monogeneic congenital disorders. The vectors most frequently used ex vivo are MLV vectors derived from MLV, whereas vectors derived from adenovirus, pox viruses, and AAV are usually used in vivo. A growing number of studies involves the use of nonviral vectors or naked DNA.

13.3.2 Gene Therapy of Monogeneic Congenital Diseases

The idea underlying gene therapy is the replacement of a defective gene by its normal, functional counterpart, for example, a mutation of the gene encoding the γc chain of the interleukin-2 and other receptors is the cause of the congenital immune disorder SCID-X1 (Severe Combined Immunodeficiency Syndrome). Owing to this defect, immunologically relevant receptors are unable to mediate the normal differentiation and immune function of lymphoid cells such as T-cells and natural killer lymphocytes (NK). Therefore, newborn babies suffering from SCID-X1 have a very limited immune system and must live in a germ-free environment. Their life expectancy is strongly reduced. Conventional treatment, that is, bone marrow transplantation, can provide a cure to a certain extent, but involves a high risk if no HLA haploidentical donor is available. For the latter situation, gene therapy within the framework of a clinical study was considered in France [20].

In this study, autologous CD34-positive bone marrow stem cells were retrovirally modified to express the functional γc chain gene. T-cells and other hematopoietic cells derived form corrected stem cells were shown to repopulate the hematopoietic cell compartment, and over a period of up to 3 years, 11 treated patients, mostly newborns, displayed a functional and nearly normal immune system. This represents the first reproducible cure of a disease by gene therapy.

A leukemia-like lympho-proliferative disease was diagnosed roughly three years after treatment of two obviously cured patients. Treatment had been started at the age of a few months. Subsequent analysis revealed that the leukemia-like disease was indeed caused by the MLV vector; the disease mechanism is termed "insertional oncogenesis" resulting from insertional mutagenesis of the proto-oncogene *LMO2*. According to the current knowledge, up to 50 cells with an integration in LMO2 may have been administered together with the approximately 10^8 genetically modified CD34-positive bone marrow cells. Owing to the expression vector integration, the transcription of the *LMO2* gene was deregulated and activated. Under normal circumstances, the body can cope with individual cells presenting preneoplastic changes like the one described. In the two treated children, however, further genetic changes must have accumulated to finally result in leukemia. Contributing factors discussed include the effect of the therapeutic γc chain gene, the product of which influences cell proliferation and differentiation, and other so far unknown genetic changes that may have occurred during the massive in vivo cell replication. In SCID-X1, the T-cell compartment is completely depleted, and is replenished after gene therapy by differentiation and replication of a few genetically corrected blood stem cells. During this process, genetic aberrations may occur with substantial frequency. However, further analysis will be required to understand the exact cause of leukemia development in *SCID-X1* gene therapy. Since hundreds of patients treated with retrovirally modified cells in the past 10 years have not developed leukemia up to now, it is currently assumed that a practical risk of leukemia exists only in *SCID-X1* gene therapy.

Gene therapy of hemophilia B also seems promising. Here, AAV vectors encoding a smaller but functional version of the human coagulation factor IX-gene were administered by intramuscular injection. A detectable increase in factor IX plasma concentration was observed. Even repeated AAV injections were well tolerated.

13.3.3
Tumor Gene Therapy

There are various gene therapy approaches that are being developed for the treatment of cancer. They are aimed at inhibiting molecular pathways underlying malignant cell transformation. In other cases, tumor cell ablation by directly applying cell-killing mechanisms, or, more indirectly, by improving immunological defense mechanisms directed against tumor cells are attempted [21].

A number of gene therapy studies involving the adenoviral transfer of tumor suppressor genes like p53 have already been performed. This is aimed at reverting malignant cell transformation or at inducing apoptosis. However, transduction following, for example, needle inoculation into tumors has been shown to be limited to a few cells close to the needle tracks. Direct tumor cell ablation by local injection of conditionally replication-competent adenoviruses in head and neck tumors led to detectable local tumor regression by direct virus-mediated cell lysis, especially if chemotherapy was used in parallel. Here, virus replication improved transduction efficiency in vivo. For the treatment of malignant brain tumors, variant herpesviruses have been inoculated into the tumor in order to lyse the tumor cells in vivo, especially if prodrugs have been administered that are converted by the viral thymidin kinase gene to a toxic drug.

In addition, a number of clinical approaches have already been tested that led to an improvement of immune recognition of tumors [22]. They involved intratumoral injection of vectors that transfer foreign *MHC* genes, such as B7.1 or B7.2, or cytokine genes, for example, interleukin-2 or granulocyte-macrophage colony stimulating factor (GM-CSF). Here, vaccinia-derived vectors such as MVA or ALVAC have often been used. Autologous or allogeneic tumor cells were also modified ex vivo by transfer of immunostimulating genes. Promising results have been reported from a phase I study in which autologous tumor cells were adenovirally modified with the *GM-CSF* gene and rapidly reinoculated to stimulate antitumor immunity.

13.3.4
Gene Therapy of Cardiovascular Diseases

Local intramuscular injection of plasmid DNA or adenoviral vectors encoding vascular epithelial growth factor or fibroblast growth factor, both able to induce the formation of new blood vessels, have been used to improve microcirculation in ischemic tissue. Needle injection of plasmid DNA has been used in leg muscle, catheter application or needle injection was also tried in ischemic heart muscle. The formation of new blood vessels and an improvement in the microcirculation has been observed.

A narrowing of the blood vessels (restenosis) often occurs after coronary blood vessel dilatation by stent implantation. This is probably caused by the proliferation of smooth muscle cells following injuring of the blood vessel endothelium by the stent. Here, the role of adenoviral or plasmid DNA–mediated transfer of the gene encoding inducible nitroxid synthase (iNOS) is thought to result in reduced cell proliferation.

13.3.5
Preventive Vaccination and Gene Therapy of Infectious Diseases

During the past 5 to 10 years, effective medicines have been developed for the treatment of AIDS. Combinations of effective chemotherapeutics are able to inhibit various steps of the replication cycle of HIV-1. This often results in reduction of the viral load in the peripheral blood, sometimes down to a level barely detectable with modern techniques. Because of the requirement for long-term treatment and the massive adverse effects related to conventional treatment by chemotherapy, gene therapy of HIV infection could offer additional therapy options. Ex vivo retroviral transfer of HIV-inhibiting genes into peripheral blood lymphocytes or CD34-positive human cells has been attempted, so far with little success. The therapeutic molecules used include (1) decoy-RNA specifying multiple copies of the Rev- or the Tat-responsive element, so-called poly-TAR or poly-RRE sequences, (2) miniantibodies (single chain Fv, scFv) able to capture viral gene products within the cell, (3) transdominant negative mutants of viral proteins such as RevM10, or (4) ribozyme RNA that enzymatically cleaves RNA. Other genes still under development are designed to prevent entry or chromosomal integration of HIV. It remains to be shown whether such gene therapy approaches present a suitable therapeutic option compared with existing chemotherapy.

The best prevention of infectious diseases is achieved by prophylactic vaccines [23]. Clinical trials using vectored vaccines based on ALVAC or MVA have been initiated. Other clinical trials pursue the goal of developing vaccines against HIV-1, malaria, hepatitis B, tuberculosis, and influenza A virus infections [24]. Vaccination regiments using poxvirus vectors such as ALVAC or MVA in combination with naked DNA as a prime vaccine, sometimes followed by further booster injections of recombinant viral antigens, are being tested in humans. Such regiments have been shown to prevent disease progression after lentivirus infection of monkeys. This illustrates the complexity of vaccination strategies that are currently pursued in vaccine research.

13.3.6 Clinical Gene Therapy for the Treatment of Other Diseases

Clinical gene therapy can also be used for the treatment of diseases not necessarily caused by single known gene defects, if promising therapeutic genes can be reasonably applied. Patients with chronic rheumatoid arthritis, for instance, should benefit from a reduction of the inflammations in joints. Such inflammations are caused or at least maintained by a cascade of events including the overexpression and increased release of a number of inflammatory cytokines. Monoclonal antibodies that are able to reduce the local concentration of the tumor necrosis factor (TNF) have already been successfully used to treat disease. Here, clinical gene transfer approaches involve the transfer of autologous synovial cells modified ex vivo by a therapeutic gene encoding interleukin-1 receptor antagonist. Alternatively, adenoviral vectors with the same gene have been directly injected into the affected joint.

13.4 Manufacture and Regulatory Aspects

The regulation of gene therapy is very complex and differs considerably in the European Union and the United States [25]. In Part IV, Annex I of Directive 2003/63/EC (which replaces Annex I of Directive 2001/83/EC), a definition of so-called gene therapeutics is given. As gene therapy not only includes therapeutic but also preventive and diagnostic use of vectors, nucleic acids, certain microorganisms, and genetically modified cells, the term "gene transfer medicinal products" as used in the relevant European guideline "Note for guidance on the quality, preclinical and clinical aspects of gene transfer medicinal products (CPMP/BWP/3088/99)" seems more exact. An accurate listing of the medicinal products that belong to the group of gene transfer medicinal products can be found in the table contained in the guideline. The definition given in the first chapter of this article is in accordance with this guideline and is in agreement with the definition of gene therapeutic products of Directive 2003/63/EC. The annex of the latter directive contains legally binding requirements for quality and safety specifications of gene transfer products. Although targeted at product licensing, these requirements may have a bearing on their characterization before clinical use. Active ingredients of gene transfer medicinal products may include, for example, vectors, naked plasmid DNA, or certain microorganisms such as conditionally replicated adenovirus. For the ex vivo strategy, the active ingredients are the genetically modified cells.

Written approval by a competent authority in conjunction with positive appraisal by an ethics committee will in future be necessary for the initiation of clinical gene therapy trials. Respective regulatory processes are currently established in all EU member states during transformation of Directive 2001/20/EC. The manufacture of clinical samples in compliance with Good Manufacturing Practice (GMP) will become compulsory. Germ-line therapy is illegal in the European Union. The law relevant for clinical gene therapy trials and manufacture of gene transfer medicinal products in Germany is the German Drug Law (AMG) and respective decrees and operation ordinances. The law governing the physicians' profession stipulates in the "Guidelines on gene transfer into human somatic cells" ("Richtlinien zum Gentransfer in menschlichen Körperzellen") that the competent ethics committee may seek advice from the central "Commission of Somatic Gene Therapy" of the Scientific Council of the German Medical Association before coming to its vote. The Paul–Ehrlich Institut is the competent authority in Germany and offers information on current clinical trial regulations.

Gene transfer medicinal products will be licensed via the centralized procedure by the European Commission. The licensing process is coordinated by the EMEA (European Agency for the Evaluation of Medicinal Products) following submission of a licensing application. The marketing authorization is governed by Council Regulation (EC) No. 2309/93. The recommendation in favor or against marketing authorization is made on the basis of Directives 75/319/EEC and 91/507/EEC by experts of the national competent authorities who are members of the "Committee for Proprietary Medicinal Products" (CPMP).

In the United States, the Center for Biologics Evaluation and Research (CBER) of the "Food and Drug Administration" (FDA) is responsible for clinical trial approval and marketing authorization.

The assessment of the licensing application focuses on the quality, safety, efficacy, and environmental risk of a gene transfer medicinal product. The manufacturing process has to be designed and performed according to Good Manufacturing Process (GMP) regulations. Like other biologicals, gene therapy products have considerably larger size and complexity compared to chemicals, and analysis of the finished product is not sufficient to control their quality and safety. A suitable process management, in-process control of all critical parameters identified within process validation are decisive factors. Gene transfer medicinal products containing or consisting of genetically modified organisms are also subject to contained-use regulations before licensing and until these organisms are applied to humans.

From the economic point of view, procedures for the manufacture of therapeutic DNA must be scalable and efficient, and, at the same time, simple and robust. Manufacturing processes are as manifold as the gene transfer methods used in gene therapy. As an example, manufacture of plasmid DNA for naked nucleic acid transfer can be briefly described as follows [26]. The methods available for plasmid production today largely originate from lab procedures for the production of DNA for analytical purposes (mini preparations) and have been adapted to fit process scale [27]. Toxic substances and those that present a hazard to the environment, expensive ingredients, and nonscalable methods must be avoided [28]. In this context, the experience gained from industrial manufacture of raw materials with the aid of bacterial cultures and virus production for the purpose of vaccine production are useful for fermentation [29]. Suitable methods for downstream processing above all include chromatographic methods with high dynamic capacity and selectivity as well as high throughput [30, 31].

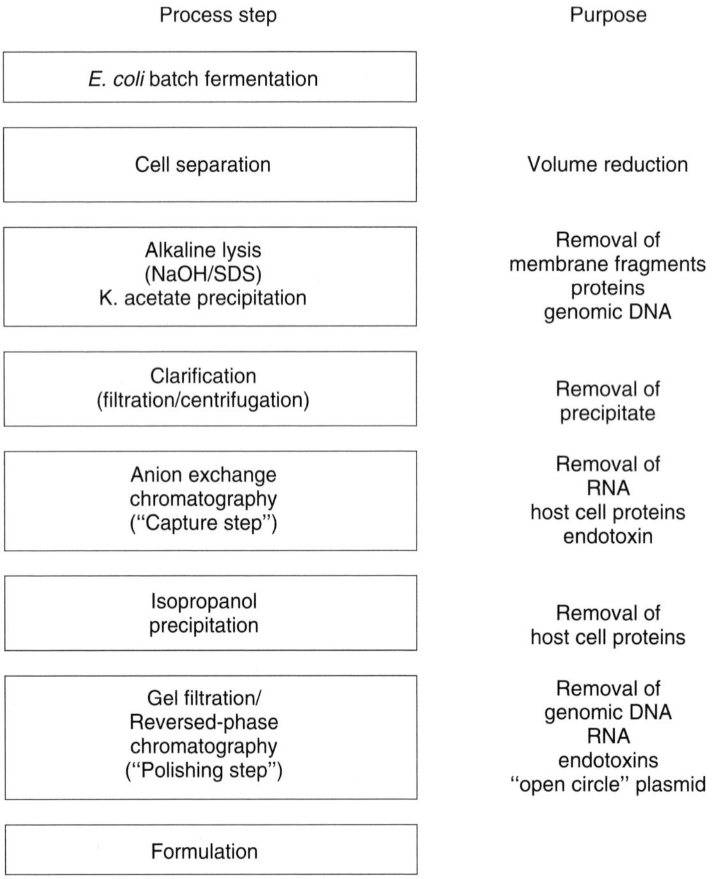

Fig. 4 Therapeutic plasmid DNA: Typical manufacturing process.

Test	Specification (Method)
Appearance	Clear colourless solution (visual inspection)
Size, restriction interfaces (identity)	Agreement with plasmid card (agarose gel electrophoresis, restriction enzyme assay)
Circular plasmid DNA (ccc)	> 95 % (Agarose gel electrophoreses, HPLC)
E. coli DNA	< 0.02 μg/μg plasmid DNA (southern blot)
Protein	Not detectable (BCA protein assay)
RNA	Not detectable (agarose gel electrophoresis)
Endotoxin	< 0.1 EU/μg plasmid DNA (LAL assay)
Sterility	No growth after 14 days (USP)
Specific activity	Conforms to reference standard (in vitro transfection)

Fig. 5 Therapeutic plasmid DNA: Typical release specifications.

In a typical procedure for the manufacture of therapeutic plasmid DNA (cf. Figs. 4 and 5), the first step is batch fermentation of *Escherichia coli* cells from a comprehensively characterized "Master Working Cell Bank" (MWCB). For this purpose, modern methods use high-density fermentation with optimized and safe *E. coli* K12 strains bearing a high number of copies of the required plasmid. The bacterial cells are harvested for further processing, resuspended in a small buffer volume, and lysed in an alkaline lysis procedure [32]. By neutralization, the plasmid DNA is renatured while a large quantity of proteins, membrane components, and genomic DNA remain denatured. After separation of the precipitate by filtration, a chromatographic step can be performed as "capture step". Because of the anionic character of the nucleic acid, anion exchange chromatography (AEX) is the method of choice. In fractionated gradient elutions, differences in the charge enable the separation from contaminated RNA. Gel filtration (GF) or reversed phase (RP) steps can be used for fine purification. For final product analysis, evidence must be provided batch-by-batch that besides the correct identity and homogeneity, critical impurities like microorganisms, host cell proteins, genomic DNA, RNA, or endotoxins have been reduced below the specified limits [33]. Removal of endotoxins is critical for in vivo gene transfer efficiency achieved with naked DNA.

Some established methods from protein chemistry can be used for processing therapeutic DNA. Parallels with the processing of proteins, however, cannot conceal the fact that nucleic acids have some very specific properties. These include the extremely high viscosity of DNA solutions, the high sensitivity of nucleic acids to gravity, the low static and dynamic capacity of their chromatographic adsorption, and the ability to penetrate filtration media with porosities well below their molecular weight ("spaghetti effect").

After first experience, plasmid concentrations of approximately 200 mg L^{-1} fermentation broth can be obtained in high-density fermentation (optical density > 50), corresponding to a yield of approximately 800 mg plasmid DNA per kilogram of dry biomass. Thus, from a fermenter of 1000 L usable volume, approximately 100 g plasmid DNA can be isolated per run in a batch fermentation at a purification yield of approximately 50%. Consequently, capacities for production of kilogram amounts can be built up with existing technologies [34].

13.5 First Experience with the Clinical Use of Gene Transfer Medicinal Products

The development of somatic gene therapy is still in its infancy. A number of theoretical risks of gene therapy have been listed, and numerous approaches and gene transfer methods are being developed in the clinic, even more in preclinical experiments.

Until today, SCID-X1 patients have apparently been cured by gene therapy using retrovirally modified bone marrow stem cells. At the same time, the occurrence of leukemia in 2 of the approximately 10 successfully treated children showed that, at this point of development, theoretical risks cannot be clearly distinguished from clinically relevant risks due to the so far insufficient clinical experience. Trends, however, show that each pathological situation will require the development of a certain adapted gene therapy approach. Thus, in the long run, gene therapy will present real therapy or prevention options, especially for a number of up-to-now insufficiently treatable or untreatable diseases.

References

1. F. Anderson, *Sci. Am.* **1995**, *273*, 96B–98B.
2. K. W. Culver, *Gene Therapy*, Mary Ann Liebert, New York, 1994.
3. R. A. Morgan, W. F. Anderson, *Annu. Rev. Biochem.* **1993**, *62*, 191–217.
4. R. C. Mulligan, *Science* **1993**, *260*, 926–932.
5. U. Gottschalk, S. Chan, *Arzneim.-Forsch./Drug Res.* **1998**, *48*, 1111–1120.
6. P. Tolstoshev, W. F. Anderson, *Genome Res. Mol. Med. Vir.* **1993**, *7*, 35–47.
7. D. Armentato, C. Sookdeo, G. White, *J. Cell Biochem.* **1994**, *18A*, 102–107.
8. R. G. Vile, S. J. Russell, *Br. Med. Bull.* **1995**, *51*, 12–15.
9. C. P. Hodgson, *Biotechnology* **1995**, *13*, 222–229.
10. K. K. Jain, *Vectors for Gene Therapy*, Scrip Reports, PJB Publications Ltd, Surrey, 1996.
11. R. J. Mannino, S. Gould-Fogerite, *Biotechniques* **1988**, *6*, 682–688.
12. S. I. Michael, D. T. Curiel, *Gene Ther.* **1994**, *1*, 223–232.
13. J. W. Wilson, M. Grossmann, J. A. Cabrera et al., *J. Biol. Chem.* **1992**, *267*, 11483–11489.
14. C. J. Buchholz, J. Stitz, K. Cichutek, *Curr. Opin. Mol. Ther.* **1999**, *5*, 613–621.
15. J. Stitz, C. J. Buchholz, K. Cichutek et al., *Virology* **2000**, *273*, 16–20.
16. J. Stitz, M. D. Muhlebach, K. Cichutek et al., *Virology* **2001**, *291*, 191–197.
17. V. Randrianarison-Jewtoukoff, M. Perricaudet, *Biologicals* **1995**, *23*, 145–147.
18. R. M. Kotin, *Hum. Gene Ther.* **1994**, *5*, 793–797.
19. S. Hacein-Bey-Abina, A. Fischer, M. Cavazzana-Calvo, *Int. J. Hematol.* **2002**, *76*, 295–298.
20. V. Randrianarison-Jewtoukoff, M. Perricaudet, *Biologicals* **1995**, *23*, 145–147.
21. R. A. Spooner, M. P. Deonarian, A. A. Epenetos, *Gene Ther.* **1995**, *2*, 1–11.
22. B. Gansbacher, K. Zier, B. Daniels, *J. Exp. Med.* **1990**, *172*, 1217–1222.
23. K. Cichutek, *Intervirology* **2000**, *43*, 331–338.
24. R. S. Nussenzweig, C. A. Long, *Science* **1994**, *265*, 1381–1384.
25. O. Cohen-Haguenauer, F. Rosenthal, B. Gansbacher et al., *Hum. Gene Ther.* **2002**, *13*, 2085–2110.
26. N. A. Horn, J. A. Meek, G. Budahazi et al., *Hum. Gene Ther.* **1995**, *6*, 565–573.
27. M. Müller, Considerations for the scale-up of plasmid DNA purification in *Nucleic Acid Isolation Methods* (Eds.: B. Bowlen, P. Dürre), American Scientific Publishers, New York, 2003.
28. M. Marquet, N. A. Horn, J. A. Meek, *Biopharm* **1995**, *10*, 26–37.
29. C. Prior, P. Bay, B. Ebert, *Pharm. Technol.* **1995**, *19*, 30–52.
30. A. P. Green, G. M. Prior, N. M. Helveston et al., *Biopharm* **1997**, *10*, 52–62.
31. G. Chandra, P. Patel, T. A. Kost et al., *Anal. Biochem.* **1992**, *203*, 169–177.
32. H. C. Birnboim, *Methods Enzymol.* **1983**, *100*, 243–249.

33. M. Marquet, N. A. Horn, J. A. Meek, *Part 1: Biopharm* **1997**, *10*(7), 42–50; *Part 2: Biopharm* **1997**, *10*(8), 40–45.
34. U. Gottschalk, The industrial perspective of somatic gene therapy in *Interdisciplinary Approaches to Gene Therapy* (Eds.: S. Müller, J. W. Simon, J. Vesting), Springer, 1997.

14
Nonviral Gene Transfer Systems in Somatic Gene Therapy

Oliver Kayser
Freie Universität Berlin, Berlin, Germany

Albrecht F. Kiderlen
Robert Koch-Institut, Berlin, Germany

14.1
Introduction

Somatic gene therapy might develop into one of the most important therapeutic strategies of the near future. Innate or acquired genetic defects are held responsible for a number of diseases such as hemophilia, cystic fibrosis (mucoviscidosis), adenosine-deaminase (ADA) deficit, and AIDS. Substituting or supplementing malfunctioning or missing genetic information by transiently or permanently inserting the appropriate gene appears to be a plausible therapeutic strategy especially from the patients' point of view. Attempts in gene therapy began in the early 1990s with great expectations that have only partially been met. No disease with a defined genetic background has so far been causally cured by gene therapy. Viral gene transfer systems have caused severe problems that could not be brought under control to date. The death of the 18-year-old Jesse Gelsinger is a tragic evidence of the basic deficits of viral transfection systems. In consequence, attention is now being focused on chemical and physical gene transfer systems. This review covers nonviral gene transfection strategies of current interest with special reference to experimental results found in vivo and for clinical trials.

14.2
What is Gene Therapy?

Gene therapy may be defined as the expression of a gene that has been introduced into a target cell or a target tissue in order to alter an existing function or to introduce a new function with the aim of curing a patient from a specific disease. In many countries this is restricted by law to somatic cells. In Germany, for example, genetic manipulation of germ cells is forbidden according to Section 5 Embryonenschutz Gesetz (EschG) and attracts a penalty of up to five years imprisonment [1]. German legislation on gene transfection as a form of therapy or medicine is still incomplete. An amendment specifically covering somatic gene therapy has not yet been passed. German Drug Regulation (Section 2 and Section 3 Deutsches Arzneimittelgesetz, AMG) defines DNA introduced for therapeutic purposes in a very general manner as a pharmaceutical product. German Gene Technology Regulation (Section 2 (3) Deutsches Gentechnikgesetz, GenTG) explicitly excludes the regulation of gene therapy [2]. Interestingly, gene therapy might be indirectly affected: theoretically *(sensu strictu)*, a patient undergoing gene therapy becomes a genetically modified organism. Consequently, his release from hospital should be subject to authorization. When discussing DNA as a pharmaceutical, a look at the European legislation, for example, at the guidelines of the European Agency for the Evaluation of Medicinal Products (EMEA) might be helpful. According to Council Regulations (EEC) No. 2309/93 – ANNEX (List A), DNA is considered a pharmaceutical product for gene therapy [3]. The necessary vector is simply defined as an additive.

Most likely, gene therapy will only work for a limited number of "suitable" diseases in a restricted commercial environment. It is very costly and affords a highly complex technology as well as an individually tailored strategy for gene delivery, depending on the relevant circumstances and aims [4]. Monogenetic diseases such as hemophilia [5], sickle cell anemia, adenosine-deaminase deficit [6], cystic fibrosis [7], or Duchenne muscular dystrophy [8] are likely first candidates, as they require the replacement or substitution of only a single gene. Trisomy 21 (Down Syndrome), on the other hand, which is also frequently mentioned in this context, is a poor candidate, as it is much more difficult to silence the additional chromosome 21 than to replace nonfunctional genes. In different forms of cancer and heart- or circulatory failures or of acquired genetic dysfunctions such as hepatitis or AIDS, two or more genes must be substituted or otherwise manipulated. Here, successful gene therapy seems unlikely at the moment owing to technical limitations in DNA transport into target cells and in tissue-specific forms of application.

Among the known innate diseases, gene therapy has been most intensively investigated in ADA deficiency and cystic fibrosis. Both reduce mean life expectancy to below 20 years, accompanied by severe symptoms that strongly affect the quality of life [6, 7].

14.3 Strategies in Gene Therapy

Independent of the respective method of gene transfer, two basic strategies for gene therapy may be discriminated (Fig. 1). Following the ex vivo strategy, cells or tissues are first removed from the patient, then exposed in vitro to the therapeutic genetic construct. If the transfection has been successful, the material is reimplanted in the patient [9, 10]. In the alternative in vivo methods, the therapeutic gene or DNA sequence is integrated into a vector and this construct is injected locally or systemically into the patient. Among other deficits, the latter method is characterized by poor tissue selectivity, rapid extracellular DNA degradation, and the danger of inducing oncogenes when using viral vectors.

Further differentiation is possible at the molecular level (Fig. 2): An intact (therapeutic) gene may simply be added to the defect one (Fig. 2a), a missing gene may be substituted

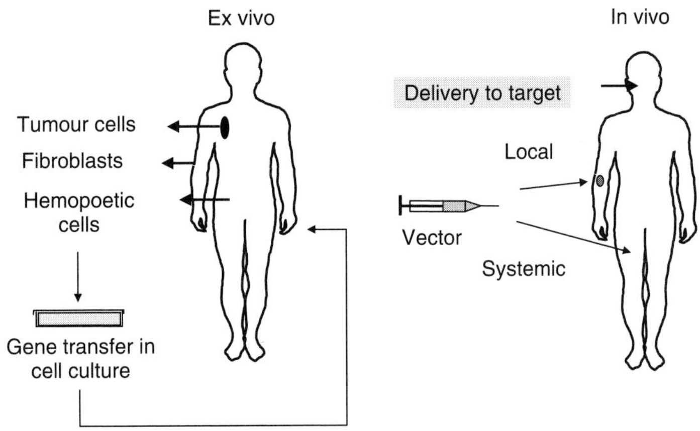

Fig. 1 Ex vivo and in vivo gene therapy strategies.

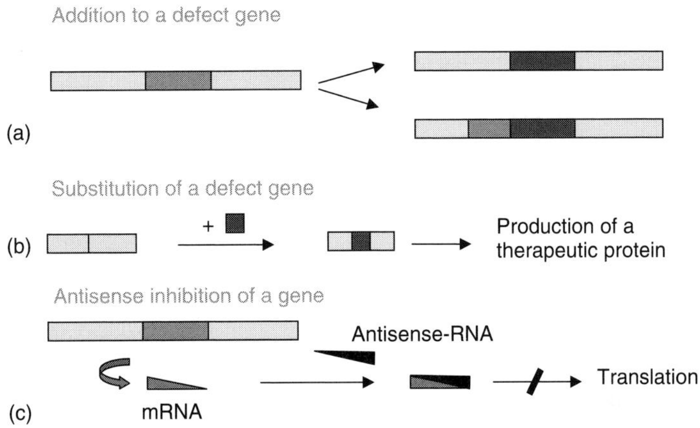

Fig. 2 Therapeutic gene functions.

(Fig. 2b), or a malfunctioning gene product may be inhibited, for example, during gene translation by giving mRNA-complimentary antisense oligonucleotides (Fig. 2c). Recently, the first antisense oligonucleotide drug, Vitravene™ (with Formivirsen as the active ingredient), has been licensed for treating CMV-retinitis in AIDS patients [11].

The description of plasmid vectors and the biotechnology necessary for their production as pharmaceuticals and licensing specifications are not major subjects of this article. These can be studied in comprehensive overviews by Hutchins [12] or Ferreira [13].

Intensive work is ongoing, both in designing new synthetic vectors and in improving DNA as a therapeutic agent. The transfection efficiency of "naked" DNA is low and it is rapidly degraded in the cytosol [14]. The expression of the therapeutic gene may be enhanced by eukaryotic promoters of viral origin such as cytomegalovirus (CMV) or simian virus [15]. The influence of the 5′UTR-site on the translation efficiency of mRNA is also being studied [16, 17]. The insertion of at least one intron into the cDNA may lead to 100-fold enhancement of mRNA [18]. The addition of a suitable terminal poly(A)-signal (bovine growth factor) has similar effects [19]. Special attention is given to the development of tissue-specific promoters, restricting transfected gene expression to the necessary therapeutic sites [20, 21]. A very elegant strategy for controlling the expression of the transfected gene is by turning it on or off with common oral drugs such as tetracycline or progesterone-antagonist. These interact with a mutated receptor acting as a transgene, thus initiating a signal transduction cascade, which finally induces the transcription (or its inhibition) of the respective gene for just as long as the drug is kept at a sufficient level [22, 23].

14.4 Gene Transfer Systems

Owing to their rapid degradation in the cytosol, genes or DNA-sequences are only rarely transfected as "naked" molecules [24]. The commonly used gene transfer methods may be segregated into biological, physical, and chemical systems (Table 1). The biological systems involving viral vectors (retro-, adeno-, or poxviruses) make up 77% of all gene transfection studies published to date [25]. However, physical and chemical methods have experienced a relative increase in recent years (12% from 1998 to 2003). This tendency reflects the often highly serious and hardly controllable side effects of viral systems. Despite recent modifications of wild-type vectors, the future of such systems for human application appears limited. One main problem of viral vectors is their strong immunogenicity. Already, the first (high dose) application may initiate an immune reaction against the given proteins, which may lead to all sorts of allergic reactions, even lethal anaphylactic shock, when repeated. Possible reversion of the virus to wild-type is another danger. The potential induction of oncogenes by retroviruses must also be mentioned [26, 27].

On this background, the development of chemical or physical gene transfer systems

Tab. 1 Gene transfer methods in somatic gene therapy [10]

	Method	*Application*	*Tissue selectivity*	*Transfection efficiency*	*Expression duration*
Chemical	Liposomes, Ca-phosphate-precipitation	Ex vivo/(in vivo)	No	Low	Transient/stabile
Physical	Microinjection, electroporation, "particle-bombardment"	Ex vivo/(in vivo)	No	Moderate	Transient
Biological	Nonviral: ligand/receptor	(Ex vivo)/in vivo	Yes	Low–moderate	Transient
	Viral: (e.g. retro-, adeno-, adeno-associated)	Ex vivo/in vivo	Some	Moderate–high	Transient/stabile

appears especially interesting, combining simple usage with maximum safety. Further requirements for an ideal gene transfer system are minimal infectivity and immunogenicity, defined chemical and physical characteristics, and the possibility of multiple dosing [27].

When looking for suitable alternatives to viral vectors, the pharmaceutical industry can offer a broad spectrum of thoroughly investigated and readily available medicinal carrier systems. As far as molecular transport and organ- or cell-specificity are concerned, the demands on modern drug delivery and gene transfer systems exhibit so much similarity, that the latter may profit significantly in the fields of biotechnology and pharmacology [28, 29]. Among the chemical vectors, liposomes, polymer nanoparticles, and polylysine particles deserve special discussion. Physical transfection systems such as electroporation and bioballistics are further examples of an efficient transfer of existing know-how in pharmaceutical technology to somatic gene therapy. It might be mentioned that gene transfer in a nonviral system must correctly be addressed as transfection, whereas in biological systems it should be referred to as transduction. Table 1 suggests a systematic arrangement of methods in gene transfer technology. In the following, the most important contemporary nonviral gene transfer systems are described and assessed.

14.5 Physical Gene Transfer Systems

14.5.1 Electroporation

During electroporation, cells or tissues are exposed to an electric field with high voltage (up to 1 kV). Short, rapid pulses cause transient membrane instability and the formation of pores with a mean lifetime of minutes [30]. Soluble DNA constructs that have been added to the culture medium or injected into the tissue may thereby enter the cell and ultimately reach the nucleus. The basic principle is also known to pharmacists as "iontophoresis" and is used for transdermal drug application [31]. This gene transfection method has so far been well accepted by patients and is safe, especially due to the low risk of infection. To date, mainly liver, muscle, and skin cells have been transformed this way, mostly via the transdermal route [32].

14.5.2 Bioballistics

Bioballistical methods are already widely used in biotechnology, for example, as "gene guns" for injecting DNA vaccines (Fig. 3). For this, linear or circular

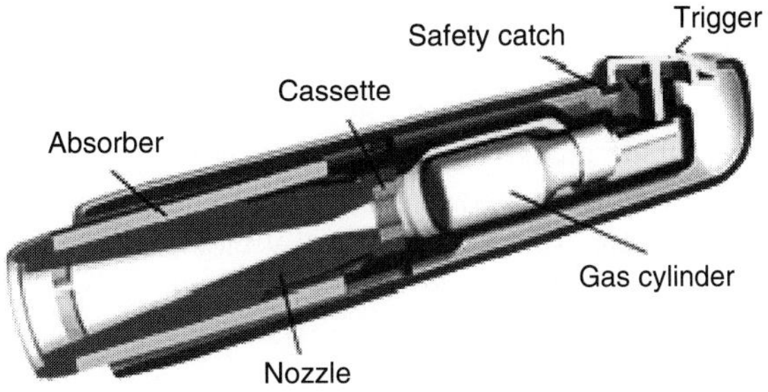

Fig. 3 Accela® gene gun (Powderject) [33].

(plasmid) DNA is adsorbed to nanometer-sized gold or tungsten particles. These are shot from a cylinder with compressed nitrogen or helium as propellant, thereby reaching speeds of up to 900 m s^{-1} [33]. When applied to the skin, the DNA/metal particles pass through dead tissue such as the *Stratum corneum*, reaching living cells. Statistically, only one of 10 000 particles reach the interior of viable cells, thereby loosing their DNA. For the DNA to then enter the nucleus, active transport mechanisms are probably necessary. Bioballistical gene transfer methods are obviously not suitable for systemic application. Further drawbacks are the need for very stable DNA and the high development costs. Their greatest advantage, on the other hand, is the fact that related technologies are already approved and commercially available such as the Accell® (Powderject) (Fig. 3) or the Helios® (BioRad) systems that are used for vaccination [34].

14.6 Chemical Vectors

A multitude of synthetic chemical vectors are being developed. Of these, cationic lipids and cationic polymers are probably the most thoroughly investigated. Already in 1987, Felgner et al. could demonstrate in vitro gene transfection using cationic lipids [35]. Transfer efficiency was systematically improved [36, 37], leading to the first clinical studies on cystic fibrosis patients [38, 39].

Irrespective of their highly variant molecular composition and 3D structure, chemical vectors such as liposomes and polymer particles have many biological and physical features in common. Positively charged amine functions are especially important as these can be loaded with the negatively charged DNA molecules [40]. Optimal interaction between plasmid-DNA and cationic additives leads to the development of colloidal, positively charged particles that may adhere to and be taken up by negatively charged cells. This is a cornerstone of chemical vector technology. Type and structure of the amine functions determine the stability of the complex, its cellular uptake in a phagosome, its release from the phagosome into the cytosol and dissociation of the DNA molecule, and even DNA-transport to the nucleus [41].

Nucleic acids as drugs are still rather unusual in pharmaceutics. They are highly negatively charged and range in sizes from 10^3 kDa (oligonucleotides) to 10^6 kDa (genes). Their targets are invariably intracellular. In contrast to most conventional pharmaceuticals, nucleic acids are too large and too strongly charged to pass cell membranes by simple passive means. Furthermore, free, that is, "naked" nucleic acids are rapidly degraded by cellular nucleases. Chemical vectors therefore have the additional job to protect the DNA they carry from enzymatic degradation. In the context of gene transfer, liposomes are also referred to as lipoplex, polymers as polyplex, and combinations as lipopolyplex particles.

14.6.1 Cationic Liposomes (Lipoplex)

Cationic liposomes were developed and already complexed with DNA over 20 years ago [35]. However, it was in the Human Genome Project that their potential as gene transfer vehicles really became apparent to geneticists and pharmacists. Cationic liposomes consist of cationic phospholipids that may be divided according to their number of

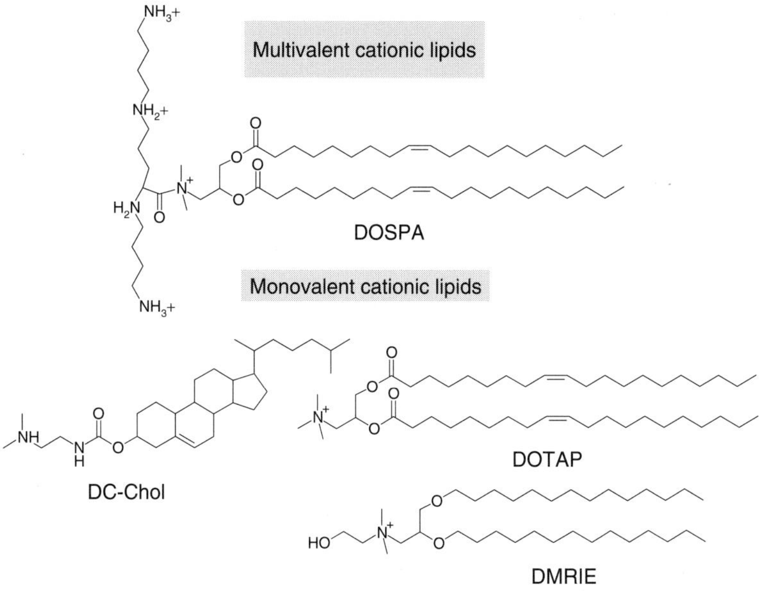

Fig. 4 Cationic lipids: DOSPA: 2,3-Dioleoyloxy-*N*-[2-(spermincarboxyamido)ethyl]-*N*,*N*-dimethyl-1-propanamminiumchloride; DOTAP: 1,2-Dioleoyloxypropyl-3-*N*,*N*,*N*-trimethylammoniumchloride; DC-chol: 3β-[*N*-(*N'*,*N'*-dimethylaminoethane)carbamoyl]-cholesterol; DMRIE: 1,2-Dimyristyloxypropyl-3-*N*,*N*-dimethylhydroxyammoniumbromide [42, 43].

tertiary amine functions into monovalent and multivalent cationic lipids [42, 43]. An overview of the most commonly used cationic lipids is given in Fig. 4. DOTAP (1,2-Dioleoyloxypropyl-3-*N*,*N*,*N*-trimethylammoniumchloride) and DC-chol (3β-[*N'*,*N'*-dimethylaminoethanecarbamoyl]-cholesterol) are possibly the most important. However, due to high toxicity, they must be mixed with helper lipids such as Dioleoylphosphatidylethanolamine (DOPE) before processing to liposomes [44]. A well-known transfection agent is Lipofectin®, which consists of equal parts of *N*-[1-(2,3-dioleoyloxy)propyl]-*N*,*N*,*N*-trimethylammonium chloride (DOTMA) and DOPE [45].

The high general toxicity of the first cationic lipids initiated the synthesis of a variety of derivatives [46]. The published in vitro data allows a first structure/activity analysis. The presence of a tertiary amine function and of 1 or 2 hydrophobic aliphatic chains are most important for producing stable liposomes and for binding negatively charged DNA. The tertiary amine functions can be combined with the hydrophobic lipid chains by either ester or ether bonds. Ester bonds are split more readily under physiological conditions than ether bonds and the lipoplex particles are consequentially metabolized more rapidly. These different metabolic characteristics are exemplified by DOTMA and DOTAP; the former is synthesized as ester, the latter as ether.

The aliphatic lipid chain may be saturated (DMRIE) or may contain double bonds (DOTAP, DOSPA). In vivo, dialkylchains with a length of 12 to 18 carbon atoms each seem to promote the best gene transfection [38]. A substitution of the alkyl chains with cholesterol, as in DC-chol, shows further advantages in the treatment of cystic fibrosis [38, 39]. Cholesterol structures have strong affinities to broncho-epithelial cells and interact less with plasma proteins [39].

Positively charged liposomes loaded with negatively charged DNA must maintain an overall positive charge in order to approach and contact target cells that are normally negatively charged. Liposomes that have been overloaded with DNA and thus receive a negative total charge exhibit low endocytosis rates and enhanced metabolic degradation [38, 39]. Proper selection of the cationic lipids and of the lipid-to-DNA ratio are decisive for the rate of complexation and the colloidal structure of the product. Finally, conformations may change in time, a process also referred to as "aging" of the DNA/lipid complex [47].

Recent studies show that intracellular stability of the lipid/DNA complex and the necessary dissociation of the DNA molecule from its vector depend both on the p*K*s of the chosen lipids and on the pH of the product [48]. Safinya et al. describe how not only the main cationic lipid but also the helper lipids influence the physical characteristics of the resulting lipid phase [49]. For instance, when formulating DOTAP liposomes with Dioleoylphosphatidylcholine (DOPC), lamellar structures are achieved. However, when exchanging DOPC for DOPE as helper lipid, inverse hexagonal micelles are produced. One may speculate whether different lipid phase characteristics lead to defined biopharmaceutical variations [50].

Cationic liposomes have also been intensively investigated in vivo. In combination with neutral helper lipids, they appear to be well tolerated even at high concentrations; they are neither immunogenic nor do they induce toxic side effects [36]. This applies to different routes of application such as intravenous, pulmonary, or nasal, none of which have been described as significantly unpleasant by the recipient patients. Nevertheless, cationic liposomes also pose a number of problems such as high serum protein binding and rapid metabolic degradation in the liver.

Liposomes are generally the most efficient at a diameter ranging from 400 to 500 nm. Larger particles are almost completely removed from the circulation by cells of the monocytic phagocytes system of liver and lung, and therefore do not reach other target areas [51, 52]. Pegylation is one means of improving pharmacokinetics. However, the useful extent of pegylation is restricted as further reduction of the zeta potential at the surface of the liposomes correlates with their reduced cellular uptake [7, 53].

Low transfection rates are a further problem. Reasons are, among others, insufficient lysosomal release of the therapeutic DNA from its carrier and/or its rapid enzymatic degradation within the lysosome or in the cytosol. The DNA that was integrated in the liposomes must be released to become effective (see Fig. 5). Studies on microinjections of lipid/DNA complexes directly into the nucleus clearly show reduced transfection rates [54]. As the nuclear pores have mean diameters of only 25 to 50 nm and passage for macromolecules is further controlled by the nuclear pore complex, passive diffusion into the karyosol is limited to particle sizes below 45 kDa [51, 55]. Conventional recombinant plasmids normally have a molecular weight ranging from 50 to 100 kDa, and therefore require active transport into the nucleus [55].

Admission of plasmid-DNA into the karyosol is steered by nuclear localization

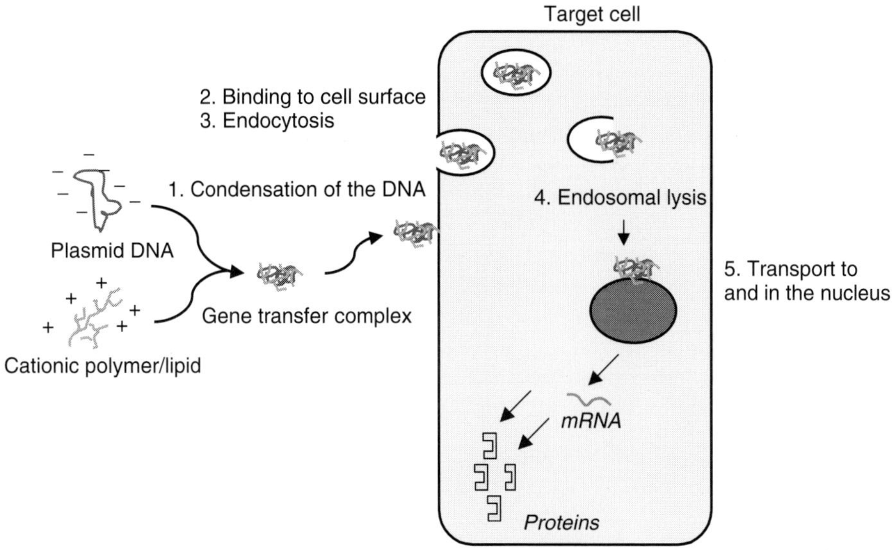

Fig. 5 Uptake, transport, and release of DNA in particulate gene transfer systems.

sequences that are assisted by importin-ß, guanin-nucleotide-binding protein (Ran), and nuclear transport factor (NTF). In the karyosol, the therapeutic DNA is transcribed by RNA polymerases into mRNA, which is transported back into the cytosol where it attaches to ribosomes for translation into protein [55, 56]. In order to improve its transport into the nucleus, the therapeutic DNA may be coupled to such nucleus localizing sequences. The latter can be found naturally in certain viruses that have developed this strategy for a most efficient nuclear invasion. For example, Rudolph et al. coupled DNA to short TAT sequences taken from the arginine-rich motif of the HIV-1-TAT protein and achieved significant enhancement of transfection. Of the 101–amino acid–long HIV-TAT sequence they synthesized, a 12–amino acid–long oligopeptide bound it to polyarginine and thus achieved a 390-fold enhanced transfection rate [57].

Unsatisfactory transfection rates and low cell- or organ specificity also initiated the development of simple liposomes to virosomes or immunoliposomes. Virosomes are liposomes that contain viral proteins or fusiogenic peptides for enhanced DNA release from the endosome and for generally improved gene transfection rates [58]. Immunoliposomes are characterized by target-specific monoclonal antibodies bound to their surface. These were first developed in the eighties for tumor-specific delivery of liposome-entrapped drugs and have since been modified for gene therapy [55, 58].

14.6.2
Polymer Particles (Polyplex)

A further class of synthetic gene vectors that has received attention in past years is cationic polymers, which condense and package DNA with high efficiency. Polymerized or oligomerized branched or nonbranched amino acid chains composed of lysine or arginine are common [59, 60]. Polyethylenimine, however, developed in 1995 by Boussif [61] and already used for

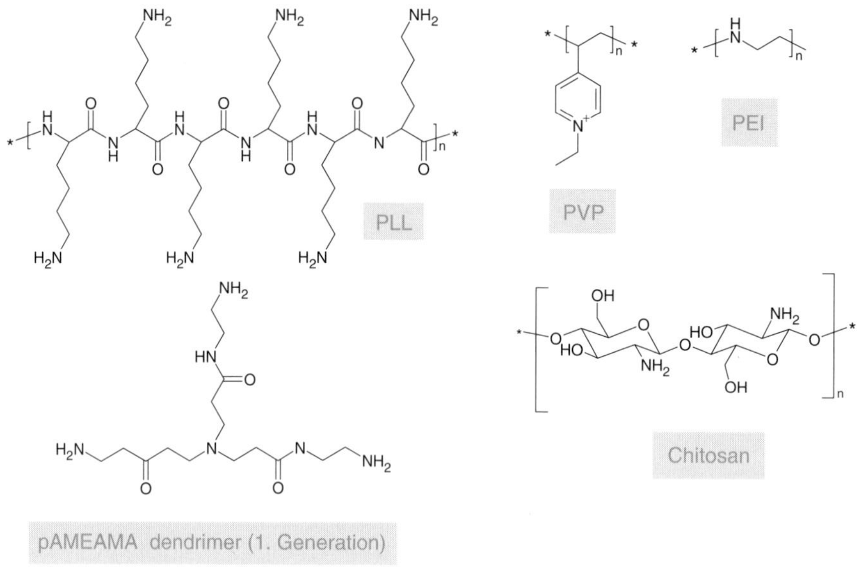

Fig. 6 Chemical structures of cationic polymers.

gene transfection experiments in vitro as well as in vivo, appears to be the most promising cationic polymer at the moment (see Fig. 6). Basically, cationic polymer gene transfer systems reveal the same pharmaceutical problems and intracellular barriers as described above for cationic liposomes.

14.6.3
Poly-L-Lysine (PLL) and Poly-L-Arginine (PLA)

Poly-L-lysine (PLL) has already been intensively used as a polymeric gene transfer system [60, 62]. It is synthesized by polymerization of the *N*-carboxyanhydrid of lysine. Arginine is polymerized in a similar manner. PLL/DNA complexes are produced by dissolving both components in aqueous media and precipitating the particulate complexes. These particles, which normally range from 400 to 500 nm, are capable of transporting nucleic acids ranging from short molecules to large artificial yeast chromosomes [62, 63]. The first in vivo studies, however, revealed substantial toxicity combined with low DNA transfection efficiency. Different chemical modifications and variations in particle size were then tested. Toxicity is substantially reduced by coating the particles with PEG derivatives [64], and the transfection rate is enhanced by attaching ligands such as transferrin, folate, or target-specific monoclonal antibodies [65]. Interestingly, pegylation also achieved significant reduction in unwanted hepatic metabolization of the particles.

Apart from lysine, polymers of other cationic amino acids such as arginine and histidine were also investigated. Conjugation of histidine to ε-L-histidine enabled the development of highly interesting PLH/DNA complexes characterized by high transfection rates [66]. One explanation for the elevated transfection efficiency may be that the highly protonated histidine structure that develops in the generally acidic (pH 6) endosomal environment may cause rapid destruction of the endosomal membrane, and thus an enhanced release of the therapeutic DNA into the cytosol.

14.6.4 Polyethyleneimine (PEI)

Linear or branched polyethyleneimines generally range in size from 1.8 to 800 kDa [60, 67, 68]. They are synthesized by cationic polymerization. Starting from 2-substituted-2-oxazoline-monomers, linear PEI with a mean molecular weight of 22 kDa, also known as ExGen 500 [69], are produced by hydrolysis. PEI/DNA complexes have already been used in many in vivo studies in animals following iv injection [70–74]; clinical studies, however, have not been reported to date. PEI/DNA complexes have repeatedly revealed very high transfection efficacy. One advantage of PEI/DNA complexes over lipoplex- or PLL particles is their intrinsic buffer capacity at lysosomal pH, leading, as described above for PLL-particles, to rapid destruction of the lysosomal membrane and DNA release [75]. This effect, also termed "proton-sponge", is brought about by the chemical structure of PEI. Polymerization produces particles with primary amines at the surface and secondary as well as tertiary amines in the interior. This causes a shift in pK_a from approximately 6.9 to 3.9. The strong protonation at a pH below 6 induces an osmotic gradient across the endosomal membrane, resulting in an influx of water, swelling, and finally disruption of the endosome with release of the PEI/DNA particles into the cytosol [75].

Though the literature reveals some discrepancies, there seems to exist an inverse relationship between the molecular weight of the PEI particles and their transfection efficiency. The interesting physicochemical characteristics of PEI-based gene transfer systems encouraged their further development, for example, to dendrimers (see below). In direct comparison, linear PEI complexes (e.g. PEI 22) seem to possess better transfection characteristics than branched (e.g. PEI 25) [70]. One explanation may be a premature dissociation and subsequent degradation of the DNA molecule.

Clinical trials with PEI complexes as gene transfer systems could so far not be undertaken because of their frequently intolerable general toxicity. Depending on the chemical structure, the lethal dose for mice ranges from 40 to 100 mg kg^{-1} body weight [76]. The main problem lies in the strong interaction between PEI complexes and erythrocytes leading to their aggregation and the danger of emboli. Pegylation of the PEI complexes may only partially help solve the problem, as a high degree of pegylation generally reduces particle uptake and thus transfection efficacy [77].

14.6.5 Dendrimers

The name "dendrimer" refers to the star- or tree-shaped, branched structures of this relatively new class of cationic gene transfer systems [8, 78–81]. They are frequently synthesized from polyamidoamines with special chemical or physical features. Probably best known are the "starburst" dendrimers with particle sizes ranging from 5 to 100 nm. These particles reveal a highly regular branched "dendritic" symmetry. Starburst dendrimers are three-dimensional oligomeric or polymeric compounds, which, initiated from small molecules as nuclei, are built layer-by-layer ("generations") by repeated chemical reaction cycles. This allows an exquisite steering of the final size, three-dimensional form, and surface chemistry of a starburst polymer by the individual selection of components and binding procedures for each generation [82].

The physical and chemical characteristics of dendrimers are mainly the result of the number and type of amine functions on the particle surface, but the secondary and tertiary amines in the inside also affect their biological features. Despite their high molecular weights, dendrimers are soluble in water. They complex DNA with great efficiency, thus giving excellent vehicles for gene transfer. Their high transfer efficiency, however, is probably less owing to high DNA adsorption rates but rather to protonation of the amine functions after endosomal uptake. As described above for the PEI complexes, this induces an osmotic gradient, leading to osmotic lysis of the organelle and enhanced release of the complex into the cytosol [82, 83]. Dendrimer/DNA complexes have proven their gene transfer efficacy in in vivo studies [79–81]. However, as already exhibited by the PEI complexes, they show strong, undesired interaction with erythrocytes, causing hemolysis. Again, the free primary amine functions on the particle surface are held responsible. Depending on the type of dendrimer and the target cell, cationic dendrimers also reveal general cytotoxicity at concentrations ranging from 50 to 300 $\mu g\ mL^{-1}$ [84]. On the other hand, no dendrimer has to date been reported to induce tumors or to substantially affect the immune system [85].

14.6.6 Chitosan

Chitosan is a fiber produced by hydrolyzing chitin, mostly from crustaceans [86]. Owing to its free amine functions, chitosan may also be protonated ($pK_a = 5.6$). In in vitro studies with Hela-cells, chitosan/DNA complexes showed a gene-transfection potency similar to that reported for PEI/DNA complexes. Plain chitosan, however, is almost insoluble in water at neutral pH (but soluble at acidic pH). For this reason, trimethylated, quaternary chitosan derivatives have been produced that are sufficiently soluble under physiological conditions and easily complex DNA molecules. In in vitro experiments with COS-1 and CaCo-2 cells, these innovative chitosan derivatives proved superior to nontreated chitosan polymers, particularly as they showed no unspecific cytotoxicity [87].

14.6.7 Poly(2-dimethylamino)ethylmethacrylate

Methacrylate polymers are used in pharmaceutical technology for microencapsulation. They are synthesized by polymerization of monomeric dimethylaminoacrylic acid to poly(2-dimethylamino)ethylmethacrylate (pDMAEMA), which is both simple and cheap. Their low general toxicity makes these polymers interesting also as gene transfer vehicles [88]. In in vitro experiments with OVCAR-3 and COS-7 cells, some pDMAEMA/DNA particles showed high transfection rates [88, 89]. This proved to be highly dependent on their size and charge. In HEPES-buffer (pH 7.4) pDMAEMA particles exhibit a positive zeta potential of around 25 mV and an average size of 100 to 200 nm. Following endosomal uptake, again the outer primary amine functions are protonated, leading to osmotic lysis of the endosome and release of the pDMAEMA/DNA particles into the cytosol.

Studies of DNA absorption onto pDMAEMA particles show that linear DNA (e.g. antisense oligonucleotides) is adsorbed more strongly than circular plasmid DNA [91]. However, this stronger adsorption has a negative influence on the gene transfer efficiency, as the DNA molecule is less likely to dissociate from the pDMAEMA particle in the cytosol. For this reason, circular DNA is preferred. As expected, DNA that is adsorbed to pDMAEMA is protected from degradation by DNAse I [88].

14.7 Outlook

The nonviral gene transfection systems introduced in this review bear significant advantages over the viral systems, but have serious drawbacks as well. One fact in favor is that many such systems are already well established in classical areas of pharmaceutical technology. Their production methods have already been optimized and safety aspects have been investigated in detail. Simple transfer of knowledge may substantially reduce developing costs. Nonviral systems are noninfectious. They allow significantly higher DNA-loading rates than viral systems, which reach their limits around 30 kb (Herpes virus). Nonviral systems are only weakly immunogenic and therefore allow – in stark contrast to viral systems – multiple application.

The main drawback of nonviral systems is that they normally only lead to the transient expression of the therapeutic gene as it is not permanently integrated into the host genome. In consequence, the therapeutic gene transfer must be regularly repeated, possibly over a long period of time. Further disadvantages are insufficient cell- or tissue specificity and low DNA transfer rates from the cytosol to the nucleus. Taken together, the gene-transfection performances of nonviral systems are even weaker than those of viral systems.

To date, it is still not possible to make a clear decision between viral and nonviral gene transfer systems. Mixed systems, hybrid vectors, which can be envisaged as "denucleated" viruses, are in the pipeline. A combination of viral surface proteins and liposomes or the integration of therapeutic DNA into artificial cells or viruses are further innovative ideas for improving somatic gene therapy. Most important is the rapid progress in three fields: in cell biology, unspecific and specific intracellular trafficking of macromolecules still raises questions; in biochemistry, further DNA carriers must be brought forward for testing; and pharmaceutical technology must supply improved and cell/tissue-specific drug delivery systems. Brought together in a rational form, such progress should make somatic gene therapy possible for selected disease forms in the near future.

References

1. German Regulation for the Protection of Human Embryos (Embryoschutz Gesetz, EschG) as proclaimed 13th December 1990, *BGBl* Part I, 1990, p. 2747.
2. German Regulation for Gene Technology (Gentechnik Gesetz, GenTG) as proclaimed 16th December 1993, BGBl Part I, p. 2066. Last altered (2. GenTG-Änderung) 8th August 2002, BGBl Part I, 2002, 59, pp. 3220–3244.
3. EMEA Council Regulation 2309/93 as proclaimed 22nd July 1993, (OJ EC L 214).
4. G. M. Rubanyi, *Mol. Aspects Med.* **2001**, *22*, 113–142.
5. H. S. Kingdon, R. L. Lundblad, *Biotechnol. Appl. Biochem.* **2002**, *35*(2), 141–148.

6. R. M. Blaese, K. W. Culver, A. D. Miller, et al., *Science* **1995**, *270*, 475–480.
7. T. R. Flotte, *Curr. Opin. Mol. Ther.* **1999**, *1*, 510–516.
8. W. D. Biggar, H. J. Klamut, P. C. Demacio, et al., *Clin. Orthop.* **2002**, *401*, 88–106.
9. C. Bordignon, L. D. Notarangelo, N. Nobili, et al., *Science* **1995**, *270*(5235), 470–475.
10. U. Gottschalk, S. Chan, *Arzneimittelforschung* **1998**, *48*, 1111–1120.
11. R. M. Orr, *Curr. Opin. Mol. Ther.* **2001**, *3*, 288–294.
12. B. Hutchins, *Curr. Opin. Mol. Ther.* **2000**, *2*, 131–135.
13. G. N. M. Ferreira, G. A. Monteiro, M. F. Duarte, et al., *Trends Biotechnol.* **2000**, *18*, 380–938.
14. D. M. Feltquate, *J. Cell. Biochem. Suppl.* **1998**, *30–31*, 304–311.
15. L. Qin, Y. Ding, D. R. Pahud, et al., *Hum. Gene Ther.* **1997**, *8*, 2019–2029.
16. M. Kozak, *J. Biol. Chem.* **1991**, *266*, 19867–19870.
17. M. Kozak, *Annu. Rev. Cell. Biol.* **1992**, *8*, 197–225.
18. M. T. Huang, C. M. Gorman, *Nucleic Acids Res.* **1990**, *18*, 937–947.
19. N. S. Yew, D. M. Wysokenski, K. X. Wang, et al., *Hum. Gene Ther.* **1997**, *8*, 575–584.
20. M. E. Coleman, F. DeMayo, K. C. Yin, et al., *J. Biol. Chem.* **1995**, *270*, 12109–12116.
21. K. Anwer, M. Shi, M. F. French, et al., *Hum. Gene Ther.* **1998**, *9*, 659–670.
22. M. Gossen, S. Freundlieb, G. Bender, et al., *Science* **1995**, *268*, 1766–1769.
23. Y. Wang, B. W. O'malley, S. Y. Tsai, *Methods Mol. Biol.* **1997**, *63*, 401–413.
24. J. A. Wolff, R. W. Malone, P. Williams, et al., *Science* **1990**, *247*, 1465–1468.
25. URL: http://www.wiley.co.uk/genmed/clinical/ (1st June 2003).
26. G. Romano, P. P. Claudio, H. E. Kaiser, et al., *In Vivo* **1998**, *12*, 59–67.
27. A. Fischer, *Cell Mol. Biol. (Noisy-le-grand)* **2001**, *47*, 1269–1275.
28. H. Ma, S. L. Diamond, *Curr. Pharm. Biotechnol.* **2001**, *2*, 1–17.
29. R. I. Mahato, L. C. Smith, A. Rolland, *Adv. Gen.* **1999**, *41*, 95–156.
30. J. Gehl, *Acta Physiol. Scand.* **2003**, *177*, 437–447.
31. N. Kanikkannan, *BioDrugs* **2002**, *16*, 339–47.
32. S. Satkauskas, M. F. Bureau, M. Puc, et al., *Mol. Ther.* **2002**, *5*, 133–140.
33. T. L. Burkoth, B. J. Bellhouse, G. Hewson, et al., *Crit. Rev. Ther. Drug Carrier Syst.* **1999**, *16*, 331–384.
34. A. D. Barrett, *Curr. Opin. Investig. Drugs* **2002**, *3*, 992–995.
35. P. L. Felgner, T. R. Gadek, M. Holm, et al., *Proc. Natl. Acad. Sci. U. S. A.* **1987**, *84*, 7413–7417.
36. N. S. Templeton, *Expert Opin. Biol. Ther.* **2003**, *3*, 57–69.
37. D. Niculescu-duvaz, J. Heyes, C. J. Springer, *Curr. Med. Chem.* **2003**, *10*, 1233–1261.
38. E. R. Lee, J. Marshall, C. S. Siegel, et al., *Hum. Gene Ther.* **1996**, *7*, 1701–1717.
39. C. Wheeler, P. L. Felgner, Y. J. Tsai, et al., *Proc. Natl. Acad. Sci. U. S. A.* **1996**, *93*, 11454–11459.
40. I. Koltover, K. Wagner, C. R. Safinya, *Proc. Natl. Acad. Sci. U. S. A.* **2000**, *97*, 14046–14051.
41. A. J. Lin, N. L. Slack, A. Ahmad, et al., *J. Drug Target.* **2000**, *8*, 13–27.
42. Z. Li, Z. Ma, *Curr. Gene Ther.* **2001**, *1*, 201–226.
43. S. Simoes, P. Pires, N. Düzgünes, et al., *Curr. Opin. Mol. Ther.* **1999**, *1*, 147–157.
44. I. Koltover, T. Salditt, J. O. Radler, et al., *Science* **1998**, *281*, 78–81.
45. P. L. Felgner, Y. J. Tsai, L. Sukhu, et al., *Ann. N. Y. Acad. Sci.* **1995**, *772*, 126–139.
46. B. J. Stevenson, A. D. Carothers, W. A. Wallace, et al., *Gene Ther.* **1997**, *4*, 210–218.
47. E. Tomlinson, A. P. Rolland, **1996**, *39*, 357–372.
48. D. V. Schaffer, N. A. Fidelman, N. Dan, et al., *Biotechnol. Bioeng.* **2000**, *67*, 598–606.
49. C. R. Safinya, *Curr. Opin. Struct. Biol.* **2001**, *11*, 440–448.
50. J. O. Rädler, I. Koltover, T. Salditt, et al., *Science* **1997**, *275*, 810–814.
51. R. Wattiaux, N. Laurent, S. Wattiaux-de Coninck, et al., *Adv. Drug Deliv. Rev.* **2000**, *41*, 201–208.
52. G. Osaka, K. Carey, A. Cuthbertson, et al., *J. Pharm. Sci.* **1996**, *6*, 612–618.
53. R. Z. Yu, R. S. Geary, J. M. Leeds, et al., *Pharm. Res.* **1999**, *16*, 1309–1315.
54. J. Vacik, B. S. Dean, W. E. Zimmer, et al., *Gene Ther.* **1998**, *6*, 1006–1014.
55. M. Johnson-Saliba, D. A. Jans, *Curr. Drug Targets* **2001**, *2*, 371–399.
56. C. K. Chan, D. A. Jans, *Immunol. Cell Biol.* **2002**, *80*, 19–130.

57. C. Rudolph, C. Plank, J. Lausier, et al., *J. Biol. Chem.* **2003**, *278*, 11411–11418.
58. N. Shi, R. J. Boado, W. M. Pardridge, *Pharm. Res.* **2001**, *18*, 1091–1095.
59. W. Zauner, M. Ogris, E. Wagner, *Adv. Drug Deliv. Rev.* **1998**, *30*, 115–131.
60. C. W. Pouton, P. Lucas, B. J. Thomas, et al., *J. Control. Release* **1998**, *53*, 289–299.
61. O. Boussif, F. Lezoualch, M. A. Zanta, et al., *Proc. Natl. Acad. Sci. U. S. A.* **1995**, *92*, 7297–301.
62. C. P. Lollo, M. G. Banaszczyk, P. M. Mullen, et al., *Methods Mol. Med.* **2002**, *69*, 1–13.
63. P. Marschall, N. Malik, Z. Larin, *Gene Ther.* **1999**, *6*, 1634–1637.
64. M. Mannisto, S. Vanderkerken, V. Toncheva, et al., *J. Control. Release* **2002**, *83*, 169–182.
65. E. Wagner, M. Ogris, W. Zauner, *Adv. Drug Deliv. Rev.* **1998**, *30*, 97–113.
66. P. Midoux, E. Lecam, D. Coulaud, et al., *Somat. Cell. Mol. Genet.* **2002**, *27*, 27–47.
67. B. Brissault, A. Kichler, C. Guis, et al., *Bioconjugate Chem.* **2003**, *14*, 581–587.
68. D. A. Tomalia, G. R. Killat in *Encyclopedia of Polymer Science and Engineering* Vol. 1 (Ed.: J. I. Kroschwitz), 2nd ed., Wiley, New York, 1985, pp. 680–739.
69. J. L. Merlin, A. N'Doye, T. Bouriez, et al., *Drug News Perspect.* **2002**, *15*, 445–451.
70. L. Wightman, R. Kircheis, V. Rossler, et al., *J. Gene Med.* **2001**, *3*, 362–372.
71. C. Plank, M. X. Tang, A. R. Wolfe, et al., *Hum. Gene Ther.* **1999**, *10*, 319–332.
72. A. Bragonzi, G. Dina, A. Villa, et al., *Gene Ther.* **2000**, *7*, 1753–1760.
73. I. Chemin, D. Moradpour, S. Wieland, et al., *J. Viral. Hepat.* **1998**, *5*, 369–375.
74. S. M. Zou, P. Erbacher, J. S. Remy, et al., *J. Gene Med.* **2000**, *2*, 128–134.
75. J. Behr, *Chimia* **1997**, *51*, 34–36.
76. A. G. Schatzlein, *Anti-Cancer Drugs* **2001**, *12*, 275–304.
77. H. Petersen, P. M. Fechner, A. L. Martin, et al., *Bioconjugate Chem.* **2002**, *13*, 845–854.
78. M. J. Cloninger, *Curr. Opin. Chem. Biol.* **2002**, *6*, 742–748.
79. S. E. Stiriba, H. Frey, R. Haag, *Angew. Chem., Int. Ed. Engl.* **2002**, *41*, 1329–1334.
80. C. L. Gebhart, A. V. Kabanov, *J. Control. Release* **2001**, *73*, 401–416.
81. G. De jong, A. Telenius, S. Vanderbyl, et al., *Chromosome Res.* **2001**, *9*, 475–485.
82. G. M. Dykes, *J. Chem. Technol. Biotechnol.* **2001**, *76*, 903–918.
83. G. R. Newkome, C. N. Moorefield, F. Vogtle, *Dendrimers and Dendrons: Concepts, Syntheses, Applications*, Wiley-VCH, New York, 2001.
84. N. Malik, R. Wiwattanapatepee, R. Klopsch, et al., *J. Control. Release* **2000**, *65*, 133–148.
85. J. F. Kukowska-Latallo, A. U. Bielinska, J. Johnson, et al., *Proc. Natl. Acad. Sci. U. S. A.* **1996**, *93*, 4897–4902.
86. L. Römpp in *Biotechnologie und Gentechnik* (Eds.: W. D. Decjwer, A. Pühler, E. Schmid), 2nd ed., Georg Thieme Verlag, Stuttgart, New York, 1999.
87. M. Thanou, B. I. Florea, M. Geldof, et al., *Biomaterials* **2002**, *23*, 153–159.
88. J. H. Van steenis, E. M. Van maarseveen, F. J. Verbaan, et al., *J. Control. Release* **2003**, *87*, 167–176.
89. N. J. Zuidam, G. Posthuma, E. T. J. De vries, et al., *J. Drug Target.* **2000**, *8*, 51–66.
90. C. Arigita, N. J. Zuidam, D. J. Crommelin, et al., *Pharm. Res.* **1999**, *16*, 1534–1541.
91. F. Verbaan, Colloidal Gene Delivery Systems – *In vivo* Fate of poly(2-(dimethyl amino) ethyl methacrylate)-based Transfection Complexes, Ph.D. Thesis, Utrecht University, Holland, 2002, pp. 14–15.

15
Xenotransplantation in Pharmaceutical Biotechnology

Gregory J. Brunn
Transplantation Biology and the Departments of Pharmacology and Experimental Therapeutics, Mayo Clinic, Rochester, MN, USA

Jeffrey L. Platt
Surgery, Immunology, and Pediatrics, Mayo Clinic, Rochester, MN, USA

No area of medicine has generated more excitement or controversy than the field of transplantation. Organ allotransplantation allows "curative" treatments for failure of the heart, kidney, liver, and lungs by replacing these diseased organs with physiologically normal ones. Replacement of beta cells via pancreatic islet or whole pancreas transplantation offers curative treatment to patients with diabetes. The main limitation to applying transplantation for the treatment of diseases is a shortage of human donors. This shortage limits the clinical application of organ transplantation to approximately 5% of the number of transplants that would be performed were the supply of organs unlimited [1, 2]. Possible solutions to this limitation have garnered considerable interest and include the use of artificial organs, "engineered tissues," stem cell transplants, and xenotransplants. Although some newer technologies have excited interest, xenotransplants of the heart, lung, kidney, and pancreatic islets are known to function well enough to sustain life. Enthusiasm for xenotransplantation also stems from the possibility that animal tissues and organs might be less susceptible to disease recurrence compared to allotransplants. Advances in cellular and molecular biology and in genetics open possibilities for use of cells, tissues, and organs to address the complications of disease, not only by replacement of abnormal cells and tissues but also by the use of transplanted tissues to impart novel physiological functions. In this regard and for some purposes, xenografts may be an ideal vehicle for introducing a novel gene or biochemical process that could be of value to the transplant recipient.

If interest in xenotransplantation is substantial, the hurdles to its application are equally so. For the past three decades, the first and preeminent obstacle to transplanting organs and tissues between species has been the immune reaction of the host against the graft. A second, and still theoretical, hurdle is the possibility that beyond the immune barrier, there might be physiologic limitations to the survival or function of a xenograft and the possibility that a xenotransplant might engender medical complications for the xenogeneic host. A third hurdle is the possibility that a xenograft might transfer infectious agents from the donor to the host, and that from the host such agents might spread to other members of society. This communication will consider the current state of efforts to overcome the various hurdles to xenotransplantation and will evaluate how genetic engineering might be applied to this end.

15.1 The Pig as a Source of Tissues and Organs for Clinical Xenotransplantation

Although it might be intuitive that the best source of xenogeneic tissues for clinical transplantation is nonhuman primates, it is the pig that is the focus of most efforts in this field. The reasons for favoring the pig as a xenotransplant source include the availability of pigs in large numbers, the ease with which the pig can be bred, the limited risk of zoonotic disease engendered by the use of pigs, and the possibility of introducing new genes into the germline of the pig.

Genetic engineering of pigs using transgenic techniques and nuclear transfer has certain advantages over conventional gene therapy (Table 1). Introducing genetic material directly into the porcine germ cell obviates the need for a vehicle, which may vary in reliability of gene delivery and may introduce secondary unintended consequences due to the vector itself. Second, the genetic material introduced into the germline can be expressed constitutively in all cells, especially in stem cells, and passed on to subsequent generations. Third, with the use of transgenic techniques, only the donor is manipulated; in conventional gene therapy, both the donor and the recipient may be affected.

Recent advances in cloning pigs [3–5] through nuclear transfer also allows "knocking out" genes. Besides the advantage of gene knockouts, nuclear transfer can be done with cultured somatic cells obviating the need for embryonic stem cells.

15.2 The Biologic and Immunologic Responses to Xenotransplantation

All xenografts elicit an immune response, including antibodies, cell-mediated immunity, natural killer (NK) cells, and inflammation [6]. However, the fate of xenografts confronted with these responses is dictated in part by the way in which the graft receives its vascular supply (Fig. 1). Isolated cells, such as hepatocytes, and "free" tissues, such as pancreatic islets and skin, derive their vascular supply through the ingrowth of host blood vessels. The process of neovascularization, as such, might be impaired in a xenograft by incompatibility of donor growth factors with the host microvasculature. To the extent that neovascularization or graft function depends upon hormones and cytokines of host origin, the function of the xenograft might also be impaired. As the host microcirculation is established, however, a xenogeneic tissue may be relatively protected from attack by host immune elements. Whole-organ grafts provide their own microcirculation and growth factors, and, as a result, incompatibility between the donor

Tab. 1 Genetic engineering in xenotransplantation: conventional gene therapy versus transgenic therapy versus cloning

	Conventional gene therapy	*Conventional transgenic techniques*	*Cloning*
Delivery	Vector or vehicle required	Injection of genetic material directly into pronuclei of fertilized egg	Transfection of cultured somatic cells
Expression	Dependent on ability of each cell to take up genetic material Requires treatment for every transplant or recipient	Genetic material introduced into the germline, leading to expression in a line of animals	Genetic material introduced into the germline, leading to expression in a line of animals
	May require repeated treatment	One manipulation	One manipulation
Immunogenicity	Delivery vehicle or transgene may be immunogenic	The transgene may elicit immune response	The transgene may elicit immune response
Target of genetic manipulation	The recipient and the graft may be transduced	Genetic manipulation of the donor only	Genetic manipulation of the donor only
Genetic manipulation	Gene addition Dominant negative	Gene addition Dominant negative	Gene addition Dominant negative Gene knockout

and the recipient is less likely to have an impact on cellular function. On the other hand, because the circulation is of donor origin, the immune, inflammatory, and coagulation systems of the recipient can act directly on donor cells, sometimes with dramatic and devastating consequences.

15.3 Hyperacute Rejection

An organ transplanted from a pig into a primate such as a human is subject to hyperacute rejection. Hyperacute rejection begins immediately upon reperfusion of the graft and destroys the graft within minutes to a few hours. Hyperacute rejection is characterized histologically by interstitial hemorrhage and thrombosis, the thrombi consisting mainly of platelets [7]. Research over the past decade has clarified the molecular basis for the hyperacute rejection of pig organs by primates [8, 9], and this knowledge has led to the development of new and incisive therapeutic approaches to averting this problem. Hyperacute rejection was once considered the most daunting hurdle to clinical application of xenotransplantation; however, hyperacute rejection can now be prevented in nearly every case.

Hyperacute rejection of porcine organ xenografts by primates is initiated by the binding of xenoreactive natural antibodies to the graft [7, 10–12]. Xenoreactive natural antibodies are present in the circulation without a known history of sensitization [13]. Contrary to expectations, xenoreactive antibodies

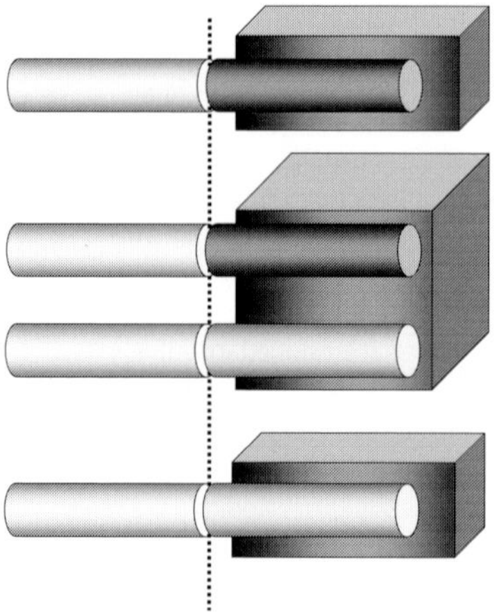

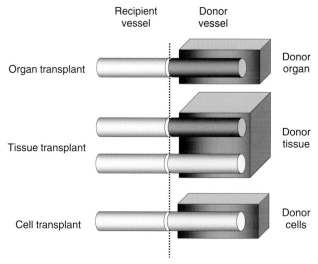

Fig. 1 Mechanisms of xenograft vascularization. Organ xenografts receive recipient blood exclusively through the donor blood vessels (top). Free tissue xenografts (e.g. pancreatic islets and skin) are vascularized partly by the ingrowth of recipient blood vessels and partly by spontaneous anastomosis of donor and recipient capillaries (middle). Cellular xenografts (e.g. hepatocytes and bone marrow cells) are vascularized by the ingrowth of recipient blood vessels (bottom).

are predominantly directed against only one antigen, a saccharide consisting of terminal Galα1,3Gal [14–17]. The importance of Galα1,3Gal as the primary antigenic barrier to xenotransplantation was demonstrated recently by experiments in which anti-Galα1,3Gal antibodies were specifically depleted from baboons using immunoaffinity columns before transplantation of pig organs [18]. Antibody binding to the newly transplanted organs was largely curtailed, and hyperacute rejection did not occur.

Although the identification of the relevant antigen for pig-to-primate xenotransplantation allows specific depletion of the offending antibodies, more enduring and less intrusive forms of therapy would be preferred. One approach to overcoming the antibody–antigen reaction is to develop lines of pigs with low levels of antigen expression [19]. Various genetic approaches aimed at "remodeling" the antigenicity of donor tissues by reducing Galα1,3Gal expression are under investigation. These approaches can be separated into three categories: (1) interference with the function of α1,3galactosyltransferase (α1,3GT), the enzyme that catalyzes the synthesis of the Galα1,3Gal moiety; (2) expression of α galactosidase, which cleaves αgalactosyl residues; and (3) deletion of the gene encoding α1,3GT from the pig genome to prevent synthesis of the saccharide. The utility of genetic modification of pig tissues to reduce Galα1,3Gal expression has recently been demonstrated by Sharma and colleagues [20], who generated transgenic pigs expressing the H-transferase. Transgenic pigs expressing H-transferase express H antigen at the terminus of some sugar chains instead of Galα1,3Gal. Another genetic approach to modify expression of Galα1-3Gal was proposed by Osman et al. [21]. Expression of α-galactosidase, which cleaves α-galactosyl residues [16], in conjunction with other galactosyltransferases, significantly reduces expression of Galα1-3Gal [21]. A third strategy to modify Galα1,3Gal was demonstrated by Miyagawa and coworkers, who generated transgenic pigs expressing the human β-1,4-*N*-acetylglucosaminyltransferase III gene (*GnT-III*) [22]. This enzyme catalyzes transfer of *N*-acetylglucosamine to maturing mannose-modified proteins as they pass through the Golgi apparatus and leads to diminished αGalα1,3Gal expression both by competing with α1,3GT and by preventing subsequent modifications of α1,3GT by insertion of an *N*-acetylglucosamine onto the growing mannose chain. The net result was diminished natural antibody reactivity on transgenic pig tissue when transplanted into primates. Although these advances illustrate the utility of genetic modification of pig donor tissue, the main limitation of these genetic approaches is that residual Galα1,3Gal may be sufficient to allow rejection reactions to occur [23].

The most obvious approach to developing xenograft donors with diminished reactivity with host antibodies would be to genetically target or "knock out" the enzyme α1,3-galactosyltransferase. Embryonic stem cells were used to knock this gene out in mice [24], demonstrating that removal of this enzyme is not lethal. The intense interest in generating pigs that lack the α*1,3GT* gene has fueled a race to delete this gene from the pig genome, and recently several groups have succeeded in this endeavor. Prather and colleagues [25] and Ayares and coworkers [26] used similar strategies to disrupt one allele of the α*1,3GT* gene (GGTA1) in pigs by first targeting the gene for disruption in fetal porcine fibroblasts. Selected clones were used as nuclear donors for enucleated pig oocytes and were the resulting embryos implanted into surrogate gilts. Both approaches yielded live, healthy piglet clones in which one copy of GGTA1 had been disrupted. These achievements demonstrated that nuclear transfer technology could be applied to pig embryos, which are notoriously fragile and difficult to manipulate. Recently, the generation of cloned pigs harboring a functional knockout of both alleles of α1,3GT was reported [27]. This first success required some serendipity in that the α1,3GT-deficient pigs were found not to be homozygous knockouts, but rather functional knockouts that arose from a process in which the knockout of one allele paired with a spontaneous single base change in the remaining *GGTA1* gene resulted in an inactivating amino acid substitution in α1,3GT. Irrespective of how the inactivation of α1,3GT was achieved in these animals, it may now be feasible to make an incisive determination as to the utility of these genetically modified pig tissues for avoiding hyperacute rejection, and possibly other hurdles to using xenotransplantation. A cautionary note has recently come to light, however, indicating that pig cells that lack both copies of the *GGTA1* gene may still synthesize the Galα1,3Gal antigen, albeit at very low levels [28]. While the generation of α1,3Gal knockout pigs may help overcome an important hurdle to xenotransplantation by preventing hyperacute rejection, it may not avert other potent xenogeneic immune responses, as will be discussed below.

15.4 Complement Activation

A second and essential step in the development of hyperacute rejection is activation of the complement system of the recipient on donor blood vessels [11]. Complement activation is triggered by the binding of complement-fixing xenoreactive antibodies to graft endothelium, and to a smaller extent perhaps, by reperfusion injury. Regardless of the mechanism leading to complement activation, a xenograft is extraordinarily sensitive to complement-mediated injury because of multiple defects in the regulation of complement (Fig. 2) [29–31]. Under normal circumstances, the complement cascade is regulated or inhibited by various proteins in the plasma and on the surface of cells. These proteins protect normal cells from suffering inadvertent injury during the activation of complement. The proteins that regulate the complement cascade function in a species-restricted fashion; that is, complement regulatory proteins inhibit homologous complement far more effectively than heterologous complement [30, 32]. Accordingly, the complement regulatory proteins expressed in a xenograft are ineffective at controlling the complement cascade of the recipient, and the graft is subject to severe complement-mediated injury [29].

To address this problem, lines of animals have been developed that are transgenic for human complement regulatory proteins and that are able to control activation of complement in the xenograft (Fig. 2) [30, 33, 34]. Animals transgenic for human decay-accelerating factor (hDAF), which regulates complement at the level of C3, together with CD59, which regulates

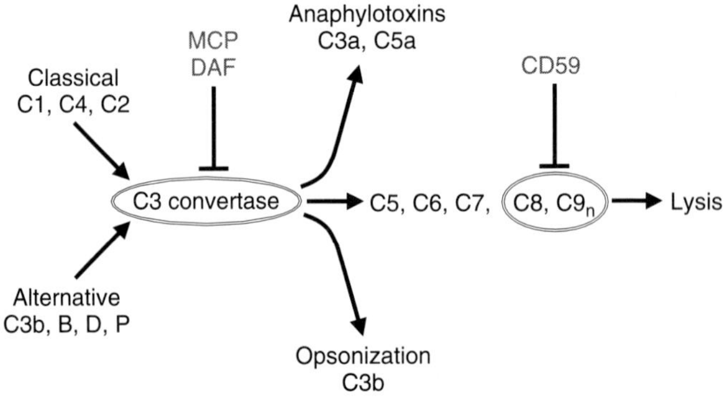

Fig. 2 Regulation of the complement system. The complement cascade, which can be activated via the classical or alternative pathway, is regulated under normal circumstances by various proteins in the plasma and on the cell surface. Three of the cell surface complement regulatory proteins are shown here. Decay-accelerating factor (DAF) and membrane cofactor protein (MCP) regulate complement activation by dissociating or promoting the degradation of C3 convertase. CD59, also known as protectin, prevents the functions of terminal complement complexes by inhibiting C8 and C9. An organ graft transplanted into a xenogeneic recipient is especially sensitive to complement-mediated injury because DAF, MCP, and CD59 expressed on the xenograft endothelium cannot effectively regulate the complement system of the recipient.

complement at the level of C8 and C9 [34], or CD46, which controls complement activation at the level of C3 and C4 [35], have demonstrated that the expression of even low levels of hDAF and CD59 or CD46 in porcine-to-primate xenografts is sufficient to allow a xenograft to avoid hyperacute rejection [36, 37]. These results, and the dramatic prolongation of xenograft survival achieved by expressing higher levels of hDAF factor in the pig [38], underscore the importance of complement regulation as a determinant of xenograft outcomes.

One of the major obstacles in testing the effects of transgenes in pig organs has been the difficulty in generating transgenic pigs. Recent work by Lavitrano, et al. [39] may accelerate the rate at which transgenic pigs may be generated and tested in transplant models. These investigators used sperm-mediated gene transfer to incorporate the *hDAF* gene into pigs and obtained a high efficiency of transgenesis (80% of pigs incorporated hDAF into the genome) and hDAF expression (43% of the transgenic pigs). The transgenic hDAF was functional in vitro, and transmitted to progeny as expected. This method, in theory, could be used to introduce multiple genes at once, or a tailor-made set of human genes that may be useful for transplant-mediated genetic therapies, as mentioned earlier.

15.5 Acute Vascular Rejection

If hyperacute rejection of a xenograft is averted, a xenograft is subject to the development of acute vascular rejection, so named because of its resemblance to acute vascular rejection of allografts [40, 41]. Acute vascular rejection (sometimes called delayed xenograft rejection) may begin within 24 hours of reperfusion and lead to graft destruction over the following days and weeks [40, 42, 43]. Although the factors important in the pathogenesis of acute vascular rejection are incompletely understood, there is growing evidence that acute vascular rejection is triggered at least in part by the binding of xenoreactive antibodies to the graft. The importance of xenoreactive antibodies in triggering acute vascular rejection is suggested by three lines of evidence: (1) antidonor antibodies are present in the circulation of recipients whose grafts are subject to acute vascular rejection [11, 40, 44, 45], (2) depletion of antidonor antibodies delays or prevents acute vascular rejection [46], and (3) administration of antidonor antibodies leads to the development of acute vascular rejection [47]. Recent studies suggest that among the antibodies that provoke acute vascular rejection are those directed against Galα1-3Gal [48, 49]. This problem thus constitutes further impetus for the continued development of α1,3Gal-deficient pigs. Regardless of which elements of the immune system trigger acute vascular rejection, it is commonly thought that this type of rejection, and especially the intravascular coagulation characteristically associated with it, are caused by the activation of endothelial cells in the transplant [40, 50, 51]. Activated endothelial cells express procoagulant molecules, such as tissue factor (TF), and proinflammatory molecules, such as *E*-selectin and cytokines [30]. The pathogenesis of acute vascular rejection is summarized in Fig. 3.

Although various therapeutic manipulations have proven successful in preventing hyperacute rejection, acute vascular rejection poses a more difficult problem, in part, because therapies are needed on an ongoing basis. For this reason, genetic

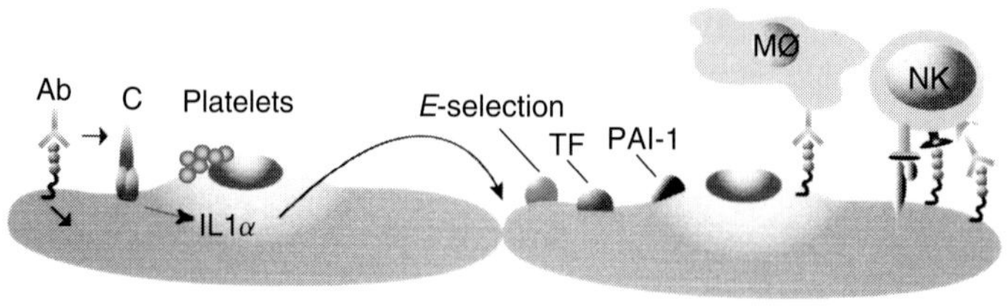

Fig. 3 Pathogenesis of acute vascular rejection. Activation of endothelium by xenoreactive antibodies (Ab), complement (C), platelets, and perhaps by inflammatory cells (natural killer (NK) cells and macrophages (M⊘) leads to the expression of new pathophysiologic properties. These new properties, such as the synthesis of tissue factor (TF) and plasminogen activator inhibitor type 1 (PAI-1), promote coagulation; the synthesis of *E*-selectin and cytokines such as IL1α promote inflammation. These changes in turn cause thrombosis, ischemia, and endothelial injury, the hallmarks of acute vascular rejection. (Adapted from *Nature* 1998: **392**(Suppl.) 11–17, with permission.) (See Color Plate p. xxiii).

modification of the donor may prove more important for dealing with acute vascular rejection than with hyperacute rejection. The various possible approaches for combating acute vascular rejection are listed in Table 2. Among these approaches, the reduction of Galα1,3Gal in xenotransplant donors may be an important part of the overall strategy, to the extent that Galα1,3Gal proves to be an important antigenic target in acute vascular xenograft rejection. Preliminary studies suggest that the level of antibody binding needed to initiate acute vascular rejection is considerably lower than the level needed to initiate hyperacute rejection [23]. Accordingly, the antigen expression would have to be reduced very significantly to achieve therapeutic benefit for acute vascular rejection. The availability of α1,3galactosyltransferase-deficient pigs should prove to be an ideal model to test this concept. In addition to lowering antigen expression, it is likely that expression of human complement regulatory proteins will be helpful in preventing acute vascular rejection. Preliminary studies suggest that interfering with the antigen–antibody reaction and controlling the complement cascade may be sufficient to prevent acute vascular rejection for at least some period of time [46]. These goals were accomplished by using animals transgenic for hDAF factor and CD59 as a source of organs, and baboons depleted of immunoglobulin as recipients. Cozzi and associates [38] achieved prolonged survival of xenografts, presumably preventing acute vascular rejection, by using transgenic pigs expressing high levels of DAF and cynomolgus monkeys treated with very high doses of cyclophosphamide. The immunosuppression perhaps prevented the synthesis of antidonor antibodies.

Work in rodents points to the potential involvement of NK cells and macrophages in mediating acute vascular rejection. However, the ability of immunoglobulin manipulation to prevent acute vascular rejection suggests that the involvement of NK cells and macrophages might be less important than in vitro studies and studies in rodents have suggested [33, 51]. On the other hand, NK cells might exacerbate the injury triggered by xenoreactive antibodies, as human NK cells have been shown to activate porcine endothelial cells in vitro [52–54].

Tab. 2 Therapeutic strategies for acute vascular xenograft rejection

Possible mechanism targeted	***Manipulation of***	
	Recipient	***Donor***
Antibody–antigen interaction	Specific depletion of xenoreactive antibodies Prevention of xenoreactive antibody synthesis (e.g. cyclophosphamide, leflunomide)	Generating transgenic pigs with low levels of antigen Generation of pig clones lacking antigen
Complement activation	Systemic anticomplement therapy (e.g. CVF, sCR1, gamma globulin)	Generation of donor pigs transgenic for human complement regulatory proteins
Endothelial cell activation	Administration of anti-inflammatory agents	Inhibition of NFκB function Introduction of protective genes
Molecular incompatibilities	Administration of inhibitors (e.g. inhibitors of complement or coagulation)	Introduction of compatible molecules

15.6 Accommodation

Fortunately, the presence of antidonor antibodies in the circulation of a graft recipient does not inevitably trigger acute vascular rejection. If antidonor antibodies are temporarily depleted from a recipient, an organ transplant can be established so that rejection does not ensue when the antidonor antibodies are returned to the circulation [55]. This phenomenon is referred to as "accommodation" [30]. Accommodation may reflect a change in the antibodies, in the antigen, or in the susceptibility of the organ to rejection. If accommodation can be established, it may be especially important in xenotransplantation because it would obviate the need for ongoing interventions to inhibit antibody binding to the graft. One potential approach to accommodation may be the use of genetic engineering to reduce the susceptibility of an organ transplant to acute vascular rejection and the endothelial cell activation associated with it [51]. Unfortunately, successful intervention at the level of such effector mechanisms is yet to be achieved. However, disruption of antibody–antigen interaction has brought about accommodation in human subjects [50, 55].

15.7 Cellular Mediated Immune Responses

Organ transplants and cellular and free tissue transplants are subject to cellular rejection. In allotransplantation, cellular rejection is controlled by conventional immunosuppressive therapy, but there is concern that, for several reasons, cellular rejection may be especially severe in xenotransplants. First, the great variety of antigenic proteins in a xenograft may lead to recruitment of a diverse set of "xenoreactive" T-cells. Second, the binding of xenoreactive antibodies and activation of the complement system may lead to amplification of elicited immune responses [56]. For example, deposition of complement in a graft may cause activation of antigen-presenting cells, in turn stimulating T-cell responses. Still another factor that might amplify the elicited immune response to a xenotransplant involves "immunoregulation," which ordinarily would circumscribe cellular immune responses, but may fail or be deficient across species. Such failure could reflect limitations in the recognition of xenogeneic cells or incompatibility of relevant growth factors, as but two examples.

Induction of immunologic tolerance has been an erstwhile goal of transplant surgeons and physicians. Especially in the case of xenotransplantation, if the current immunosuppressive regimens are not sufficient, induction of immunologic tolerance may be required. At least three approaches are being pursued: (1) the generation of mixed hematopoietic chimerism, (2) the establishment of microchimerism by various means, and (3) thymic transplantation [57–59]. The development of mixed hematopoietic chimerism through the introduction of donor bone marrow [60] has worked very well across rodent species [61, 62], although success may be limited by xenoreactive antibodies and the engraftment impaired by incompatibility of host growth factors or microenvironment [63]. Fortunately, there is evidence that these problems can be overcome [58]. Various approaches to peripheral tolerance, such as the blockade of costimulation by administration of a fusion protein consisting of a soluble form of the CTLA-4 molecule and immunoglobulin (CTLA-4-Ig), are being pursued.

Still another factor in the cellular response to a xenotransplant involves the action of NK cells. Natural killer cell functions can be amplified by cell surface receptors that recognize Galα1,3Gal [64]. Natural killer cell functions are downregulated by receptors that recognize homologous major histocompatibility complex (MHC) class I [65, 66]. Human NK cells may be especially active against xenogeneic cells because of stimulation on the one hand and failure of downregulation on the other. The possible involvement of NK cells in xenograft rejection might be addressed by generation of transgenic pigs expressing on the cell surface MHC-like molecules that will more effectively recognize corresponding receptors on NK cells and that will downregulate the function of NK cells.

How a xenogeneic donor could be modified genetically to enhance the development of tolerance or to limit elicited immune responses is still uncertain. Clearly, efforts to control the natural immune barriers to xenotransplantation may contribute to limiting the elicited immune response. To the extent that recipient T-cells recognize donor cells directly, that is, the T-cell receptors of the recipient recognizing native MHC antigens on donor cells, a xenogeneic donor might be engineered in such a way to reduce corecognition (through CD4 and CD8) or costimulation (through CD28 or other T-cell surface molecules) or to express inhibitory molecules such as CD59 or Fas ligand. These approaches and the expression of inhibitory molecules, which are being considered as approaches to gene therapy in allotransplantation, may well prove more effective in xenotransplantation because inhibitory genes can be introduced as transgenes and thereby expressed in all relevant cells in the graft. Another useful and perhaps necessary approach will involve genetic modifications to allow the survival and function of donor bone marrow cells.

15.8
Physiologic Hurdles to Xenotransplantation

Progress in addressing some of the immunological obstacles to xenotransplantation has brought into focus the question of the extent to which a xenotransplant would function optimally in a foreign host. A recent demonstration that the porcine kidney and the porcine lung can replace the most important functions of the primate kidney and primate lung are encouraging [67, 68]. Subtle defects in physiology across species may nevertheless exist. Organs such as the liver, which secrete a variety of proteins and which depend on complex enzymatic cascades, may prove incompatible with a primate host. Accordingly, one important application of genetic engineering in xenotransplantation may be the amplification or modulation of xenograft function to allow for more complete establishment of physiologic function or to overcome critical defects. For example, recent studies by Akhter and associates [69] and Kypson and coauthors [70] aimed at improving the function of cardiac allografts by manipulation of β-adrenergic signaling, and this technique might be adapted to the xenotransplant to improve cardiac function. On the other hand, most cellular processes and biochemical cascades are intrinsically regulated to meet the overall physiologic needs of the whole individual. The key question then is, which of the many potential defects actually need to be repaired.

Another potential hurdle to the clinical application of xenotransplantation is the possibility that the xenograft may disturb normal metabolic and physiologic functions in the recipient. For example, Lawson and coworkers [71, 72] have shown that porcine thrombomodulin fails to interact adequately with human thrombin and protein C to generate activated protein C. This defect could lead to a prothrombotic diathesis because of failure of generation of activated protein C. Of even greater concern is the possibility that the transplantation of an organ, such as the liver, could add prothrombotic or proinflammatory products into the blood of the recipient. Although perhaps a great many physiologic defects can be detected at the molecular level, the critical question will be, which of these defects is important at the whole-organ level or with respect to the well being of the recipient, and which must be repaired by pharmaceutical or genetic means.

15.9
Zoonosis

The increasing success of experimental xenotransplants and therapeutic trials bring to the fore the question of zoonosis, that is, infectious disease introduced from the graft into the recipient. The transfer of infectious agents from the graft to the recipient is a well-known complication of allotransplantation. To the extent that infection of the recipient in this way increases the risks of transplantation, the risk can generally be estimated and a decision made on the basis of the risk versus the potential benefits conferred by the transplant. The concern about zoonosis in xenotransplantation is not so much the risk to the recipient of the transplant, but the risk that an infectious agent will be transferred from the recipient to the population at large. Fortunately, all of the microbial agents known to infect the pig can be detected by screening and potentially eliminated from a population of xenotransplant donors. There is concern, however, that the pig may harbor endogenous retroviruses, which are inherited with genomic DNA and which might become activated and transferred to the cells of the recipient. For example, Patience and coauthors [73] recently reported that a C-type retrovirus endogenous to the pig could be activated in pig cells, leading to the release of particles that can infect human cell lines. Whether this virus or other endogenous viruses can actually infect across species and whether such infection would lead to disease are unknown, but remain a subject of current epidemiologic investigation. If cross-species infection does prove to be an important issue, genetic therapies might also be used to address this problem. The simplest genetic therapy would involve breeding out the organism, but this approach might fail if the organism were widespread or integrated at multiple loci. Some genetic therapies have been developed to potentially control human immunodeficiency viruses [74]. Although these therapies have generally failed because it has been difficult or impossible to gain expression of the transferred genes in stem cells and at levels sufficient to deal with high viral loads, the application of such therapies might be much easier in xenotransplantation because the therapeutic genes could be delivered through the germline. Ultimately, if elimination of endogenous retroviruses were necessary, it could potentially be accomplished by gene targeting and cloning, as discussed above.

15.10 A Scenario for the Clinical Application of Xenotransplantation

Successful application of xenotransplantation in the clinical arena requires insights into not only immunology but also physiology and infectious disease, all of which have been discussed briefly here in the context of genetic therapy. In recent years, important advances have been made in elucidating the immunologic hurdles of pig-to-primate transplantation. Although this scientific progress is important and exciting, xenotransplantation will likely enter the clinical arena through a step-by-step process. A first step, free tissue xenografting, is in limited clinical trials already [75–77], and preliminary evidence is encouraging as porcine free tissue xenografts appear to endure in a human recipient [77]. One immediate application of free tissue xenografting would be treatment of cirrhosis caused by hepatitis virus, using targeted infusion of porcine hepatocytes [6]. The promise of this approach is enhanced because (1) pig hepatocytes are resistant to viral reinfection, (2) rat models of cirrhotic liver failure indicate that porcine hepatocyte xenotransplants may endure and function well [78], and (3) predicted demand for hepatic transplantation due to hepatitis C-induced cirrhotic liver disease is likely to worsen the already acute shortage of livers available for transplant. Another potential extension of free tissue xenografting is the transplantation of xenogeneic islets of langerhans for the treatment of type 1 diabetes. Xenogeneic islet transplants may prove to be less liable to destruction by the autoimmune processes that underlies this disease. Temporary or "bridge" organ transplantation will probably follow free tissue xenografting. Bridge transplants will not address the problem of the shortage of human organs, but incisive analysis of the outcomes of these transplants will provide important information about the remaining immunologic hurdles and the potential physiologic and infectious considerations. With this information, further therapies including genetic engineering may allow the use of porcine organs as permanent replacements. Even then, one can envision ongoing efforts to apply genetic therapies that will optimize graft function and limit the complications of transplantation.

While it may be that the use of pigs as a source of organs and tissues for transplantation is not far off, exciting advances in tissue engineering, stem cell technology, and in vitro organogenesis may broaden the use of animals in human medicine. Adult and embryonic stem cell culture has given rise to organ-specific tissues with functional characteristics of the corresponding organs [79]. Although these cultures are unlikely to yield fully developed functional organs for transplantation into humans, pigs or other animals could be used as recipients of these culture-initiated tissues and allow completion of development. These organs, grown and maintained in animals, may then be available for transplantation on an "as needed" basis. Pigs may thus serve as xenograft "recipients" prior to becoming organ "donors." The recent success in cloning of animals, including pigs, raises the possibility of transferring nuclei from a human patient's cells into enucleated stem cells of an animal, and then growing the cells in animals to generate differentiated human tissue that is autologous with the patient. The lessons learned from genetic manipulation of animals in the quest to make animal organs suitable for transplantation into humans may find their best application in generating animals suitable for use as biological reactors to grow human organs suitable for transplantation into humans.

Acknowledgments

Supported by grants from the Heart, Lung, and Blood Institute of the National Institutes of Health.

References

1. 2001 Annual Report of the U.S. Organ Procurement and Transplantation Network and the Scientific Registry for Transplant Recipients: Transplant Data 1991–2000, Department of Health and Human Services, Health Resources and Services Administration, Office of Special Programs, Division of Transplantation, Rockville, MD; United Network for Organ Sharing, Richmond, VA; University Renal Research and Education Association, Ann Arbor, MI. Retrieved December 9, 2002, from the World Wide Web: http://www.optn.org//data/annualReport.asp.
2. R. W. Evans in *Xenotransplantation* (Ed.: J. L. Platt), ASM Press, Washington, DC, 2001, pp. 29–51.
3. A. Onishi, M. Iwamoto, T. Akita et al., *Science* **2000**, *289*, 1188–1190.
4. I. A. Polejaeva, S. Chen, T. D. Vaught et al., *Nature* **2000**, *407*, 86–90.
5. J. Betthauser, E. Forsberg, M. Augenstein et al., *Nat. Biotechnol.* **2000**, *18*, 1055–1059.
6. M. Cascalho, J. L. Platt, *Immunity* **2001**, *14*, 437–446.
7. J. L. Platt, *Hyperacute Xenograft Rejection*, Austin: R.G. Landes, 1995.
8. J. L. Platt in *Samter's Immunologic Diseases* (Eds.: K. F. Austen, M. M. Frank et al.), Lippincott Williams & Wilkins, Philadelphia, 2001, pp. 1132–1146.
9. J. L. Platt, *Transplantation* **2000**, *69*, 1034–1035.
10. D. K. C. Cooper, P. A. Human, G. Lexer et al., *J. Heart Transplant.* **1988**, *7*, 238–246.
11. J. L. Platt, R. J. Fischel, A. J. Matas et al., *Transplantation* **1991**, *52*, 214–220.
12. J. L. Platt, M. A. Turman, H. J. Noreen et al., *Transplantation* **1990**, *49*, 1000–1001.
13. R. Y. Calne, *Transplant. Proc.* **1970**, *2*, 550–556.
14. M. S. Sandrin, H. A. Vaughan, P. L. Dabkowski et al., *Proc. Natl. Acad. Sci. U. S. A.* **1993**, *90*, 11391–11395.
15. A. H. Good, D. K. C. Cooper, A. J. Malcolm et al., *Transplant. Proc.* **1992**, *24*, 559–562.
16. B. H. Collins, W. Parker, J. L. Platt, *Xenotransplantation* **1994**, *1*, 36–46.
17. W. Parker, D. Bruno, Z. E. Holzknecht et al., *J. Immunol.* **1994**, *153*, 3791–3803.

18. S. S. Lin, D. L. Kooyman, L. J. Daniels et al., *Transplant. Immunol.* **1997**, *5*, 212–218.
19. C. G. Alvarado, A. H. Cotterell, K. R. McCurry et al., *Transplantation* **1995**, *59*, 1589–1596.
20. A. Sharma, J. F. Okabe, P. Birch et al., *Proc. Natl. Acad. Sci U. S. A.* **1996**, *93*, 7190–7195.
21. N. Osman, I. F. McKenzie, K. Ostenried et al., *Proc. Natl. Acad. Sci. U. S. A.* **1997**, *94*, 14677–14682.
22. S. Miyagawa, H. Murakami, Y. Takahagi et al., *J. Biol. Chem.* **2001**, *276*, 39310–39319.
23. W. Parker, S. S. Lin, J. L. Platt, *Transplantation* **2001**, *71*, 313–319.
24. A. D. Thall, P. Malý, J. B. Lowe, *J. Biol. Chem.* **1995**, *270*, 21437–21440.
25. L. Lai, D. Kolber-Simonds, K. W. Park et al., *Science* **2002**, *295*, 1089–1092.
26. Y. Dai, T. D. Vaught, J. Boone et al., *Nat. Biotechnol.* **2002**, *20*, 251–255.
27. C. J. Phelps, C. Koike, T. D. Vaught et al., *Science* **2003**, *299*, 411–414.
28. A. Sharma, B. Naziruddin, C. Cui et al., *Transplantation* **2003**, *75*, 430–436.
29. S. Miyagawa, H. Hirose, R. Shirakura et al., *Transplantation* **1988**, *46*, 825–830.
30. J. L. Platt, G. M. Vercellotti, A. P. Dalmasso et al., *Immunol. Today* **1990**, *11*, 450–456.
31. A. P. Dalmasso, *Immunopharmacology* **1992**, *24*, 149–160.
32. A. P. Dalmasso, G. M. Vercellotti, J. L. Platt et al., *Transplantation* **1991**, *52*, 530–533.
33. D. White, J. Wallwork, *Lancet* **1993**, *342*, 879–880.
34. J. L. Platt, J. S. Logan, *Transplant. Rev.* **1996**, *10*, 69–77.
35. L. E. Diamond, C. M. Quinn, M. J. Martin et al., *Transplantation* **2001**, *71*, 132–142.
36. K. R. McCurry, D. L. Kooyman, C. G. Alvarado et al., *Nat. Med.* **1995**, *1*, 423–427.
37. G. W. Byrne, K. R. McCurry, M. J. Martin et al., *Transplantation* **1997**, *63*, 149–155.
38. E. Cozzi, N. Yannoutsos, G. A. Langford et al. in *Xenotransplantation: The Transplantation of Organs and Tissues Between Species* (Eds.: D. K. C. Cooper, E. Kemp, J. L. Platt et al.), Springer, Berlin, 1997, pp. 665–682.
39. M. Lavitrano, M. L. Bacci, M. Forni et al., *Proc. Natl. Acad. Sci. U. S. A.* **2002**, *99*, 14230–14235.
40. J. R. Leventhal, A. J. Matas, L. H. Sun et al., *Transplantation* **1993**, *56*, 1–8.
41. K. A. Porter in *Pathology of the Kidney* (Ed.: R. H. Heptinstall), Little Brown & Company, Boston, 1992, pp. 1799–1933, Vol. 3.
42. J. C. Magee, B. H. Collins, R. C. Harland et al., *J. Clin. Investig.* **1995**, *96*, 2404–2412.
43. S. S. Lin, J. L. Platt, *J. Heart Lung Transplant.* **1996**, *15*, 547–555.
44. J. J. McPaul, P. Stastny, R. B. Freeman, *J. Clin. Investig.* **1981**, *67*, 1405–1414.
45. L. C. Paul, F. H. J. Claas, L. A. Van Es et al., *N. Engl. J. Med* **1979**, *300*, 1258–1260.
46. S. S. Lin, B. C. Weidner, G. W. Byrne et al., *J. Clin. Investig.* **1998**, *101*, 1745–1756.
47. R. J. Perper, J. S. Najarian, *Transplantation* **1967**, *5*, 514–533.
48. A. H. Cotterell, B. H. Collins, W. Parker et al., *Transplantation* **1995**, *60*, 861–868.
49. K. R. McCurry, W. Parker, A. H. Cotterell et al., *Hum. Immunol.* **1997**, *58*, 91–105.
50. W. Parker, S. Saadi, S. S. Lin et al., *Immunol. Today* **1996**, *17*, 373–378.
51. F. H. Bach, H. Winkler, C. Ferran et al., *Immunol. Today* **1996**, *17*, 379–384.
52. D. J. Goodman, M. Von Albertini, A. Willson et al., *Transplantation* **1996**, *61*, 763–771.
53. A. M. Malyguine, S. Saadi, J. L. Platt et al., *Transplantation* **1996**, *61*, 161–164.
54. A. M. Malyguine, S. Saadi, R. A. Holzknecht et al., *J. Immunol.* **1997**, *159*, 4659–4664.
55. M. W. Chopek, R. L. Simmons, J. L. Platt, *Transplant. Proc.* **1987**, *19*, 4553–4557.
56. N. S. Ihrcke, L. E. Wrenshall, B. J. Lindman et al., *Immunol. Today* **1993**, *14*, 500–505.
57. T. E. Starzl, A. J. Demetris, N. Murase et al., *Immunol. Today* **1993**, *14*, 326–332.
58. D. H. Sachs, T. Sablinski, *Xenotransplantation* **1995**, *2*, 234–239.
59. Y. Zhao, K. Swenson, J. J. Sergio et al., *Nat. Med.* **1996**, *2*, 1211–1216.
60. M. Sykes, L. A. Lee, D. H. Sachs, *Immunol. Rev.* **1994**, *141*, 245–276.
61. H. Li, C. Ricordi, A. J. Demetris et al., *Transplantation* **1994**, *57*, 592–598.
62. I. Aksentijevich, D. H. Sachs, M. Sykes, *Transplantation* **1992**, *53*, 1108–1114.
63. H. A. Gritsch, R. M. Glaser, D. W. Emery et al., *Transplantation* **1994**, *57*, 906–917.
64. L. Inverardi, B. Clissi, A. L. Stolzer et al., *Transplant. Proc.* **1996**, *28*, 552.
65. H.-G. Ljunggren, K. Karre, *Immunol. Today* **1990**, *11*, 237–244.
66. L. L. Lanier, J. H. Phillips, *Immunol. Today* **1996**, *17*, 86–91.

67. J. H. Lawson, J. L. Platt in *Dialysis and Transplantation: A Companion to Brenner and Rector's The Kidney* (Eds.: W. Owen, B. Pereira, M. Sayegh), W. B. Saunders Co, Philadelphia, 2000, pp. 653–660.
68. C. W. Daggett, M. Yeatman, A. J. Lodge et al., *J. Thorac. Cardiovasc. Surg.* **1998**, *115*, 19–27.
69. S. A. Akhter, C. A. Skaer, A. P. Kypson et al., *Proc. Natl. Acad. Sci. U. S. A.* **1997**, *94*, 12100–12105.
70. A. P. Kypson, K. Peppel, S. A. Akhter et al., *J. Thorac. Cardiovasc. Surg.* **1998**, *115*, 623–630.
71. J. H. Lawson, J. L. Platt, *Transplantation* **1996**, *62*, 303–310.
72. J. H. Lawson, R. D. Sorrell, J. L. Platt, *Circulation* **1997**, *96*, 565.
73. C. Patience, Y. Takeuchi, R. A. Weiss, *Nat. Med.* **1997**, *3*, 282–286.
74. B. A. Sullenger, H. F. Gallardo, G. E. Ungers et al., *Cell* **1990**, *63*, 601–608.
75. C. G. Groth, O. Korsgren, A. Tibell et al., *Lancet* **1994**, *344*, 1402–1404.
76. R. S. Chari, B. H. Collins, J. C. Magee et al., *N. Engl. J. Med.* **1994**, *331*, 234–237.
77. T. Deacon, J. Schumacher, J. Dinsmore et al., *Nat. Med.* **1997**, *3*, 350–353.
78. H. Nagata, M. Ito, J. Cai et al., *Gastroenterology* **2003**, *124*, 422–431.
79. J. A. Thomson, J. Itskovitz-Eldor, S. S. Shapiro et al., *Science* **1998**, *282*, 1145–1147.

16
Sculpturing the Architecture of Mineralized Tissues: Tissue Engineering of Bone from Soluble Signals to Smart Biomimetic Matrices

Ugo Ripamonti, Lentsha Nathaniel Ramoshebi, Janet Patton, June Teare, Thato Matsaba and Louise Renton
Bone Research Unit of the Medical Research Council and the University of the Witwatersrand, 2193 Parktown, Johannesburg, South Africa

16.1 Introduction

The study of the molecular and cellular biology of bone morphogenetic and osteogenic proteins (BMPs/OPs), morphogens endowed with the striking prerogative of initiating de novo bone formation by induction, has profoundly modified our understanding of cell differentiation and the induction of tissue morphogenesis [1–14]. Indeed, in vivo studies over the past 12 years have revealed how osteoblastic differentiation and the induction of bone formation are controlled via the deployment of a set of specific soluble signals, the BMPs/OPs, members of the transforming growth factor-β (TGF-β) supergene family. To induce bone formation, however, the soluble signals require to be reconstituted with an insoluble substratum that triggers the bone differentiation cascade, as shown in nonhuman and human primates [15–19].

The capability of the pharmaceutical industry to develop peptides and proteins or morphogens, defined as form-generating substances [20] capable of imparting specific different pathways to responding cells initiating the cascade of pattern formation and the attainment of tissue form and function, has increased markedly in the twenty-first century. Morphogens of the transforming growth factor-β (TGF-β) superfamily have now become available in commercially viable quantities produced as rationally designed gene products in addition to naturally occurring ones in purified form and in large scale for use in clinical contexts [17, 21, 22]. However, even if morphogens' availability by recombinant DNA technology produced by the pharmacological industry is cost-effective, significant challenges to morphogens' delivery still limit their utilization as therapeutic agents. Morphogens of the TGF-β superfamily are autocrine- and paracrine-soluble signals that have widespread pleiotropic functions in vivo [3, 5, 7, 8]. Following their direct administration into the blood stream, it is highly improbable that tissue engineers will be able to direct such morphogens to specific receptors on target cells in sufficient quantities to evoke a desired therapeutic response without promoting potentially deleterious side effects in several tissues and organs of the mammalian body.

This monograph describes the soluble signals that initiate bone formation by induction and outlines novel concepts of bone tissue engineering as evaluated in primate models. The monograph further reports the apparent redundancy in endochondral bone formation by molecularly different members of the TGF-β superfamily. Our studies reported below indicate that in the adult primate – and in the adult primate only – recombinantly produced or naturally derived TGF-β isoforms are powerful inducers of endochondral bone formation, with a specific activity equal to, if not higher than, identical doses of recombinant hBMPs/OPs. We will further highlight the use of novel biomimetic matrices to deliver the biological activity of the osteogenic members of the TGF-β superfamily. Experimental studies in ovariectomized (OVX) primates of the species *Papio ursinus* will describe the local delivery of naturally derived and recombinantly produced hBMPs/OPs using a reconstituted basement membrane gel (Matrigel®) to treat systemic bone loss in nonhuman primates and thus by extension to human primates affected by osteoporosis. Lastly, we describe the development and use of biomimetic matrices with site-specific geometric modifications and endowed with the striking prerogative of initiating de novo bone formation by induction in heterotopic extraskeletal sites of primates even in the absence of exogenously applied osteogenic gene products of the TGF-β superfamily [16, 23–25].

16.2 Osteogenic Soluble Signals of the TGF-β Superfamily

Bone is in both a soluble and a solid state, and there is a continuum between the soluble and insoluble states regulated by signals in solution interacting with the insoluble extracellular matrix [7, 10]. Nature relies on common yet limited molecular mechanisms tailored to provide the emergence of specialized tissues and organs. The TGF-β superfamily is indeed an elegant example of nature's parsimony in programming multiple specialized pleiotropic functions deploying molecular isoforms with minor variations in amino acid motifs within highly conserved carboxy-terminal regions.

In preclinical and clinical contexts, tissue regeneration in postnatal life recapitulates events that occur in the normal course of embryonic development and morphogenesis [5, 8, 10]. Both embryonic development and tissue regeneration are equally regulated by a selected few, highly conserved families of morphogens, the soluble signals of the TGF-β superfamily. Among the many tissues in the body, bone has considerable potential for repair and regeneration and could well be considered a prototype for tissue repair and regeneration in molecular terms [3–5, 14, 16, 25].

Bone morphogenetic proteins/osteogenic proteins (BMPs/OPs), members of the TGF-β superfamily, are soluble mediators of tissue morphogenesis and powerful regulators of cartilage and bone differentiation in embryonic development and regeneration in postnatal life [1, 3–5, 14, 16, 25]. A striking prerogative of the osteogenic members of the TGF-β superfamily, whether naturally derived or produced by DNA recombinant technologies, is their ability to induce de novo endochondral bone formation in extraskeletal heterotopic sites in postnatal life as a recapitulation of events that occur in the normal course of embryonic development [3, 5, 12, 14, 25]. To induce endochondral bone differentiation to be exploited in preclinical and clinical contexts, the osteoinductive soluble signals require the reconstitution with an insoluble signal or substratum that triggers the bone differentiation cascade [7–9].

The menu for enunciating the rules that sculpt the architecture of corticocancellous structures of the bone and regulate bone regeneration and bone tissue engineering in clinical contexts lists complex interactions between soluble and insoluble signals [7]. Tissue engineering in clinical contexts requires three key components: an osteoinductive signal; an insoluble substratum that delivers the signal and acts as a scaffold for new bone formation; and host cells capable of differentiation into bone cells in response to the osteoinductive signal. The signals responsible for osteoinduction are conferred by the osteogenic members of the TGF-β superfamily [10, 15, 25].

The reconstitution of doses of recombinant human osteogenic protein-1 (rhOP-1), with insoluble substrata such as the inactive insoluble collagenous bone matrix additionally prepared from xenogeneic sources, restores the biological activity and results in the long-term efficacy of single applications of gamma-irradiated hOP-1 delivered by xenogeneic bovine collagenous matrices in regenerating large defects of membranous bone prepared in the calvarium of the adult primate *Papio ursinus* [15, 26]. The operational reconstitution of the soluble morphogenetic signal (hOP-1) with an insoluble substratum (the collagenous insoluble bone matrix) underscores the critical role of the insoluble signal of the collagenous matrix for the induction of tissue morphogenesis and regeneration [10, 15, 27]. These findings obtained in the adult primate indicate that a single application of gamma-irradiated hOP-1 combined with the gamma-irradiated xenogeneic bovine collagenous bone matrix carrier is effective in regenerating and maintaining the architecture of the induced bone at doses of 0.5- and 2.5-mg hOP-1 per gram of carrier matrix (Fig. 1) [15]. Information concerning the efficacy and safety of gamma-irradiated osteogenic devices in nonhuman primates is an important prerequisite for clinical applications.

The fact that a single BMP/OP initiates bone formation by induction does not preclude the requirement for interactions with other morphogens deployed synchronously and synergistically during the cascade of bone formation by induction, which may proceed via the combined action of several BMPs/OPs resident within the natural milieu of the extracellular matrix of bone [15, 25]. Partially purified preparations from bone matrix are known to contain, in addition to specific BMPs/OPs, several other proteins and some as yet poorly characterized mitogens [28]. Indeed, 90 days after implantation, regenerated tissue induced by 2.5 mg of partially purified BMPs/OPs combined with gamma-irradiated matrix, had mineralized bone and osteoid volumes comparable to specimens induced by 0.5-mg hOP-1 devices (Fig. 1) [15].

Partially purified preparations from bone matrix obtained using chromatographic procedures as described [15, 29–31] are known to contain BMP-2, BMP-3, and OP-1 but not detectable TGF-βs [32]. In a clinical trial in humans, osteogenic devices prepared by partially purified BMPs/OPs reconstituted with

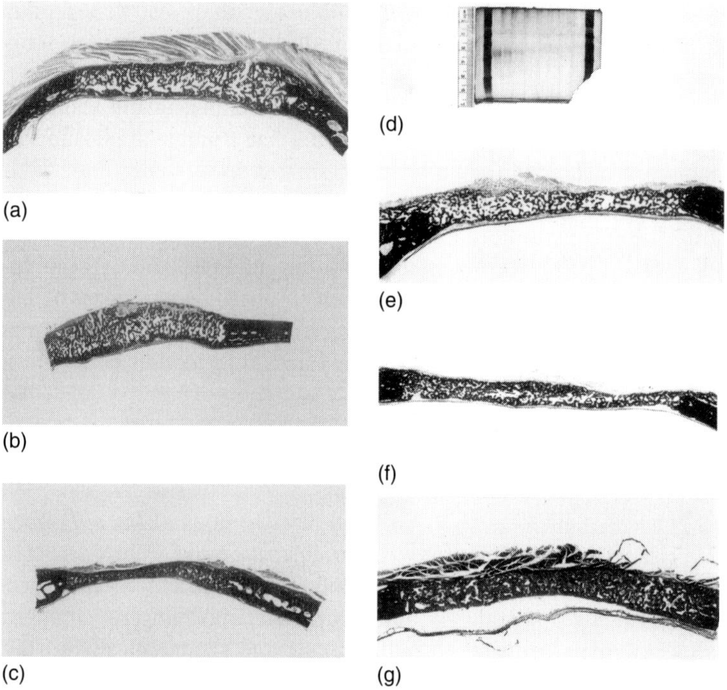

Fig. 1 Low-power photomicrographs of specimens of calvarial defects implanted with bone morphogenetic/osteogenic proteins (BMPs/OPs). (a) 0.1 mg of hOP-1 delivered by xenogeneic bovine collagenous matrix 90 days after implantation. (b and c) 2.5- and 0.5-mg hOP-1 delivered by gamma-irradiated bovine collagenous matrix harvested on day (b) 90 and (c) 365 after implantation showing complete regeneration of the defects with thick trabeculae of newly formed mineralized bone surfaced by continuous osteoid seams. Original magnification: (a) ×3, (b and c) ×2. (d) Sodium dodecyl sulfate-polyacrylamide gel electrophoresis showing purification of osteogenin (BMP-3) to homogeneity from baboon bone matrix. (e, f, and g): 0.5 (e) and 2.5 (f and g) mg of naturally derived BMPs/OPs delivered by gamma-irradiated bovine collagenous bone matrix: complete regeneration of the calvarial defects 90 days after implantation. Original magnification ×3. Undecalcied sections cut at 7 μm stained with Goldner's trichrome.

human bone matrix as described [31, 33] were combined with sterile saline and applied to mandibular defects as a paste. Histological examination on undecalcified sections prepared from bioptic material obtained three months after implantation showed that successfully implanted BMPs/OPs devices induced mineralized bone trabeculae with copious osteoid seams lined by contiguous osteoblasts. Additionally, bone deposition directly onto nonvital matrix provided unequivocal evidence of osteoinduction [31] (Fig. 2).

To date, more than 40 related proteins with BMP/OP-like sequences and activities have been sequenced and cloned, but little is known about their interactions during the cascade of bone formation by induction or about the biological and therapeutic significance of

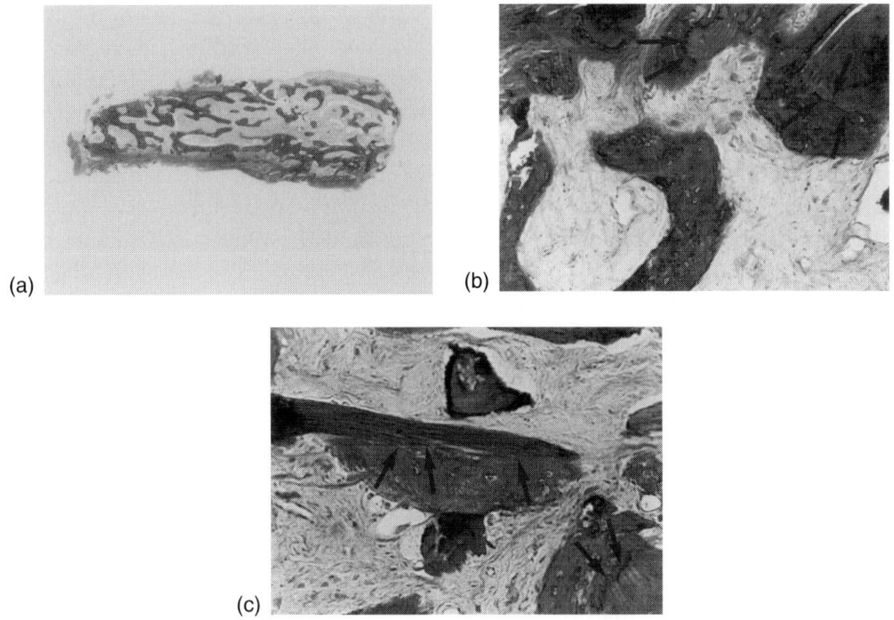

Fig. 2 Photomicrographs of tissue induction and morphogenesis in bioptic material 90 days after implantation of naturally derived BMPs/OPs purified from bovine bone matrix in human mandibular defects. (a) Trabeculae of newly formed mineralized bone covered by continuous osteoid seams within highly vascular stroma. (b) and (c) High-power views showing cellular mineralized bone surfaced by osteoid seams. Newly formed and mineralized bone directly opposing the implanted collagenous matrix carrier (arrows) confirms bone formation by induction. Undecalcified sections at 7 μm stained with Goldner's trichrome. Original magnification: (a) ×14; (b) ×40; and (c) ×50. (See Color Plate p. xxiii).

this apparent redundancy. Recombinantly produced hBMP-2, hBMP-4, and hOP-1 singly initiate bone formation by induction in vivo [1, 34–36]. It is likely that the endogenous mechanisms of bone repair and regeneration in postnatal life necessitates the deployment and concerted actions of several of the BMPs/OPs present within the extracellular matrix of bone [15, 25]. Whether the biological activity of partially purified BMPs/OPs is the result of the sum of a plurality of BMP/OP activities or of a truly synergistic interaction among BMP/OP family members deserves appropriate investigation [15, 25].

In addition to bone induction in postfetal life, the BMPs/OPs are involved in inductive events that control pattern formation during morphogenesis and organogenesis in such disparate tissues and organs as the kidney, eye, nervous system, lung, teeth, skin, and heart [37]. These strikingly pleiotropic effects of BMPs/OPs may spring from minor amino acid sequence variations in the carboxyl-terminal region of the proteins [38] as well as in the transduction of distinct signal pathways by individual Smad proteins after transmembrane serine-threonine kinase receptor activation [39, 40].

In vitro studies indicate that both hOP-1 and hBMP-2 modulate messenger RNA (mRNA) expression of related BMP/OP family members and in vivo studies are now mandatory to identify therapeutic approaches on the basis of the information of gene regulation by hBMPs/OPs [41, 42]. Ultimately, it will be necessary to gain insights into the distinct spatial and temporal patterns of expression of BMPs/OPs and TGF-ß family members elicited by a single application of recombinant hBMPs/OPs, including hOP-1.

To investigate expression patterns of gene products following single applications of a recombinant hBMP/OP, doses of hOP-1 (0.1, 0.5, and 2.5 mg) were implanted in heterotopic extraskeletal sites of the *rectus abdominis* muscle and orthotopically in calvarial defects of adult primate *Papio ursinus.* mRNA expression of OP-1, TGF-ß, BMP-3, and collagens Type II and IV showed upregulation of the genes in ossicles generated by the higher dose of the hOP-1 as evaluated on day 15 and 30, though with significant differences between tissues generated in orthotopic and in heterotopic sites [Thato Matsaba, unpublished data]. In vivo studies in primate models should now be used to design therapeutic approaches on the basis of the information of gene regulation by hOP-1 [25].

The processes of tissue morphogenesis and regeneration of the complex tissue morphologies of the periodontal tissues rest on the sequential expression of BMP/OP proteins during regenerative events and also on the expression and synthesis of related morphogenetic gene products of the large TGF-ß superfamily [43]. The initiating events in periodontal regeneration are transitory and lead to sequential molecular and cellular outcomes stimulating subsequent events such as chemotaxis, differentiation, proliferation, and angiogenesis, leading ultimately to the morphogenesis and remodeling of the periodontal tissues [43–45].

The pleiotropic functions of the BMPs/OPs have been further shown after their implantation in furcation defects prepared in adult primates *Papio ursinus* [43, 44, 46, 47]. Undecalcified semi-thin sections of furcation defects of the chacma baboon *Papio ursinus* treated with doses of highly purified bone-derived BMPs/OPs showed not only alveolar bone but also cementogenesis and periodontal ligament regeneration, the essential ingredients to engineer periodontal tissue regeneration [44, 46, 48] (Fig. 3).

Single applications of relatively low doses of hOP-1 (0.1- and 0.5-mg hOP-1 per gram of collagenous matrix as carrier) preferentially induced cementogenesis as evaluated 60 days after implantation in surgically induced furcation defects in *Papio ursinus* [47]. This seemingly specific cementogenic function of hOP-1 suggested that a structure/activity profile could reside within BMP/OP family members to control tissue morphogenesis and regeneration of disparate tissues and organs [43, 44, 47, 48].

More challenging was the demonstration of cementogenesis and alveolar bone regeneration in periodontally induced furcation defects with root surfaces chronically exposed to periodontal pathogens [43]. A pathogenic human strain of *Porphyromonas gingivalis* was inoculated into the furcation areas of the first and second mandibular molars of four adult chacma baboons twice a month for 12 months. Chronic periodontitis was induced in all four animals as assessed by probing periodontal pocket depths, intraoral radiographs, and microbiological analyses that confirmed the presence of *Porphyromonas gingivalis* [43]. Two months after scaling, root planing, and a plaque-control regimen with clinical resolution of gingivitis, mucoperiosteal flaps were elevated to expose Class II furcation

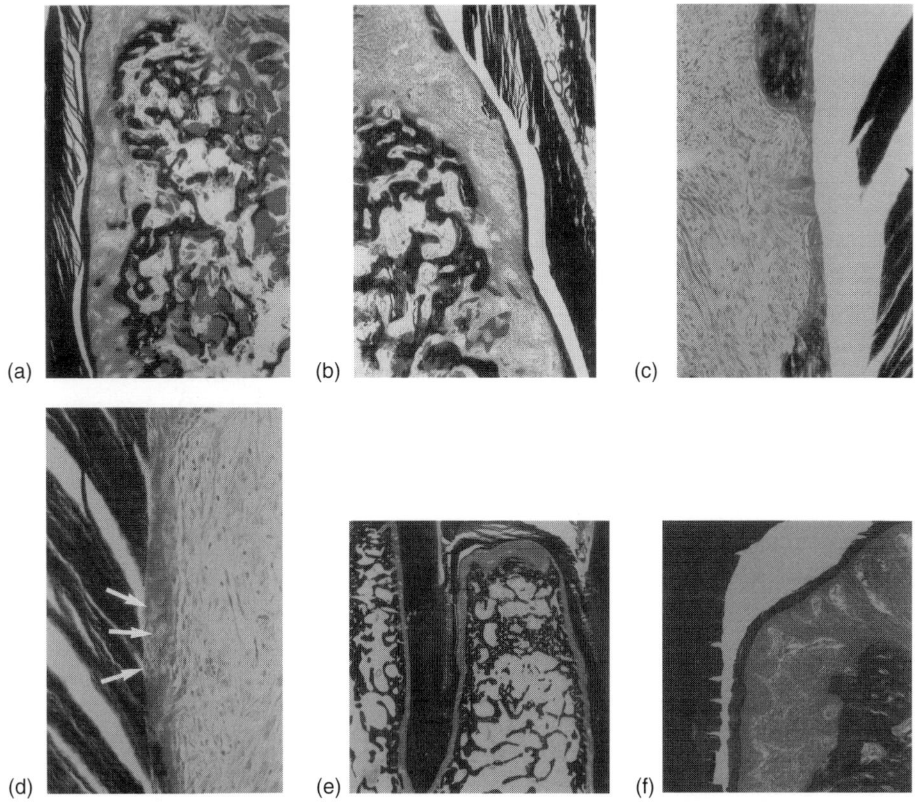

Fig. 3 Photomicrographs of periodontal tissue engineering and morphogenesis by BMPs/OPs in the primate *Papio ursinus*. (a and b) Furcation defects 60 days after implantation of 250 μg of naturally derived BMPs/OPs showing regeneration of cementum, periodontal ligament fibers, and mineralized alveolar bone surfaced by continuous osteoid seams. (c and d) High-power views: the undecalcified sections cut at 7 μm permit one to identify the newly deposited cementum (orange red) as yet to be mineralized, mineralized cementum (in blue), and cementogenesis with foci of nascent mineralization in pale blue (arrows) within cementoid collagenic material (in red). Note in (c) the generation of Sharpey's fibers within the newly formed cementoid. (e and f) Photomicrographs of periodontitis-induced furcation defects treated with 2.5 mg of hOP-1 per gram of bovine-insoluble collagenous bone matrix as carrier. Complete regeneration with *restitutio ad integrum* of the periodontal tissues with newly induced cementum, periodontal ligament, and alveolar bone with Sharpey's fibers coursing from the regenerated alveolar bone to (f) the newly formed cementum. Original magnification: (a and b) ×15; (c and d) ×100; (e) ×6; and (f) ×40. Undecalcified sections cut at 4 μm stained with Goldner's trichrome.

defects of the affected molars filled with granulation tissue. After root planing and debridement, furcation defects were implanted with 0.5 and 2.5 mg of gamma-irradiated hOP-1 per gram of xenogenic bovine-insoluble collagenous bone matrix as carrier.

Serial undecalcified sections prepared six months after surgery showed regeneration of alveolar bone and induction of cementogenesis with Sharpey's fibers uniting the regenerated bone to the newly induced cementum with 0.5- and 2.5-mg hOP-1 indicating an additional specific use of hOP-1 for tissue engineering and morphogenesis in clinical context [43] (Fig. 3).

The study also demonstrates that a single recombinant morphogen, originally isolated as osteogenic protein, induces a cascade of pleiotropic molecular and morphological events leading to the regeneration of the complex morphologies of the periodontal tissues, including alveolar bone, cementum, and the assembly of a functionally oriented periodontal ligament system [43].

16.3 Site–tissue Specificity of Bone Induction by TGF-β Isoforms in the Primate

In the bona fide heterotopic assay for bone induction in the subcutaneous site of rodents [49, 50], the TGF-β isoforms, either purified from natural sources or expressed by DNA recombinant technologies, do not initiate endochondral

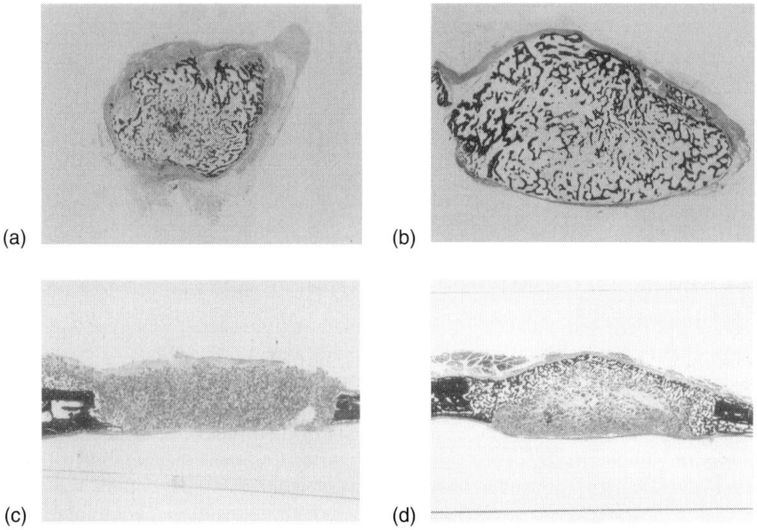

Fig. 4 Tissue morphogenesis and site–tissue-specific osteoinductivity of recombinant human-transforming growth factor-β2 (hTGF-β2) in the adult primate *Papio ursinus*. (a and b) Endochondral bone induction and tissue morphogenesis by hTGF-β2 implanted in the *rectus abdominis* muscle and harvested (a) 30 and (b) 90 days after heterotopic implantation. Heterotopic bone induction by a single administration of (a) 5- and (b) 25-µg hTGF-β2 delivered by 100 mg of guanidinium-inactivated collagenous matrix. (c and d) Calvarial specimens harvested from the same animals as shown in (a and b). (c) Lack of bone formation in a calvarial defect 30 days after implantation of 10-µg hTGF-β2 delivered by collagenous bone matrix. (d) Osteogenesis, albeit limited, is found in a specimen treated with 100-µg hTGF-β2 with bone formation only pericranially 90 days after implantation. Note the delicate trabeculae of newly formed bone facing scattered remnants of collagenous matrix particles, embedded in a loose and highly vascular connective tissue matrix. Original magnification: (a and b) ×4.5; (c and d) ×3. Undecalcified sections cut at 4 µm stained with Goldner's trichrome. (See Color Plate p. xxiv).

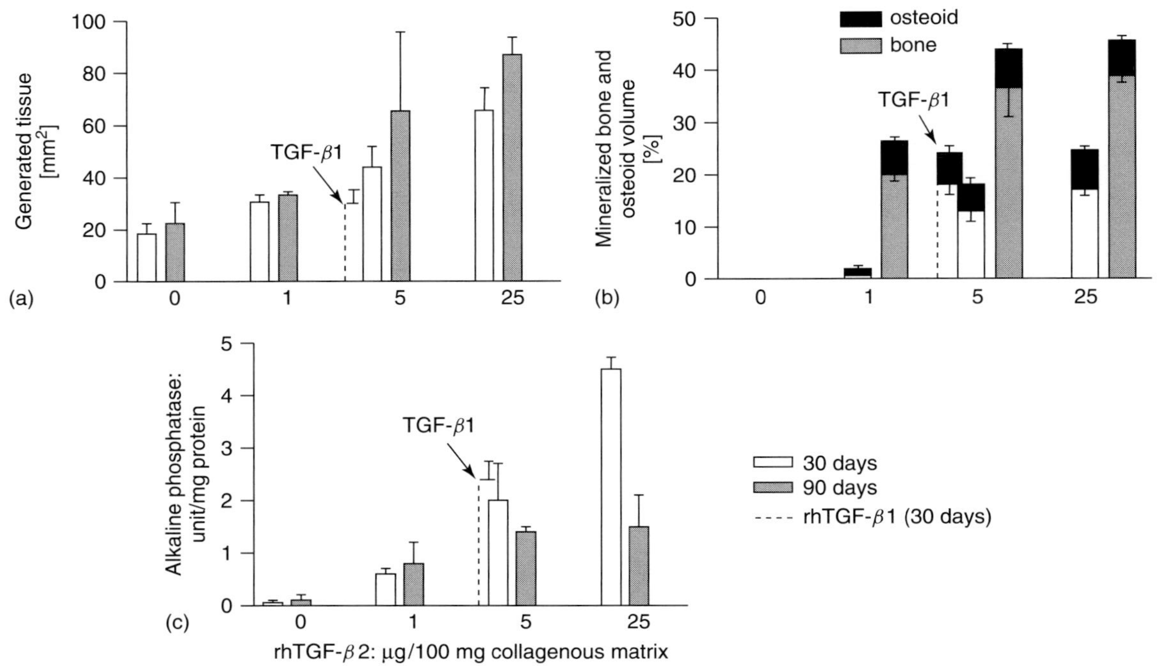

Fig. 5 Effect of transforming growth factor-β (hTGF-β) isoforms on key parameters of heterotopic bone formation by induction on day 30 and 90. (a) Computerized analysis of newly generated tissue (in mm^2) induced by doses of hTGF-β1 and hTGF-β2 delivered by inactivated collagenous matrix. (b) Histomorphometric results of induced mineralized bone and osteoid volumes (in %) in the newly formed ossicles. (c) Effects of hTGF-β1 and hTGF-β2 on tissue alkaline phosphatase activity in the newly generated heterotopic ossicles. The alkaline activity of the supernatant after homogenization of implants was determined with 0.1-M *p*-nitrophenylphosphate as substrate at pH 9.3 and 37 °C for 30 min. One unit of enzyme liberates one micromole of *p*-nitrophenylphosphate under the assay conditions used. Alkaline phosphatase is expressed as units of activity per milligram of protein [29, 50].

bone formation [51, 52]. Strikingly, and in contrast, the mammalian TGF-β isoforms have shown a marked site- and tissue-specific endochondral osteoinductivity, yet remarkably in the primate only [53–57] (Figs. 4 and 5). In the higher vertebrates such as the primate species, the presence of several related but different molecular forms with osteogenic activity highlights the biological significance of this apparent redundancy and indicates multiple interactions during both embryonic development and bone regeneration in postnatal life [25, 54]. Indeed, a potent and accelerated synergistic interaction in endochondral bone formation has been shown with the binary application of human recombinant or native TGF-β1 with hOP-1 in both heterotopic and orthotopic sites of primates (Fig. 6) [25, 54, 55, 58].

The TGF-β isoforms are powerful inducers of endochondral bone when implanted in the *rectus abdominis* muscle of the primate *Papio ursinus* at doses of 1, 5, and 25 μg per 100 mg of collagenous matrix as carrier, yielding large corticalized ossicles by day 90 (Figs. 4 and 5). Endochondral bone initiated by TGF-β isoforms expresses mRNA of bone induction markers including BMP-3 and OP-1 [55–57].

A significant striking result is that the bone-inductive activity of TGF-β isoforms in the primate is site- and tissue-specific with rather substantial endochondral bone induction in the *rectus abdominis* muscle but absent osteoinductivity in orthotopic sites on day 30 and limited osteogenesis in orthotopic sites on day 90 (Fig. 4).

The observed site and tissue specificity of induction in the nonhuman primate *Papio ursinus* and thus by extension to *homo sapiens* may be due to the presence or absence of multiple variable-responding cells, the expression of inhibitory binding proteins or the influence of the downstream antagonists of the TGF-β signaling, Smad6 and Smad7 [59–61]. Indeed, current results of mRNA studies on tissues generated by TGF-β isoforms in heterotopic and orthotopic sites demonstrated robust expression of the TGF-β self-regulatory proteins Smad6 and Smad7 orthotopically, but only modest expression in tissue from heterotopic sites (unpublished observation). These findings represent one possible explanation for the poor osteoinductivity of TGF-β isoforms observed in nonhealing calvarial defects and indicate that overexpression of Smad6 and Smad7 downregulate the osteoinductivity of the TGF-β isoforms when deployed orthotopically [61].

Conceivably, the rapid induction of endochondral bone by hTGF-β isoforms could be utilized for the generation of large ossicles in the *rectus abdominis* muscle of human patients. Thirty days after heterotopic implantation, generated ossicles could be harvested and morsellized fragments transplanted into bony defects affecting the same patient in an autogenous fashion to treat defects either in the axial or craniofacial skeleton including periodontal osseous defects. The rapidity of tissue morphogenesis and induction of bone formation complete with mineralization of the outer cortex of the ossicles and bone marrow formation by day 30 is of particular importance for repair and regeneration of bone in the elderly, where repair phenomena are temporally delayed and healing progresses slower than in younger patients. Potentially, fragments of autogenously induced bone could be morsellized from ossicles induced in the *rectus abdominis* after the binary application of hOP-1 and relatively low doses of a TGF-β isoform, a synergistic strategy known to yield massive mineralized ossicles with large seams of osteoid populated by contiguous

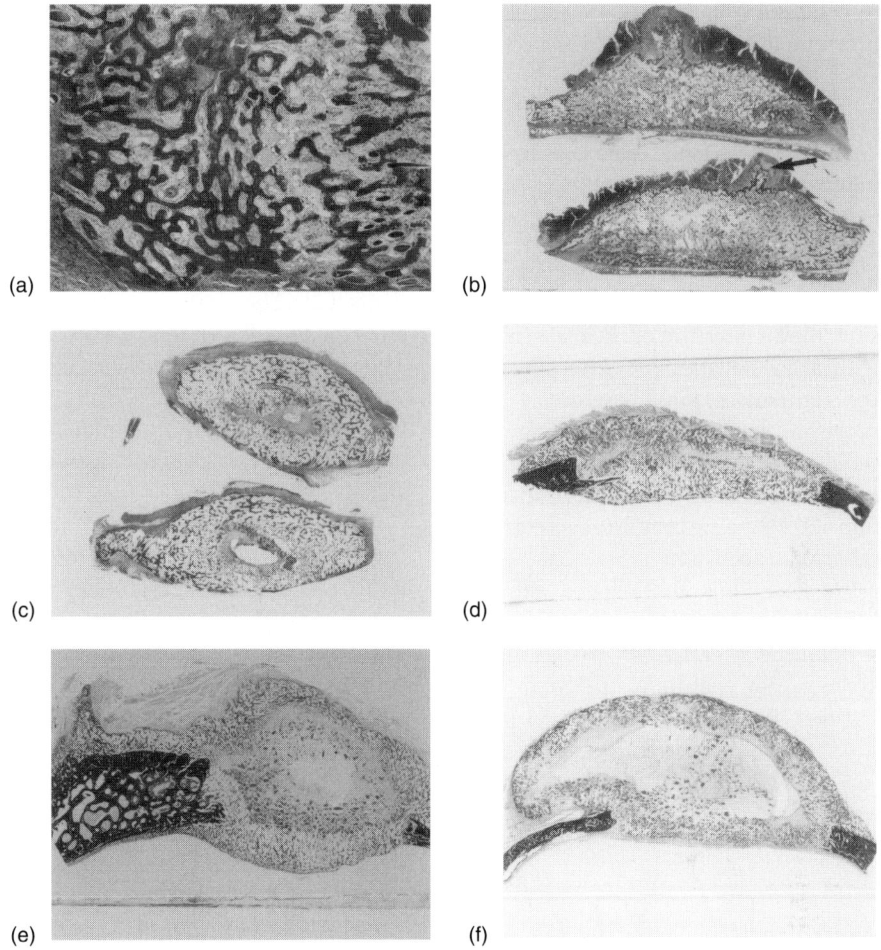

Fig. 6 Synergistic tissue morphogenesis and heterotopic bone induction by the combinatorial action of recombinant human osteogenic protein-1 (hOP-1) and transforming growth factor-β1 (hTGF-β1). (a) Rapid and extensive induction of mineralized bone in a specimen generated by 25-μg hOP-1 combined with 0.5-μg hTGF-β1 on day 15. Mineralized trabeculae of newly formed bone are covered by osteoid seams populated by contiguous osteoblasts. (b and c) Photomicrographs of massive ossicles that had formed between the muscle fibers and the posterior fascia of the *rectus abdominis* using binary applications of 25- and 125-μg hOP-1 interposed with 5-μg hTGF-β1 on day 30. Corticalization of the large heterotopic ossicles with displacement of the *rectus abdominis* muscle and extensive bone marrow formation permeating trabeculae of newly formed bone. Arrow in (b) points to a large area of chondrogenesis protruding within the rectus abdominis muscle. (d, e, and f) Low-power photomicrographs of calvarial defects treated by binary applications of 100-μg hOP-1 and 5 μg of naturally derived TGF-β1 purified from porcine platelets as described [55] and harvested on day 30. The calvarial specimens show extensive bone differentiation with pronounced vascular tissue invasion and displacement of the calvarial profile 30 days after implantation of the binary morphogen combinations. Original magnification: (a) ×30; (b, c) ×3.5; (d, e, and f) ×3. Undecalcified sections cut at 4 μm and stained with Goldner's trichrome. (See Color Plate p. xxv).

osteoblastic cells by day 15 and 30 after heterotopic implantation [25, 54, 55, 58].

16.4 Treatment of Systemic Bone Loss by Local Induction of Bone

The biosynthesis and assembly of extracellular matrix with angiogenesis and vascular invasion is a prerequisite to restore the architecture of skeletal structures with a constellation of extracellular matrix components [5]. Therefore, a critical provision for tissue engineering is the sculpturing of the optimal extracellular matrix scaffolding [5] for the transformation of responding cells into secretory bone cells and osteoblasts.

We have thus investigated extracellular matrix components reconstituted as a biomimetic carrier matrix as delivery systems for naturally derived, highly purified BMPs/OPs and hOP-1 in the rodent bioassay [62]. Matrigel®, a soluble extract of the Engelbreth-Holm-Swarm tumor [63] gels at room temperature to form a reconstituted basement membrane [63]. It contains laminin, type IV collagen, entactin, nidogen, heparan sulphate proteoglycan, and additional growth factors [63], and was tested to deliver the BMPs/OPs after subcutaneous implantation in rodents [62]. The matrix proteins laminin and type IV collagen bind BMPs/OPs [64]. The combination of the Matrigel® carrier with morphogens successfully induced bone formation, indicating that Matrigel® is an effective carrier of osteogenic soluble signals [62].

We have used the Matrigel® matrix reconstituted with BMPs/OPs for a novel local treatment of systemic bone loss in our cohort of OVX primates *Papio ursinus* [62]. The bone mineral density of the lumbar vertebrae of the OVX primates was significantly affected by estrogen depletion [62]. Histomorphometric data on iliac crest biopsies showed that bone was permanently lost 36 months after ovariectomy. Injections of 0.5 mg of naturally derived BMPs/OPs into the lumbar vertebrae 3 and 4 using an X-ray image intensifier for guidance was successfully performed for the first time in primates [62] (Fig. 7). This novel method may prove to be valuable for the treatment of systemic bone loss by localized injections of osteogenic proteins of the TGF-β superfamily in clinical contexts.

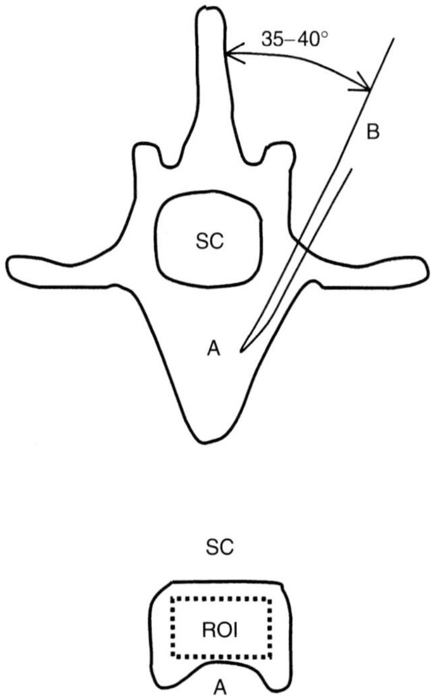

Fig. 7 Schematic representation of (a) the primate *Papio ursinus* lumbar vertebra and (b) the bone marrow biopsy needle positioning during local administration of naturally derived bone morphogenetic proteins. Insertion of the needle was monitored on X-ray image intensifier. Bottom panel: sagittal representation of (a) the lumbar vertebral body and its region of interest (ROI); SC: spinal canal.

16.5 Geometric Induction of Bone Formation

Biomimetic matrices endowed with intrinsic osteoinductivity, that is, capable of initiating de novo bone formation in heterotopic sites of primates even in the absence of exogenously applied BMPs/OPs, have been developed in our laboratories [23, 24] (Fig. 8). Sintered highly crystalline hydroxyapatites induce bone formation in adult primates via

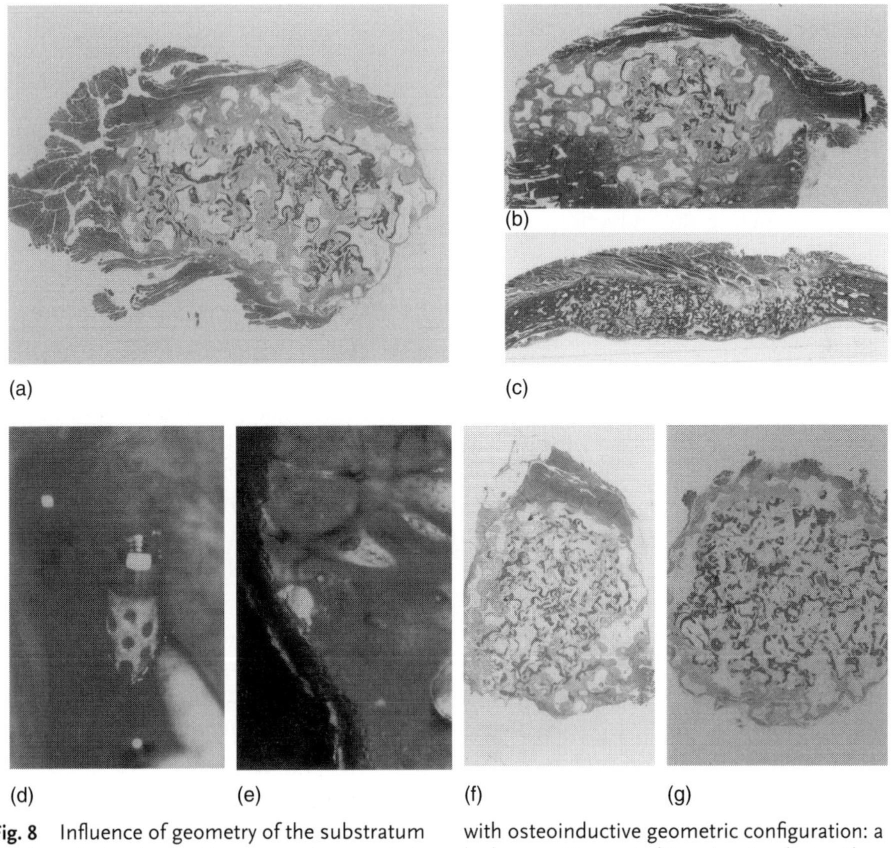

Fig. 8 Influence of geometry of the substratum on tissue induction and bone morphogenesis in highly crystalline sintered hydroxyapatites. (a and b) Intrinsic and spontaneous bone induction within the porous spaces of hydroxyapatite biomatrices implanted heterotopically in the *rectus abdominis* of an adult primate without the addition of exogenously applied BMPs/OPs. (c) Low-power view of a sintered hydroxyapatite specimen 90 days after implantation showing complete bone growth across the porous spaces of the sintered hydroxyapatite disc implanted in a calvarial defect of the adult primate *Papio ursinus*. (d) Preclinical application of implants with osteoinductive geometric configuration: a hydroxyapatite-coated titanium implant with a series of concavities prepared on the coated surface is implanted in the edentulous ridge of an adult primate. (e) High-power view of an undecalcified section showing a concavity region and the osteointegration to the coating of hydroxyapatite 60 days after implantation. (f and g) Bone induction by hTGF-β2 in sintered porous hydroxyapatites pretreated with 1-(f) and 25-µg (g) hTGF-β2 and harvested on day 90. (a, b, c, f, and g) Decalcified sections cut at 4 µm. Original magnification: (a) $\times 8$; (b) $\times 6$; (c) $\times 3 \times 4$; (e) $\times 120$; (f and g) $\times 6$.

intrinsic osteoinductivity regulated by the geometry of the substratum [23, 24] (Fig. 8).

Current experiments in our laboratories have confirmed that the geometry of biomimetic matrices is not the only driving force in osteoinduction since the structure of the insoluble signal dramatically influences and regulates gene expression, with the induction of bone as a secondary response [25, 65]. Soluble signals induce morphogenesis; physical forces imparted by the geometric topography of the insoluble signal dictate biological patterns, constructing the induction of bone and regulating the expression of selected mRNA of gene products of the TGF-β superfamily as a function of the structure [66].

Our molecular, biochemical, and morphological data have indicated that the specific geometric configuration in the form of concavities within highly crystalline hydroxyapatite biomimetic matrices is the driving molecular and morphogenetic microenvironment, conducive and inducive to a specific sequence of events leading to bone formation by induction [23, 24]. The specific geometry of the biomimetic matrices initiates a bone-inductive microenvironment by providing geometrical structures biologically and architecturally conducive and inducive to optimal sequestration and synthesis of osteogenic members of the TGF-β superfamily [23–25] and particularly capable of stimulating angiogenesis, a prerequisite for osteogenesis. Angiogenesis may indeed provide a temporally regulated flow of cell populations capable of expression of the osteogenic phenotype.

We have recently investigated whether the BMPs/OPs shown to be present by immunolocalization in the concavities are adsorbed onto the sintered biomimetic matrices from the circulation or rather are locally produced after local expression and synthesis by transformed cellular elements resident within the concavity microenvironment [23, 65]. We now propose the following cascade of molecular and morphogenetic events culminating in the induction of bone in heterotopic sites of primates and initiating within concavities of the *smart* biomimetic matrices:

1. Vascular invasion and capillary sprouting within the invading tissue with capillary elongation in close contact with the implanted hydroxyapatite biomatrix.
2. Attachment to and differentiation of mesenchymal cells at the hydroxyapatite/soft tissue interface of the concavities. Expression of TGF-β and BMPs/OPs family member genes in differentiating osteoblast-like cells resident within the concavities of the *smart* biomimetic matrices as shown by Northern blot analyses of tissue harvested from the concavities of the substratum [65].
3. Synthesis of specific TGF-β superfamily member proteins as markers of bone formation by induction from resident transformed osteoblast-like cells onto the sintered crystalline hydroxyapatite as shown by immunolocalization of OP-1 and BMP-3 within the cellular cytoplasm and at the interface of the hydroxyapatite biomatrix with the invading mesenchymal tissue [23, 24, 65].
4. *Intrinsic* osteoinduction with further differentiation of osteoblastic cell lines, which is dependent upon a critical threshold of endogenously produced BMPs/OPs initiating bone formation by induction as a secondary response [65].

16.6 Conclusion

Tissue regeneration in postnatal life recapitulates events that occur in the normal course of embryonic development and morphogenesis. Both embryonic development and tissue regeneration are equally regulated by a selected few and highly conserved families of morphogens. This plurality of gene products are members of the TGF-β superfamily. The initiation of bone formation during embryonic development and postnatal osteogenesis involves a complex cascade of molecular and morphogenetic processes that ultimately lead to the architectural sculpture of precisely organized multicellular structures. In the primate only, heterotopic bone induction is initiated by naturally derived BMPs/OPs and TGF-βs, recombinant hBMPs/OPs and hTGF-βs, and sintered hydroxyapatites biomimetic matrices with a specific geometric configuration. Bone tissue develops as a mosaic structure in which members of the TGF-β superfamily singly, synergistically, and synchronously initiate and maintain the developing morphological structures and play different roles at different time points of the morphogenetic cascade. Osteogenic members of the TGF-β superfamily are sculpturing tissue constructs, helping to engineer skeletal tissue regeneration in molecular terms: morphogens exploited in embryonic development are reexploited and redeployed in postnatal tissue regeneration.

Biomimetic biomaterial matrices are now designed to obtain specific biological responses so much so that the use of biomaterials capable of initiating bone formation via osteoinductivity even in the absence of exogenously applied BMPs/OPs is fast altering the horizons of therapeutic bone regeneration. Our results have indicated that the geometry of the substratum is not the only driving force since the structure of the insoluble signal dramatically influences and regulates gene expression and the induction of bone as a secondary response. Soluble signals induce morphogenesis; physical forces imparted by the geometric topography of the insoluble signal dictate biological patterns, constructing the induction of bone and regulating the expression of osteogenic gene transcripts and their translation products initiating bone formation as a function of the structure [66–68].

Acknowledgments

This work is supported by grants from the South African Medical Research Council, the University of the Witwatersrand, Johannesburg, the National Research Foundation and by ad hoc grants from the Bone Research Unit. We thank the Central Animal Services of the University for the continuous help with primate experimentation.

References

1. J. M. Wozney, V. Rosen, A. J. Celeste et al., *Science* **1988**, *242*, 1528–1534.
2. J. M. Wozney, *Mol. Reprod. Dev.* **1992**, *32*, 160–167.
3. A. H. Reddi, *Curr. Opin. Cell Biol.* **1992**, *4*, 850–855.
4. A. H. Reddi, *Cytokine Growth Factor Rev.* **1997**, *8*, 11–20.
5. A. H. Reddi, *J. Cell Biochem.* **1994**, *56*, 192–195.
6. A. H. Reddi, *Curr. Opin. Cell Biol.* **1994**, *4*, 737–744.
7. A. H. Reddi, *Cell* **1997**, 159–161.
8. A. H. Reddi, *Nat. Biotechnol.* **1998**, *16*, 247–252.

9. A. H. Reddi, *Biochem. Soc. Trans.* **2000**, *28*, 345–349.
10. A. H. Reddi, *Tissue Eng.* **2000**, *6*, 351–359.
11. U. Ripamonti, A. H. Reddi, *Adv. Plast. Reconstr. Surg.* **1995**, *11*, 47–65.
12. U. Ripamonti, S. Vukicevic, *South Afr. J. Sci.* **1995**, *91*, 277–280.
13. U. Ripamonti, *South Afr. J. Bone Joint Surg.* **1997**, *7*, 21–32.
14. U. Ripamonti, N. Duneas, *Plast. Reconstr. Surg.* **1998**, *101*, 227–239.
15. U. Ripamonti, B. Van Den Heever, J. Crooks et al., *J. Bone Miner. Res.* **2000**, *15*, 1798–1809.
16. U. Ripamonti, L. N. Ramoshebi, T. Matsaba et al., *J. Bone Joint Surg. Am.* **2001**, *83-A*, S1-166–127.
17. G. E. Friedlaender, C. R. Perry, J. D. Cole, *J. Bone Joint Surg. Am.* **2001**, *83-A*, S151–S158.
18. E. H. Groeneveld, E. H. Burger, *Eur. J. Endocrinol.* **2000**, *142*, 9–21.
19. H. G. Moghadam, M. R. Urist, G. K. Sandor et al., *J. Craniofac. Surg.* **2001**, *12*, 119–127.
20. A. M. Turing, *Philos. Trans. R. Soc. London* **1952**, *237*, 37–72.
21. P. J. Boyne, *J. Bone Joint Surg. Am.* **2001**, *83-A*, S146–S150.
22. U. M. Wikesjo, R. G. Sorennsen, J. M. Wozney, *J. Bone Joint Surg. Am.* **2001**, *83-A*, S136–S145.
23. U. Ripamonti, J. Crooks, A. N. Kirkbride, *South Afr. J. Sci.* **1999**, *95*, 335–343.
24. U. Ripamonti, Smart biomaterials with intrinsic osteoinductivity: geometric control of bone differentiation in *Bone Engineering* (Ed.: J. E. Davis), EM2 Corporation, Toronto, Canada, 2000, pp. 215–222.
25. U. Ripamonti, Osteogenic proteins of the transforming growth factor-ß superfamily in *Encyclopedia of Hormones* (Eds.: H. L. Henry, A. W. Norman), Academic Press, 2003, pp. 80–86.
26. U. Ripamonti, B. Van Den Heever, T. K. Sampath et al., *Growth Factors* **1996**, *13*, 273–289.
27. T. K. Sampath, A. H. Reddi, *Proc. Natl. Acad. Sci. U. S. A.* **1981**, *78*, 7599–7603.
28. P. V. Haushka, A. E. Mavrakos, M. D. Iafrati et al., *J. Biol. Chem.* **1996**, *261*, 12665–12674.
29. U. Ripamonti, A. Ma, N. Cunningham et al., *Matrix* **1992**, *12*, 369–380.
30. U. Ripamonti, A. Ma, N. Cunningham et al., *Plast. Reconstr. Surg.* **1993**, *91*, 27–36.
31. C. Ferretti, U. Ripamonti, *J. Craniofac. Surg.* **2002**, *13*, 434–444.
32. A. H. Reddi, N. S. Cunningham, *Biomaterials* **1990**, *11*, 33–34.
33. U. Ripamonti, C. Ferretti, Mandibular reconstruction using naturally derived bone morphogenetic proteins: a clinical trial report in *Advances in Skeletal reconstruction Using Bone Morphogenetic Proteins* (Ed.: T. S. Lindholm), World Scientific Publishing Co., 2002, pp. 277–289.
34. E. A. Wang, V. Rosen, J. S. D'alessandro et al., *Proc. Natl. Acad. Sci. U. S. A.* **1990**, *87*, 2220–2224.
35. R. G. Hammonds, R. Schwall, A. Dudley et al., *Mol. Endocrinol.* **1991**, *5*, 149–155.
36. T. K. Sampath, J. C. Maliakal, P. V. Haushka et al., *J. Biol. Chem.* **1992**, *267*, 20352–20362.
37. G. Thomadakis, L. N. Ramoshebi, J. Crooks et al., *Eur. J. Oral Sci.* **1999**, *107*, 368–377.
38. K. Staehling-Hampton, P. D. Jackson, M. J. Clark et al., *Cell Growth Differ.* **1994**, *5*, 585–593.
39. K. Miyazono, *Bone* **1999**, *25*, 91–93.
40. K. Miyazono, P. Ten Dijke, C. H. Heldin, *Adv. Immunol.* **2000**, *75*, 115–157.
41. D. Chen, M. A. Harris, G. Rossini et al., *Calcif. Tissue Int.* **1997**, *60*, 238–290.
42. Y. Honda, R. Kniutsen, D. D. Strong et al., *Calcif. Tissue Int.* **1997**, *60*, 297–301.
43. U. Ripamonti, J. Crooks, J. Teare et al., *South Afr. J. Sci.* **2002**, *98*, 361–368.
44. U. Ripamonti, A. H. Reddi, *Crit. Rev. Oral Biol. Med.* **1997**, *8*, 154–163.
45. U. Ripamonti, J. R. Tasker, U. Ripamonti et al., *Curr. Pharm. Biotechnol.* **2000**, *1*, 47–55.
46. U. Ripamonti, M. Heliotis, van den Heever et al., *J. Periodontal Res.* **1994**, *29*, 439–445.
47. U. Ripamonti, M. Heliotis, D. C. Rueger et al., *Arch. Oral Biol.* **1996**, *41*, 121–126.
48. U. Ripamonti, Induction of cementogenesis and periodontal ligament regeneration by bone morphogenetic proteins in *Bone Morphogenetic Proteins: Biological Characteristics and Reconstructive Repair*, Chapter 17 (Ed.: T. S. Lindholm), R.G. Landes Company and Academic Press, 1996, pp. 189–198.
49. M. R. Urist, *Science* **1965**, *150*, 893–899.
50. A. H. Reddi, C. B. Huggins, *Proc. Natl. Acad. Sci.* **1972**, *69*, 1601–1605.

51. A. B. Roberts, M. B. Sporn, R. K. Assoian et al., *Proc. Natl. Acad. Sci. U. S. A.* **1986**, *83*, 4167–4171.
52. T. Matsaba, L. N. Ramoshebi, J. Crooks et al., *Growth Factors* **2001**, *19*, 73–86.
53. U. Ripamonti, C. Bosch, B. Van Den Heever et al., *J. Bone Miner. Res.* **1996**, *11*, 938–945.
54. U. Ripamonti, N. Duneas, B. Van Den Heever et al., *J. Bone Miner. Res.* **1997**, *12*, 1584–1595.
55. N. Duneas, J. Crooks, U. Ripamonti, *Growth Factors* **1998**, *15*, 259–277.
56. U. Ripamonti, J. Crooks, T. Matsaba et al., *Growth Factors* **2000**, *17*, 269–285.
57. U. Ripamonti, J. Teare, T. Matsaba et al., Site, tissue and organ specificity of endochondral bone induction and morphogenesis by TGF-beta isoforms in the primate *Papio ursinus, Proceedings of the 2001 FASEB Summer Research Conference*, Tucson, Arizona, USA, 2001.
58. U. Ripamonti, J. R. Tasker, Bone induction by transforming growth factor-ß in the primate and synergistic interaction with bone morphogenetic proteins in *Advances in Skeletal Reconstruction Using Bone Morphogenetic Proteins* (Ed.: T. S. Lindholm), World Scientific Publishing Co., Singapore, 2002, pp. 79–95.
59. T. Imamura, M. Takase, A. Nishihara et al., *Nature* **1997**, *389*, 622–626.
60. A. Nakao, M. Afrakhte, A. Morén et al., *Nature* **1997**, *389*, 631–635.
61. U. Ripamonti, J. Teare, T. Matsaba et al., Site, tissue and organ specificity of endochondral bone induction and morphogenesis by TGF-ß isoforms in the primate *Papio ursinus, FASEB Summer Conference: The TGF-ß Superfamily: Signaling and Development*, Tucson, Arizona, USA, July 7–12, 2001.
62. U. Ripamonti, B. Van Den Heever, M. Heliotis et al., *South Afr. J. Sci.* **2002**, *98*, 429–433.
63. H. K. Kleinman, M. L. McGarvey, J. R. Hassell et al., *Biochemistry* **1986**, *25*, 312–318.
64. V. M. Paralkar, A. K. N. Nandekar, R. H. Pointer et al., *J. Biol. Chem.* **1990**, *265*, 17281–17284.
65. U. Ripamonti, L. N. Ramoshebi, J. Patton et al., Soluble signals and insoluble substrata: Novel molecular cues instructing the induction of bone in *The Skeleton* (Eds.: E. J. Massaro, J. M. Rogers), Humana Press, Totowa, NJ, 2003, in press.
66. U. Ripamonti, T. Matsaba, J. Teare et al., Intrinsic osteoinductivity by smart biomimetic matrices, *Abstract in the 4th International Conference on Bone Morphogenetic Proteins*, Sacramento, USA, October 17–21, 2002.
67. U. Ripamonti, J. Teare, T. Matsaba et al., Tissue engineering by bone morphogenetic proteins, *Abstract in the 4th International Conference on Bone Morphogenetic Proteins*, Sacramento, USA, October 17–21, 2002.
68. U. Ripamonti, J. Patton, T. Matsaba et al., Endochondral bone induction by TGF-ß proteins: site and tissue specificity of induction in the primate, *Abstract in the 4th International Conference on Bone Morphogenetic Proteins*, Sacramento, USA, October 17–21, 2002.

Subject Index

d

p